Potential Theory and Stochastics in Albac

Aurel Cornea Memorial Volume

Potential Theory and Stochastics in Albac

Aurel Cornea Memorial Volume

Conference Proceedings

Albac, September 4–8, 2007

Edited by

D. Bakry
L. Beznea
N. Boboc
M. Röckner

Theta
Bucharest 2009

Editors
D. Bakry
L. Beznea
N. Boboc
M. Röckner

The Theta Foundation
Calea Griviţei 21
010702 Bucharest, Romania
E-mail: theta@theta.ro
http://www.theta.ro

CIP Description of the National Library
Descrierea CIP a Bibliotecii Naţionale a României

POTENTIAL THEORY AND STOCHASTICS.
CONFERENCE (2007 ; ALBAC)
Potential theory and stochastics in Albac : Aurel Cornea memorial volume : conference proceedings : Albac, September, 4–8, 2007 - Bucureşti : Editura Fundaţiei Theta, 2009
ISBN 978–973–87899–5–1

517.956.225

ISBN 978–973–87899–5–1

Printed on acid free paper.

Printed and bound by ARTA GRAFICĂ, Bucharest, Romania

CONTENTS

Foreword

This collection of papers is the proceedings of the *Potential Theory and Stochastics* Conference which was held in Albac, Romania, from September 4 to 8, 2007. It represents the Aurel Cornea memorial volume.

The conference in Albac was organized by the *Simion Stoilow* Institute of Mathematics of the Romanian Academy, the *1 Decembrie 1918* University of Alba Iulia and the University of Piteşti.

The meeting was dedicated to the memory of Aurel Cornea and it was attended by many of his friends, collaborators and colleagues. On the whole, forty-two participants from twelve countries attended the event. We thank them all for their contribution to the success of the meeting.

Thirty-four talks were delivered during the five days of the conference. In addition a message from Mrs. Adelheid Cornea was presented to the participants by Eleutherius Symeonidis.

The first article of the proceedings presents the life and work of Aurel Cornea and it is completed by his list of scientific publications.

The volume contains twenty refereed scientific contributions of the participants, covering the main topics of the conference: geometric aspects in potential theory; Dirichlet structures and stochastic analysis; potential theoretical methods for the analysis of infinite dimensional processes, stochastic partial differential equations; nonlinear partial differential equations and potential theory. We also refer to the list of talks given at Albac by the participants of the conference right after this introduction.

The conference was made possible by generous support from *Deutsche Forschungsgemeinschaft* and *Romanian Academy* (through Project GZ: 436 RUM 113/23/0-1), *Romanian Ministry of Education, Research and Youth* (contracts CEx06-M3-102/2006, 2-CEx06-11-10/2006, CEex-5-D11-23/2005, 2-CEx06-11-18/2006 and CNCSIS grant No. 2236/2006, code 815), *SOFTWIN Group* and *Romanian Ministry of Foreign Affairs*.

This volume has been edited with support from the *Romanian Ministry of Education, Research and Youth* (through the above mentioned contracts). Support from PN II Program of the same ministry (contract with CNCSIS code ID 209/2007) is also gratefully acknowledged.

Thanks are due to Daniel Breaz and Nicoleta Breaz, the local organizers of the meeting, and to Andrei Oprina for preparing the final form of the manuscript.

The Editors

LIST OF TALKS

Dominique Bakry: On gradient bounds for the heat kernel on the Heisenberg group

Allami Benyaiche: On the Dirichlet problem associated to a coupled PDE

Lucian Beznea: Potential analysis on Wiener spaces

Mounir Bezzarga: On a singular value problem and the boundary Harnack principle on the exterior of the unit ball

Marco Biroli: Homogenization for Riemannian Dirichlet forms on open sets with holes and Neumann boundary condition on holes

Hans-Peter Blatt: Potential theoretical aspects in constructive function theory

Jürgen Bliedtner: Remarks on the Lebesgue measure in classical potential theory

Nicu Boboc: Non-symmetric resistance forms

Anca Iuliana Bonciocat: On the geometry of discrete metric spaces

Sirkka-Liisa Eriksson: Hyperbolic harmonic functions and their functions theory

Masatoshi Fukushima: Lévy systems and Feller measures for Markov processes

Stephen J. Gardiner: Recent progress on questions of gravi-equivalence

Marius Ghergu: On a class of singular elliptic equations

Kohur GowriSankaran: A result on the boundary behavior of separately superharmonic functions on the product of Brelot spaces

Wolfhard Hansen: Convexity properties of harmonic measures

Walter Hoh: On generalized hyperbolic-like Feller processes

Niels Jacob: Some higher order differential and pseudo-differential operators associated with random fields

Klaus Janssen: Subordination of right resolvents

Kazuhiro Kuwae: On a Liouville type theorem for harmonic maps to convex spaces via Markov chains

Peter A. Loeb: Applications to stochastic processes of rich measures formed from non-standard models

Jaroslav Lukeš: Choquet and exposed sets in potential theory

Terry Lyons: On signed probability measures and some old work of Krylov

Zhi-Ming Ma: Nonlinear filtering of semi-Dirichlet processes

Yoshihiro Mizuta: Function spaces of variable exponent

Ivan Netuka: Potential theory of the farthest point distance function

Marian Panţiruc: Fourier transform on the heat semigroup on the Heisenberg group

Mihai N. Pascu: Hot spots conjecture and maximum principles

Eugen Popa: Polyinvariant elements in a semi-dynamical system

Michael Röckner: Strong solutions of stochastic porous media equations: A survey of recent results

Byron Schmuland: Reversible Fleming-Viot processes

Wilhelm Stannat: A new a priori estimate for 2D-stochastic Navier-Stokes equations w.r.t. Gaussian measures

Corneliu Udrea: Excessive functions with respect to Monge-Ampère resolvents and concave functions

Jiří Veselý: Life and work of Aurel Cornea (1933 - 2005)

LIST OF PARTICIPANTS

Miroslav BACAK, Mathematical Institute, Charles University Prague, Sokolovska 83, 18675 Prague, Czech Republic.
E-mail: bacak@karlin.mff.cuni.cz

Dominique BAKRY, Laboratoire de Statistique et Probabilités, Université Paul Sabatier, 118 route de Narbonne, 31062 Toulouse, France.
E-mail: bakry@cict.fr

Allami BENYAICHE, Faculté des Sciences de Kénitra, Université IBN Tofail, Kénitra, Morroco.
E-mail: a_benyaiche@yahoo.fr

Lucian BEZNEA, "Simion Stoilow" Institute of Mathematics of the Romanian Academy, P.O. Box 1-764, RO-014700 Bucharest, Romania.
E-mail: lucian.beznea@imar.ro

Mounir BEZZARGA, Institut Preparatoire aux Etudes d'Ingenieur de Tunis, rue Jawaher Lel Nehru, 1008 Montfleury-Tunis, Tunisia.
E-mail: mounir.bezzarga@yahoo.fr

Marco BIROLI, Department of Mathematics "F. Brioschi", Politecnico di Milano, Piazza Leonardo da Vinci 32, Milano 20133, Italy.
E-mail: marco.biroli@polimi.it

Hans-Peter BLATT, Katholische Universität Eichstätt-Ingolstadt, 85072 Eichstätt, Germany.
E-mail: hans.blatt@ku-eichstaett.de

Jürgen BLIEDTNER, Fachbereich Mathematik Johann Wolfgang Goethe-Universität, Robert-Mayer-Strasse 8, D-60054 Frankfurt/Main, Germany.
E-mail: bliedtne@math.uni-frankfurt.de

Nicu BOBOC, Faculty of Mathematics and Computer Sciences, University of Bucharest, Str. Academiei 14, RO-010014 Bucharest, Romania.
E-mail: nboboc@math.math.unibuc.ro

Anca Iuliana BONCIOCAT, "Simion Stoilow" Institute of Mathematics of the Romanian Academy, P.O. Box 1-764, RO-014700 Bucharest, Romania.
E-mail: anca@wiener.iam.uni-bonn.de

Daniel BREAZ, Department of Mathematics, "1 Decembrie 1918" University of Alba Iulia, Str. N. Iorga, No. 11-13, RO-510009 Alba Iulia, Romania.
E-mail: dbreaz@uab.ro, breazdaniel@yahoo.com

Nicoleta BREAZ, Department of Mathematics, "1 Decembrie 1918" University of Alba Iulia, Str. N. Iorga, No. 11-13, RO-510009 Alba Iulia, Romania.
E-mail: nicoletabreaz@yahoo.com

Gheorghe BUCUR, Faculty of Mathematics and Computer Sciences, University of Bucharest, Str. Academiei 14, RO-010014 Bucharest, Romania, and "Simion Stoilow" Institute of Mathematics of the Romanian Academy.

Sirkka-Liisa ERIKSSON, Department of Mathematics, Tampere University of Technology, P.O. Box 553, FI - 33101 Tampere, Finland.
E-mail: sirkka-liisa.eriksson@tut.fi

Masatoshi FUKUSHIMA, Department of Mathematics, Kansai University, Suita, Osaka 564-8680, Japan.
E-mail: fuku2@mx5.canvas.ne.jp

Stephen J. GARDINER, School of Mathematical Sciences, University College Dublin, Belfield, Dublin 4, Ireland.
E-mail: stephen.gardiner@ucd.ie

Marius GHERGU, "Simion Stoilow" Institute of Mathematics of the Romanian Academy, P.O. Box 1-764, RO-014700 Bucharest, Romania.
E-mail: marius.ghergu@imar.ro

Kohur GOWRISANKARAN, Department of Mathematics and Statistics, McGill University, 1107 Burnside Hall, 805 Sherbrooke W., Montreal, Canada.
E-mail: kohur.gowrisankaran@mcgill.ca

Marian HAIDUCU, Faculty of Mathematics and Computer Science, University of Piteşti, Str. Târgul din Vale nr. 1, Piteşti, Romania.
E-mail: marian.haiducu@yahoo.fr

Wolfhard HANSEN, Faculty of Mathematics, University of Bielefeld, P.O. Box 100131, 33501 Bielefeld, Germany.
E-mail: hansen@math.uni-bielefeld.de

Walter HOH, Faculty of Mathematics, University of Bielefeld, P.O. Box 100131, 33501 Bielefeld, Germany.
E-mail: hoh@math.uni-bielefeld.de

Niels JACOB, Department of Mathematics, Swansea University, Singleton Park, Swansea SA2 8PP, Wales, Great Britain.
E-mail: n.jacob@swansea.ac.uk

Klaus JANSSEN, Henrich Heine Institute of Mathematics, University of Düsseldorf, Universitätsstr. 1, 40225 Düsseldorf, Germany.
E-mail: janssenk@uni-duesseldorf.de

Kazuhiro KUWAE, Department of Mathematics, Faculty of Education, Kumamoto University, Kumamoto 860-8555, Japan.
E-mail:kuwae@gpo.kumamoto-u.ac.jp

Peter A. LOEB, Department of Mathematics, University of Illinois at Urbana-Champaign, 1409 W. Green Street Urbana, Illinois 61801-2975, USA.
E-mail: loeb@math.uiuc.edu

Jaroslav LUKEŠ, Mathematical Institute, Charles University Prague, Sokolovska 83, 18675 Prague, Czech Republic.
E-mail: lukes@karlin.mff.cuni.cz

Terry LYONS, Mathematical Institute, University of Oxford, 24-29 St Giles', Oxford OX1 3LB, Great Britain.
E-mail: tlyons@maths.ox.ac.uk

Zhi-Ming MA, Institute of Applied Mathematics, Academy of Mathematics and Systems Science, CAS No. 55, Zhong-guan-cun East Road, Beijing 100080, China.
E-mail: mazm@amt.ac.cn

Yoshihiro MIZUTA, Hiroshima University, Higashi-Hiroshima 739, Japan.
E-mail: mizuta@mis.hiroshima-u.ac.jp

Emil MOLDOVEANU, Faculty of Mathematics and Computer Science, University of Piteşti, Str. Târgul din Vale nr. 1, Piteşti, Romania.
E-mail: analizam@inMail24.com

Ivan NETUKA, Mathematical Institute, Charles University Prague, Sokolovska 83, 18675 Prague, Czech Republic.
E-mail: netuka@karlin.mff.cuni.cz

Andrei-George OPRINA, "Simion Stoilow" Institute of Mathematics of the Romanian Academy, P.O. Box 1-764, RO-014700 Bucharest, Romania.
E-mail: oandrei22@yahoo.com

Marian PANŢIRUC, Faculty of Mathematics, "Al. I. Cuza" University, Bd. Carol I nr. 11, 700506 Iaşi, Romania.
E-mail: pantirucmarian@yahoo.com

Mihai N. PASCU, Faculty of Mathematics and Computer Science, Transilvania University, 50 Iuliu Maniu St, Braşov 500091, Romania.
E-mail: mihai.pascu@unitbv.ro

Eugen POPA, Faculty of Mathematics, "Al. I. Cuza" University, Bd. Carol I nr 11, 700506 Iaşi, Romania.
E-mail: epopa@uaic.ro

Liliana POPA, Chair of Math. "Gh. Asachi" Technical University, Bd. Dimitrie Mangeron nr. 67, 700050 Iaşi, Romania.
E-mail: epopa@uaic.ro

Michael RÖCKNER, Faculty of Mathematics, University of Bielefeld, P.O. Box 100131, 33501 Bielefeld, Germany.
E-mail: roeckner@math.uni-bielefeld.de

Byron SCHMULAND, Department of Mathematical and Statistical Sciences, University of Alberta Edmonton, Alberta, Canada.
E-mail: schmuland2002@yahoo.ca

Wilhelm STANNAT, Fachbereich Mathematik, TU Darmstadt, Schlossgartenstr. 7, D-64289 Darmstadt, Germany.
E-mail:stannat@mathematik.tu-darmstadt.de

Eleutherius SYMEONIDIS, Katholische Universität Eichstätt-Ingolstadt, 85072 Eichstätt, Germany.
E-mail: e.symeonidis@ku-eichstaett.de

Maria TEODOREANU, Research Institute for Electrical Engineering (ICPE), Bucharest, Romania.
E-mail: maria.teodoreanu@gmail.com

Corneliu UDREA, Faculty of Mathematics and Computer Science, University of Piteşti, Str. Târgul din Vale nr. 1, Piteşti, Romania.
E-mail: corneliu_udrea@yahoo.com

Jiří VESELÝ, Mathematical Institute, Charles University Prague, Sokolovska 83, 18675 Prague, Czech Republic.
E-mail: jvesely@karlin.mff.cuni.cz

Potential Theory and Stochastics in Albac, xvii–xxvi
Aurel Cornea Memorial Volume

LIFE AND WORK OF AUREL CORNEA

(JULY 6, 1933 – SEPTEMBER 3, 2005)

JIŘÍ VESELÝ

Aurel Cornea was born in Transylvania, part of Romania, in a small village Veneţia de Jos as the first child of Nicolaie and Rozalia Cornea. His talent was apparent already in the elementary school and hence his parents moved after the World War II to the capital of Romania to secure for him a better education. Unfortunately, at the age of 14 he had an accident during one of his chemical experiments and he lost his sight. He missed one school year due to recovery but his enormous will helped him to get over and to continue in his studies. His interest shifted from chemistry to the queen of sciences – mathematics. And already in that time he made up his mind to be a mathematician as good as possible.

He entered the Bucharest University where he studied mathematics with the help of his schoolmates and colleagues; some of them became his life friends and collaborators. There are stories of eye witnesses confirming Cornea's ability to perform very complicated computations by heart which sound almost unbelievable (cf. an interesting article [O1]). He became a student of Simion Stoilow and under his supervision he received his Ph.D. in 1960 for the thesis *On the behavior of the analytic transformations at the Martin boundary*. After his studies he worked at the *Institute of Mathematics of Romanian Academy* (1955 – 1978), then at the *Faculty of Mathematics and Physics of University of Bucharest* (since 1975) and also at the *National Institute for Scientific and Technical Creation* (INCREST). At the Bucharest University he taught stochastic processes, convexity and potential theory. With small interruptions he spent in Romania first 45 years of his life.

A remark on the consequences of the political division of the world into blocks is in order: in the group of countries belonging to Warsaw treaty scientists had limited access to western sources (journals and books) as well as travels abroad were hardly possible. In case of Romania, frontiers up to 1966 were practically closed and journeys to countries outside the block were almost excluded. Even travels within the block were for some scientist difficult or impossible. As a consequence of this situation many people used those limited chances to leave the block and tried to build up new career abroad. (cf. [O1]).

In 1967 Corneliu Constantinescu and Cornea were invited by Brelot to Paris and a year later on the basis of Bauer's invitation they both spent two months in Erlangen. The same year Cornea met Adelheid Guber who just lost her husband Siegfried Guber (a pupil of Bauer) in a tragic accident in Romania. They became friends and started to

2000 *Mathematics Subject Classification.* 01A60 (31-06, 31C99).
Key words and phrases. Dirichlet problem, harmonic mapping, controled convergence.

exchange regularly letters. This exchange lasted for years (incidentally, Cornea wrote his letters himself and they were readable!). Occasionally they met, for example in Italian Stresa already in 1969 where Constantinescu and Cornea took part at the *Summer school on potential theory* organized there by CIME (and paid by NATO). In 1974 they met in Berlin and decided to join their lives. A visit of family Cornea in Romania followed, but there was a long way to realize the dream on possibility to get married.

In 1978 there was an *International Mathematical Congress of Mathematicians* at Helsinki and also (now traditional) joint *Romanian-Finnish Seminar on Complex Analysis* in Joensuu. Instead of returning back to Romania, Cornea went to Canada where he learned how to work with special computer equipment for blind people. From there he went to USA and finally to Germany. This period of his life was very difficult but there were colleagues and friends who helped him. During his stay in Frankfurt am Main he applied for a position at Eichstätt; I quote from Constantinescu's article [O1]: *He [Cornea] became a candidate for a position at "Katholishe Universität Eichstätt" and was elected. At this moment he was 47, blind, Orthodox, had a Romanian passport with a short-term German visa, and there were 47 mathematicians who were candidates for the position. His success in obtaining the chair amounted to a miracle.* The commission was impressed by Cornea's talk but in the case of specialists in the field such as Bauer it was not a surprise. But also the head of the commission and the president of the university, Professor Mosis (a theologian!) said, that even for him it was clear that the talk was the best. Shortly after receiving this position the dreamed wedding with Guber took place.

Cornea spent the rest of his life in Germany. He organized the newly created *Department of Mathematics* at Eichstätt university, lectured there on analysis and conducted research in mathematics. He never broke contacts with Romania and quite often he invited collaborators from Romania to Eichstätt. His life was happy and harmonic, devoted to his research, his family and contacts with many of his colleagues and friends. Since 1999 he was retired and within the frame of *Deutsche Akademisher Austauschdienst* he lectured (2000 – 2004) on potential theory at University of Bucharest. He substantially contributed to the creation of so called Romanian school of potential theory (and related topics).

What was astonishing for all of us was the fact that he was able to ignore his blindness. He did all: trips to mountains, biking (on tandem bike) and many other things. He was interested in music which he loved very much, at home he was able to make little repairs. He prepared texts of articles or slides for his lectures himself. While he lived in Romania he quite often explained things difficult to learn to his 17 years younger sister. Later on he did the same for his stepdaughters and he was a very good teacher. As a mathematician he was a man of great invention, with tremendous ability of concentration on problems and great sense to isolate the principal properties of mathematical objects he studied.

His unexpected death in a tragic accident was like a bolt from the blue: Full of plans, he lectured at a conference in Romania in course of *The 10th Romanian-Finnish Seminar* in August 2005. On September 3, 2005 he died during a car accident. His death was a great loss for his family, relatives and friends, but also for his beloved mathematics.

Let us turn to Cornea's mathematics. There is an article [O2] of his colleagues describing content of monographs which Cornea wrote with different coauthors. The first book which he wrote with Nicu Boboc and Constantinescu was *Potential theory on Riemannian surfaces* (Romanian) [B1]. This was followed in 1963 by *Ideale Ränder Riemannscher Flächen* [B2] written in collaboration with Constantinescu. The book was appreciated and so the authors were offered to write another book for Springer. Originally the plan of publishing house was that the book should be written in cooperation with Heinz Bauer but since Bauer was overloaded by other duties, Cornea worked on it again with Constantinescu only. The book appeared in 1972 under the title *Potential theory on harmonic spaces.* The article [O1] describes the cooperation of those two authors and friends. Later during his stay in Romania, Cornea wrote another book *Order and potential resolvent families of kernels* (with Gabriela Licea) [B4] and while he was already in Eichstätt yet another book written with Boboc and Gheorghe Bucur [B5] had appeared: *Order and convexity in potential theory: H-cones.* Many results previously published in articles by Cornea and his collaborators were included to those books. In the list of Cornea's works items [1] – [14] (except [3] and [6]) are related to [B2], items [15] – [34] (with exceptions [22], [23], [25] – [29] and [33]) are related to [B3].

Here another comment is appropriate: It was not easy to work on such monographs in Romania since up to 1954 access to results of "western" science was almost impossible. After Stalin's death it substantially changed and thus political changes influenced very much the content of the book [B3]: easier contact with foreign mathematics is recognizable. Bauer wrote in the Preface to the book: *Based on ideas of Brelot, Doob and Tautz, the theory of harmonic spaces has developed so rapidly in the last 15 years that up to now original papers, seminar and lecture notes were the only source for further studies in the field. In view of this situation, C. Constantinescu and A. Cornea who had both considerably influenced the development of the theory, undertook the difficult task to present contemporary knowledge of harmonic spaces as completely as possible.*

The result of their effort is this book. (...) With this monograph the authors do not only close the serious gap in the existing mathematical literature, the monograph certainly will also have a strong impact on future research in the field of potential theory. And soon it was obvious that he was right. At the same year when the book was published Cornea received the title docent doctor. Unfortunately the book also marks the end of his long lasting cooperation (but not friendship !) with Constantinescu who left Romania in 1972 and changed the field of his interests to measure theory.

My memories on the first meeting with Cornea are connected with Summer school in Stresa in 1969. For me and my colleagues it was the first chance to meet him and also other specialists in potential theory from France, Germany, Romania, USA and other countries. The meeting was for us of great importance not only as the first travel to a western country but also as the possibility to hear lectures on different aspects of potential theory. In addition, as a consequence of meeting Constantinescu and Cornea, we received the manuscript of [B3] before its publication.

As I already mentioned Cornea's books are characterized in the article [O2] and their content corresponds in a great extent with his articles (in the list of references approx. first 50 items). To some articles from the rest we will come back below. In spite of

Now let us assume (iii) and take a set $A \subset U$ and $y \in \Delta \cap \overline{A}$. Fix $0 < \varepsilon < +\infty$. If $\limsup_{x \to y, x \in A} k(x) \le n$ for an $n \in \mathbb{N}$, then

$$\begin{aligned} &\limsup_{x \to y, x \in A} \operatorname{dist}\big(h(x), f(y)\big) \le \\ &\le \limsup_{x \to y, x \in A} \big(\operatorname{dist}\big(h(x), f(y)\big) - \varepsilon\, k(x)\big) + \limsup_{x \to y, x \in A} \varepsilon k(x) \le \\ &\le \limsup_{x \to y, x \in A} \big(\operatorname{dist}\big(h(x), f(y)\big) - \varepsilon\, k(x)\big) + \varepsilon n \le \varepsilon n\,. \end{aligned}$$

Since it holds for any $\varepsilon > 0$ the estimate gives condition (a) from the definition of controlled convergence.

Suppose now $\limsup_{x \to y, x \in A} k(x) = +\infty$. Fix $n \in \mathbb{N}$ and choose a neighborhood $V(y)$ of y such that

$$(x \in U \cap V) \implies \big(\operatorname{dist}\big(h(x), f(y)\big) - \tfrac{1}{n}\, k(x) \le \tfrac{1}{n}\big)\,.$$

For any such x we have

$$\frac{\operatorname{dist}\big(h(x), f(y)\big)}{1 + k(x)} = \frac{\big(\operatorname{dist}\big(h(x), f(y)\big) - \tfrac{1}{n}\, k(x) - \tfrac{1}{n}\big)}{k(x) + 1} + \tfrac{1}{n} \le \tfrac{1}{n}\,,$$

and for any $z \in Y$ we get

$$\limsup_{x \to y, x \in A} \frac{\operatorname{dist}\big(h(x), z\big)}{1 + k(x)} \le \tfrac{1}{n} + \frac{\operatorname{dist}\big(f(y), z\big)}{1 + n}\,.$$

Since $n \in \mathbb{N}$ was arbitrarily chosen we receive condition (b) from the definition of the control convergence.

If the conditions of the Theorem are satisfied by h, k, f and

$$U_n := \{x \in U;\, k(x) \le n\}\,, \qquad n \in \mathbb{N}\,,$$

then $h = f$ on $\Delta \cap U_n$, f restricted on the set $\Delta \cap \overline{U_n}$ is continuous and uniquely determined on $\bigcup_{n=1}^{\infty} U_n$ by h.

Lemma: *Suppose that Y_1 and Y_2 are metric spaces and $F : Y_1 \otimes Y_2 \to Y$ is a Lipschitz mapping. Choose $\Delta_1, \Delta_2 \subset X$ and put $\Delta = \Delta_1 \cap \Delta_2$. Suppose that $h_j : U \to Y_j$ is convergent to $f_j : \Delta_j \to Y_j$ controlled by $k_j : U \to [0, +\infty)$, $j = 1, 2$. If $h := F(h_1, h_2)$ on U and $f = F(f_1, f_2)$ on Δ, then h is convergent to f controlled by $k = k_1 + k_2$.*

Cornea applied the technique of controlled convergence in several situations. In this way he obtained solution of the Dirichlet problem in harmonic spaces, theorem on Lebesgue derivative of integrable function and Fatou type and Beurling-Deny type theorems. In [63] it is shown how to construct with the help of the previous Lemma (and some others) controll functions in more complicated situations. Let us briefly formulate what control convergence gives for *harmonic mappings*.

Suppose that X is a harmonic space with countable base in which constants are harmonic: on X it is considered a sheaf $W \to \mathcal{M}(W, Y)$ of continuous mappings into Y satisfying certain suitable axioms and $U \subset X$ is given open relatively compact set, $\Delta = \partial U$.

Definition: A mapping $f : \Delta \to Y$ is called *resolutive* for the Dirichlet problem, if there is a harmonic mapping $h \in \mathcal{M}(W, Y)$ converging to f controlled by a non-negative harmonic function k on U.

Theorem: *For any resolutive function $f : \Delta \to Y$ there exists a unique solution of the Dirichlet problem on U with boundary data f.*

In the classical case it is the PWB-solution for any "classically resolutive" function f and the solution can be obtained even by considering upper and lower Perron classes for f consisting entirely from *harmonic functions.*

What to say at the end? Aurel Cornea was an excellent mathematician, good father and a good friend. He had a special sense of humor and always wanted to please friends. He was extremely brave in his fight against his fate. He never broke his contact with Romania. He is buried in Eichstätt not far from the nice flat he and his wife chose for their old age. On his grave there is a symbol

$$\Delta u = 0$$

and we can read there a part of a poem of his probably most favorite Romanian poet Mihai Eminescu (1850 – 1889):

Mai am un sigur dor (1883/4):

Cum n-oi mai fi pribeag	[As I will then no more
De-atunci înainte,	A wanderer be,
M-or troieni cu drag	Let them with fondness store
Aduceri aminte.	My memory.[2]]

References

Articles of Aurel Cornea

[1] *On the behaviour of analytic functions in the neighbourhood of the boundary of a Riemann surface,* Nagoya Math. J. **12** (1957), 55–58.

[2] *Au sujet de parties singuliéres de la frontiére idéale* (Romanian), Com. Acad. R. P. Romîne **8** (1958), 639–642.

[3] *Über eine Formel in der Extremisierungstheorie* (German), Rev. Math. Pures Appl. **3** (1958), 431–436.

[4] *Über den idealen Rand und einige seiner Anwendungen bei der Klassifikation der Riemannschen Flächen* (German) (with Corneliu Constantinescu), Nagoya Math. J. **13** (1958), 169–233.

[5] *Comportement des transformations analytiques des surfaces de Riemann sur la frontiére de Martin* (French) (with Corneliu Constantinescu), C. R. Acad. Sci. Paris **249** (1959), 355–357.

[6] *Über einige Probleme von M. Heins* (German) (with Corneliu Constantinescu), Rev. Math. Pures Appl. **4** (1959), 277–281.

[7] *Über das Verhalten der analytischen Abbildungen Riemannscher Flächen auf dem idealen Rand von Martin* (German) (with Corneliu Constantinescu), Nagoya Math. J. **17** (1960), 1–87.

[8] *Über den Martinschen idealen Rand einer Riemannschen Fläche* (German) (with Corneliu Constantinescu), Rev. Math. Pures Appl. **5** (1960), 21–25.

[9] *Sur la frontiére Martin d'une surface riemannienne* (Romanian) (with Corneliu Constantinescu), Acad. R. P. Romîne Fil. Cluj Stud. Cerc. Mat. **11** (1960), 261–265.

[2] German translation: Das Irren auf und nieder, Es ist zu Ende nun, Erinnerung wird mich betten, Und stille werd ich ruhn.

[10] *Spaces of harmonic functions on Riemann surfaces* (with Corneliu Constantinescu), Rev. Math. Pures Appl. **6** (1961), 373–390.

[11] *Le théoreme de Beurling et la frontiére idéale de Kuramochi* (French) (with Corneliu Constantinescu), C. R. Acad. Sci. Paris **254** (1962), 1732–1734.

[12] *Analytische Abbildungen Riemannscher Flächen* (German) (with Corneliu Constantinescu), Rev. Math. Pures Appl. **8** (1963), 67–72.

[13] *Normale Kompaktifizierungen Riemannscher Flächen* (German) (with Corneliu Constantinescu), Rev. Math. Pures Appl. **8** (1963), 73–75.

[14] *Frontiéres idéales des surfaces de Riemann* (French) (with Corneliu Constantinescu), Rev. Math. Pures Appl. (Bucarest) **8** (1963), 227–242.

[15] *On the Dirichlet problem in the axiomatic theory of harmonic functions* (with Nicu Boboc and Corneliu Constantinescu), Nagoya Math. J. **23** (1963), 73–96.

[16] *Sur le probléme de Dirichlet dans l'axiomatique des functions harmoniques* (French) (with Nicu Boboc and Corneliu Constantinescu), C. R. Acad. Sci. Paris **256** (1963), 2754–2757.

[17] *On the axiomatic of harmonic functions I, II* (with Corneliu Constantinescu), Ann. Inst. Fourier (Grenoble) **13** (1963), 373-388; ibid. **13** (1963), 389–394.

[18] *Compactifications of harmonic spaces* (with Corneliu Constantinescu), Nagoya Math. J. **25** (1965), 1–57.

[19] *Axiomatic theory of harmonic functions. Balayage* (with Nicu Boboc and Corneliu Constantinescu), Ann. Inst. Fourier (Grenoble) **15** (1965), 37–70.

[20] *Cônes des fonctions continues sur un espace compact* (French) (with Nicu Boboc), C. R. Acad. Sci. Paris **261** (1965), 2564–2567.

[21] *Axiomatic theory of harmonic functions. Non-negative superharmonic functions* (with Nicu Boboc and Corneliu Constantinescu), Ann. Inst. Fourier (Grenoble) **15** (1965), 283–312.

[22] *General optional sampling of supermartingales* (with Gabriela Licea), Rev. Roumaine Math. Pures Appl. **10** (1965), 1349–1367.

[23] *Behaviour of harmonic functions at a nonregular boundary point* (with Nicu Boboc), Bull. Math. Soc. Sci. Math. R. S. Roumanie **10** (1966), 63-74 (1967).

[24] *Sur la dénombrabilité á l'infini d'un espace harmonique de Brelot* (French), C. R. Acad. Sci. Paris Sér. A-B **264** (1967), A190–A191.

[25] *Convex cones of lower semicontinuous functions on compact spaces* (with Nicu Boboc), Rev. Roumaine Math. Pures Appl. **12** (1967), 471–525.

[26] *Comportement des balayées des mesures ponctuelles. Comportement des solutions du probléme de Dirichlet aux points irréguliers* (French) (with Nicu Boboc), C. R. Acad. Sci. Paris Sér. A-B **264** (1967), A995–A997.

[27] *Semigroups of transitions on harmonic spaces* (with Nicu Boboc, and Corneliu Constantinescu), Rev. Roumaine Math. Pures Appl. **12** (1967), 763–805.

[28] *Nonnegative hyperharmonic functions. Balayage and natural order* (with Nicu Boboc), Rev. Roumaine Math. Pures Appl. **13** (1968), 609–618.

[29] *Balayage des mésures par rapport á un cône de fonctions inférieurement semi-continues sur un espace localement compact* (with Nicu Boboc), 1968 Séminaire de Théorie du Potentiel, Dirigé par M. Brelot, G. Choquet et J. Deny: 1966/67, Exp. 22 11 pp. Secrétariat mathématique, Paris.

[30] *Espaces harmoniques. Axiome* D *et théoreme de convergence* (French) (with Nicu Boboc), Rev. Roumaine Math. Pures Appl. **13** (1968), 933–947.

[31] *Examples in the theory of harmonic spaces* (with Corneliu Constantinescu), (1968), Seminar über Potentialtheorie, 161–171 Springer, Berlin.

[32] *Weak compact sets in vector lattices and convergence theorems in harmonic spaces*, (1968) Seminar über Potentialtheorie, 173–180, Springer, Berlin.

[33] *Une démonstration unifiée des théoremes de sections de P. A. Meyer* (French) (with Gabriela Licea), Z. Wahrscheinlichkeitstheorie und Verw. Gebiete **10** (1968), 198–202.

[34] *Nuclearity on harmonic spaces* (with Corneliu Constantinescu), Tôhoku Math. J. **21** (1969), 558–572.

[35] *Cônes convexes ordonnés. Représentations intégrales* (French) (with Nicu Boboc), C. R. Acad. Sci. Paris Sér. A-B 271 (1970), A880–A883.

[36] *Cônes convexes ordonnés. H-cônes et biadjoints des H-cônes* (French) (with Nicu Boboc), C. R. Acad. Sci. Paris Sér. A-B 270 (1970), A1679–A1682.

[37] *Cônes convexes ordonnés. H-cônes et adjoints de H-cônes* (French) (with Nicu Boboc), C. R. Acad. Sci. Paris Sér. A-B 270 (1970), A596–A599.

[38] *Cones of potentials on topological spaces* (with Nicu Boboc and Gheorghe Bucur), Rev. Roumaine Math. Pures Appl. **18** (1973), 815–865.

[39] *H-cones and potential theory. Collection of articles dedicated to Marcel Brelot on the occasion of his 70th birthday* (with Nicu Boboc and Gheorghe Bucur), Ann. Inst. Fourier (Grenoble) **25** (1975), 71–108.

[40] *Sur l'existence des valeurs propres pour les applications compactes* (French) (with Ion Ichim), Rev. Roumaine Math. Pures Appl. **22** (1977), 599–606.

[41] *Natural topologies on H-cones. Weak completeness* (with Nicu Boboc and Gheorghe Bucur), Rev. Roumaine Math. Pures Appl. **24** (1979), 1013–1026.

[42] *Hilbertian and lattice-theoretical methods in potential theory* (with Nicu Boboc and Gheorghe Bucur), Romanian-Finnish Seminar on Complex Analysis (Proc., Bucharest, 1976), 517–549, Lecture Notes in Math., **743**, Springer, Berlin (1979).

[43] *Bases and essential bases in H-cones. Functional analysis: surveys and recent results, II* (with Herbert Höllein), (Proc. Second Conf. Functional Anal., Univ. Paderborn, Paderborn, 1979), 69–86, Notas Mat., 68, North-Holland, Amsterdam-New York (1980).

[44] *Carrier theory and negligible sets on a standard H-cone of functions* (with Nicu Boboc and Gheorghe Bucur), Rev. Roumaine Math. Pures Appl. **25** (1980), 163–197.

[45] *Autodual H-cones* (with Nicu Boboc and Gheorghe Bucur), Potential theory, Copenhagen 1979 (Proc. Colloq., Copenhagen, 1979), 64–77, Lecture Notes in Math., **787**, Springer, Berlin (1980).

[46] *Autodual H-cones* (with Nicu Boboc and Gheorghe Bucur), Hokkaido Math. J. **10** (1981), Special Issue, 37–67.

[47] *Morphisms and contractions on standard H-cones* (with H. Höllein), Hokkaido Math. J. **10** (1981), Special Issue, 157–212.

[48] *An approximation theorem for cones of potentials* (with Rainer Wittmann), Analysis **3** (1983), 317–335.

[49] *Geordnete konvexe Kegel in der Potentialtheorie* (German) On the work of Leonhard Euler (Berlin, 1983), 157–170, Birkhäuser, Basel (1984).

[50] *Continuity of réduites and balayaged functions*, Théorie du potentiel (Orsay, 1983), 173–182, Lecture Notes in Math., 1096, Springer, Berlin (1984).

[51] *Order continuity of the greatest lower bound of two functionals* (with Sirkka-Liisa Eriksson), Analysis **7** (1987), 173–184.

[52] *On semiregular points of the Martin boundary* (with Peter A. Loeb), Proc. Amer. Math. Soc. **103** (1988), 117–124.

[53] *A characterization of the fine sheaf property*, Potential theory (Prague, 1987), 63–69, Plenum, New York (1988).

[54] *A convergence property for conditional expectation* (with Peter A. Loeb), Ann. Probab. **17** (1989), 353–356.

[55] *An identity theorem for logarithmic potentials*, Osaka J. Math. **28** (1991), 829–836.

[56] *Harmonic approximation on arcs* (with Thomas Bagby and Paul M. Gauthier), Constr. Approx. **9** (1993), 501–507.

[57] *Finiteness principle and Harnack principle*, ICPT '91 (Amersfoort, 1991), 203–216, Kluwer Acad. Publ., Dordrecht (1994).

[58] *Résolution du probléme de Dirichlet et comportement des solutions á la frontiére á l'aide des fonctions de contrôle* (French), C. R. Acad. Sci. Paris Sér. I Math. **320** (1995), 159–164.

[59] *Martin compactification for discrete potential theory and the mean value property* (with Jiří Veselý), Potential Anal. **4** (1995), 547–569.

[60] *Solution of the Dirichlet problem on Riemannian manifolds by integration on the unit sphere in the tangent space*, Rev. Roumaine Math. Pures Appl. **47** (2002), 645–660 (2003).

[61] *Applications of controlled convergence in analysis*, Analysis and topology, 257–275, World Sci. Publishing, River Edge, NJ (1998).

[62] *Approximation of measurable processes with continuous and with simple processes* (with Gabriela Licea), An. Univ. of Bucharest Mat. **52** (2003), 149–154.

[63] *Controlled convergence: boundary behaviour and the Dirichlet problem*, Rev. Roumaine Math. Pures Appl. **51** (2006), 597-606.

[64] *Compact excessive functions and Markov processes: a general case and applications* (with Lucian Beznea and Michael Röckner), RIMS Kôkyûroku Bessatsu **B6**, 31-37, Kyoto (2008).

[65] *Potential theory on Wiener space revisited and the corresponding study for infinite dimensional Levy processes* (with Lucian Beznea and Michael Röckner). To appear.

Books

[B1] *Potential theory on Riemannian surfaces* (Romanian), (with Nicu Boboc and Corneliu Constantinescu), 160 pp., Seminar S. Stoilow, 1959-1960, Editura Academiei R. P. Romîne, Bucharest, 1962.

[B2] *Ideale Ränder Riemannscher Flächen* (German) (with Corneliu Constantinescu), Ergebnisse der Mathematik und ihrer Grenzgebiete, **32** Springer-Verlag, Berlin-Göttingen-Heidelberg, 1963.

[B3] *Potential theory on harmonic spaces* (with Corneliu Constantinescu), with a preface by H. Bauer. Die Grundlehren der mathematischen Wissenschaften, **158**. Springer-Verlag, New York-Heidelberg, 1972.

[B4] *Order and potential resolvent families of kernels* (with Gabriela Licea), Lecture Notes in Mathematics, **494**. Springer-Verlag, Berlin-New York, 1975.

[B5] *Order and convexity in potential theory: H-cones* (with Nicu Boboc and Gheorghe Bucur), Lecture Notes in Mathematics, **853**. Springer, Berlin, 1981.

Quoted articles

[O1] Constantinescu, C.: *Aurel Cornea (1933–2005)*, Revue Roumaine de Mathématiques Pures et Appliquées **LI**, 2006, 533 - 539.

[O2] Beznea, L., Boboc, N., Bucur, Gh.: *Aurel Cornea, the mathematician*, Revue Roumaine de Mathématiques Pures et Appliquées **LI**, 541 - 547.

Mathematical Institute of Charles University, Sokolovská 83, 186 75 Praha 8 - Karlín 9, Czech Republic

E-mail address: `jvesely@karlin.mff.cuni.cz`

Potential Theory and Stochastics in Albac, 1–10
Aurel Cornea Memorial Volume

SUBELLIPTIC LI-YAU ESTIMATES ON THREE DIMENSIONAL MODEL SPACES

DOMINIQUE BAKRY, FABRICE BAUDOIN, MICHEL BONNEFONT, AND BIN QIAN

ABSTRACT. We describe three elementary models in three dimensional subelliptic geometry which correspond to the three models of the Riemannian geometry (spheres, Euclidean spaces and Hyperbolic spaces) which are respectively the $SU(2)$, Heisenberg and $SL(2)$ groups. On those models, we prove parabolic Li-Yau inequalities on positive solutions of the heat equation. We use for that the Γ_2 techniques that we adapt to those elementary model spaces. The important feature developed here is that although the usual notion of Ricci curvature is meaningless (or more precisely leads to bounds of the form $-\infty$ for the Ricci curvature), we describe a parameter ρ which plays the same rôle as the lower bound on the Ricci curvature, and from which one deduces the same kind of results as one does in Riemannian geometry, like heat kernel upper bounds, Sobolev inequalities and diameter estimates.

1. FRAMEWORK AND INTRODUCTION

The estimation of heat kernel measures is a topic which had been under thorough investigation for the last thirty years at least, see [12, 8]. Among the many techniques developed for that, the famous Li-Yau parabolic inequality [12] is a very powerful tool, which relies in Riemannian geometry bounds on the gradient on heat kernels to lower bounds on the Ricci curvature. More precisely, in the simplest form, it asserts that, if E is a smooth Riemannian manifold with dimension n and non negative Ricci curvature, then if f is any positive solution of the heat equation

$$\partial_t f = \Delta f,$$

where Δ is the Laplace Beltrami operator of E, then, if $u = \log f$

$$\partial_t u \geq |\nabla u|^2 - \frac{n}{2t}.$$

This is a very precise and powerful estimate. For the model case, which is here the Euclidean space $E = \mathbb{R}^n$ and when f is the heat kernel (that is the solution of the heat equation starting at time $t = 0$ from a Dirac mass), then this inequality is in fact an equality.

From this inequality, one may easily deduce Harnack inequalities and hence precise bounds on the heat kernel.

Many generalizations of this inequality have been developed, all of them including lower bounds on the Ricci tensor. In particular, it works for a general elliptic operator L under the assumption that it satisfies a curvature-dimension inequality $CD(\rho, n)$,

---

2000 *Mathematics Subject Classification*. 35H20, 58J35.

Key words and phrases. Li-Yau inequality, subellipticity, gradient bounds, heat kernel, P_2-techniques.

which is the furthermost generalization on the notion of lower bound on the Ricci curvature, see [6, 4].

In the non elliptic case, things appear to be infinitely more complicated. In particular, most of the hypoelliptic systems do not satisfy any $CD(\rho,n)$ inequality (any reasonable notion of lower bound on the Ricci tensor leads to the value $-\infty$). Nevertheless, some Li-Yau inequalities may be obtained, see [9].

In what follows, we shall use the Γ_2 techniques developed in [4] to produce these Li-Yau bounds. The method developed here works quite well on the simple models developed here (Heisenberg groups, $SU(2)$, $SL(2)$), but could be easily generalized to a larger class of hypoelliptic operators. We shall not try to present here the most general results, but concentrate for simplicity on the three model cases mentioned above. In fact, they should be thought of as the analogous of the model spaces of Riemmanian geometry (Euclidean spaces, Spheres and Hyperbolic spaces).

In all what follows, given an elliptic second order operator L on a smooth manifold, with no constant term, one defines

$$\Gamma(f,g)=\frac{1}{2}(L(fg)-fLg-gLf)$$

which stands for $\nabla f\cdot\nabla g$ in the Riemannian case, and the curvature dimension inequality is defined from the operator Γ_2

$$\Gamma_2(f,f)=\frac{1}{2}(\Gamma(f,f)-2\Gamma(f,Lf).$$

Then, L is said to satisfy a $CD(\rho,n)$ inequality if, for any smooth function f, one has

$$\Gamma_2(f,f)\geq\rho\Gamma(f,f)+\frac{1}{n}(Lf)^2.$$

The parabolic Li-Yau inequality is then described in terms of the quantity $|\nabla f|^2=\Gamma(f,f)$ and the parameters ρ and n. For the Laplace Beltrami operator $L=\Delta$ on a smooth Riemannian manifold, this amounts to say that the dimension is at most n and that the Ricci curvature is bounded below by ρ. In the hypoelliptic models that we describe below, however, no such inequality holds (the best possible constant ρ is $-\infty$), but we shall produce some analogous of the Li-Yau inequality through a parameter ρ which therefore plays the rôle of a substitute for the Ricci curvature.

In what follows we consider a three-dimensional Lie group $\mathbf{G}$ with Lie algebra $\mathfrak{g}$ and we assume that there is a basis $\{X,Y,Z\}$ of $\mathfrak{g}$ such that

$$[X,Y]=Z$$
$$[X,Z]=-\rho Y$$
$$[Y,Z]=\rho X$$

where $\rho\in\mathbb{R}$.

Example 1.1 ($\mathbf{SU}(2)$, $\rho=1$). *The Lie group* $\mathbf{SU}(2)$ *is the group of* 2×2, *complex, unitary matrices of determinant* 1. *Its Lie algebra* $\mathfrak{su}(2)$ *consists of* 2×2, *complex, skew-adjoint matrices of trace* 0. *A basis of* $\mathfrak{su}(2)$ *is formed by the Pauli matrices:*

$$X=\frac{1}{2}\begin{pmatrix}0&1\\-1&0\end{pmatrix},\ Y=\frac{1}{2}\begin{pmatrix}0&i\\i&0\end{pmatrix},Z=\frac{1}{2}\begin{pmatrix}i&0\\0&-i\end{pmatrix},$$

for which the following relationships hold

$$[X,Y]=Z,\quad [X,Z]=-Y,\quad [Y,Z]=X.$$

Example 1.2 (Heisenberg group, $\rho=0$). *The Heisenberg group* $\mathbb{H}$ *is the group of* 3×3 *matrices:*

$$\begin{pmatrix} 1 & x & z \\ 0 & 1 & y \\ 0 & 0 & 1 \end{pmatrix},\quad x,y,z\in\mathbb{R}.$$

The Lie algebra of $\mathbb{H}$ *is spanned by the matrices*

$$X=\begin{pmatrix} 0 & 1 & 0 \\ 0 & 0 & 0 \\ 0 & 0 & 0 \end{pmatrix},\ Y=\begin{pmatrix} 0 & 0 & 0 \\ 0 & 0 & 1 \\ 0 & 0 & 0 \end{pmatrix}\ \textit{and}\ Z=\begin{pmatrix} 0 & 0 & 1 \\ 0 & 0 & 0 \\ 0 & 0 & 0 \end{pmatrix},$$

for which the following equalities hold

$$[X,Y]=Z,\ [X,Z]=[Y,Z]=0.$$

Example 1.3 ($\mathbf{SL}(2)$, $\rho=-1$). *The Lie group* $\mathbf{SL}(2)$ *is the group of* 2×2*, real matrices of determinant* 1*. Its Lie algebra* $\mathfrak{sl}(2)$ *consists of* 2×2 *matrices of trace* 0*. A basis of* $\mathfrak{sl}(2)$ *is formed by the matrices:*

$$X=\frac{1}{2}\begin{pmatrix} 1 & 0 \\ 0 & -1 \end{pmatrix},\ Y=\frac{1}{2}\begin{pmatrix} 0 & 1 \\ 1 & 0 \end{pmatrix},\ Z=\frac{1}{2}\begin{pmatrix} 0 & 1 \\ -1 & 0 \end{pmatrix},$$

for which the following relationships hold

$$[X,Y]=Z,\quad [X,Z]=Y,\quad [Y,Z]=-X.$$

We consider on the Lie group $\mathbf{G}$ the subelliptic, left-invariant, second order differential operator

$$L=X^2+Y^2,$$

as well as the heat semigroup

$$P_t=e^{tL}.$$

We also set

$$\Gamma(f,f)=\frac{1}{2}(Lf^2-2fLf)=(Xf)^2+(Yf)^2,$$

and

$$\Gamma_2=\frac{1}{2}(L\Gamma(f,f)-2\Gamma(f,Lf)).$$

In the present setting,

$$\begin{aligned}\Gamma_2(f,f)=&(X^2f)^2+(Y^2f)^2+\frac{1}{2}\left((XY+YX)f\right)^2\\ &+\frac{1}{2}(Zf)^2+\rho\Gamma(f,f)-2(Xf)(YZf)+2(Yf)(XZf).\end{aligned}\tag{1.1}$$

The mixed term $-2(Xf)(YZf)+2(Yf)(XZf)$ prevents to find any lower bound on this quantity involving $\Gamma(f,f)$ and $(Lf)^2$ only, whence the absence of any $CD(\rho,n)$ inequality.

2. Li-Yau type estimates for the heat semigroup

The classical method of Li and Yau [12] consists in applying the maximum principle to a carefully chosen expression. The method developed in [4] is quite different. Considering a positive solution of the heat equation $\partial_t f = Lf$, and denoting $f \mapsto P_t f$ the associated heat kernel, one writes $u = \log f$ and look at the expression

$$\Phi(s) = P_s(f(t-s)\Gamma(u(t-s), u(t-s))),$$

defined for $0 < s < t$. Then, one obtains through the $CD(\rho, n)$ inequality a differential inequality

$$\Phi'(s) \geq (A\Phi(s) + B)^2 + C,$$

where A, B, C are expressions which are constant in t but may depend on the function f. Then, the parabolic Li-Yau inequality is obtained as a consequence of this differential inequality.

Here, we shall develop this method a bit further, looking at more complicated quantities like

$$P_s(f(t-s)(a(s)\Gamma(u(t-s), u(t-s)) + b(s)(Zu(t-s))^2)),$$

and try to get some differential inequality on it. The computations developed here are not restricted to Lie group, since we only use an generalized $CD(\rho, n)$ inequality. There are many hypoelliptic systems that may be treated under the same lines. The reason why we restrict ourselves to those model cases described previously are mainly for pedagogical reasons.

We have the following inequality, which is our technical starting point:

Proposition 2.1. *Let $f : \mathbf{G} \to \mathbb{R}$ be positive. Let $t > 0$, for all $x \in \mathbf{G}$ and $s \in [0, t]$, consider the expressions*

$$\Phi_1(s) = P_s((P_{t-s}f)\Gamma(\ln P_{t-s}f))(x)$$

and

$$\Phi_2(s) = P_s((P_{t-s}f)(Z\ln P_{t-s}f)^2)(x).$$

Then, for every differentiable, non-negative and decreasing function $b : [0, t] \to \mathbb{R}$,

$$\left(-b'\Phi_1 + b\Phi_2\right)'(s) \geq -b'(s)\left(\left(\frac{b''(s)}{b'(s)} + 2\frac{b'(s)}{b(s)} + 2\rho\right) LP_t f(x) - \frac{1}{4}\left(\frac{b''(s)}{b'(s)} + 2\frac{b'(s)}{b(s)} + 2\rho\right)^2 P_t f(x)\right).$$

Proof. We fix a positive function f, $t > 0$ and we perform all the following computations at a given point x.

With the same notations as Proposition 2.1, straightforward (but quite tedious) computations show that

$$\Phi_1'(t) = 2P_s((P_{t-s}f)\Gamma_2(\ln P_{t-s}f))$$

and

$$\Phi_2'(s) = 2P_s((P_{t-s}f)\Gamma(Z\ln P_{t-s}f)).$$

For the last equality we use the crucial facts that

$$[L, Z] = 0$$

and

$$X(f)Z(f)[X,Z](f)+Y(f)Z(f)[Y,Z](f)=0.$$

Now, thanks to the Cauchy-Schwarz inequality, the expression (1.1), shows that for every $\lambda>0$, and every smooth function g,

$$\Gamma_2(g)\geq\frac{1}{2}(Lg)^2+\frac{1}{2}(Zg)^2+\left(\rho-\frac{1}{\lambda}\right)\Gamma(g)-\lambda\Gamma(Zg).$$

We therefore obtain the following differential inequality

$$\Phi_1'(s)\geq P_s((P_{t-s}f)(L\ln P_{t-s}f)^2)+\Phi_2(s)+\left(2\rho-\frac{2}{\lambda}\right)\Phi_1(s)-\lambda\Phi_2'(s).$$

We now have that for every $\gamma\in\mathbb{R}$,

$$(L\ln P_{t-s}f)^2\geq 2\gamma L\ln P_{t-s}f-\gamma^2,$$

and

$$L\ln P_{t-s}f=\frac{LP_{t-s}f}{P_{t-s}f}-\frac{\Gamma(P_{t-s}f)}{(P_{t-s}f)^2}.$$

Thus, for every $\lambda>0$ and every $\gamma\in\mathbb{R}$,

$$\Phi_1'(s)\geq\left(2\rho-\frac{2}{\lambda}-2\gamma\right)\Phi_1(s)+\Phi_2(s)-\lambda\Phi_2'(s)+2\gamma LP_tf-\gamma^2P_tf.$$

Now for two functions a and b defined on the time interval $[0,t)$ with a positive, we have

$$(a\Phi_1+b\Phi_2)'\geq\left(a'+(2\rho-\frac{2}{\lambda}-2\gamma)a\right)\Phi_1+(a+b')\Phi_2+(-a\lambda+b)\Phi_2'+2a\gamma LP_tf-a\gamma^2P_tf.$$

So, if b is a positive decreasing function on the time interval $[0,t)$, by choosing in the previous inequality

$$a=-b',$$

$$\lambda=-\frac{b}{b'},$$

and

$$\gamma=\frac{1}{2}\left(\frac{b''}{b'}+2\frac{b'}{b}+2\rho\right),$$

we get the desired result. □

As a first corollary, by using the function

$$b(s)=(t-s)^{\alpha},\quad \alpha>2$$

and integrating from 0 to t, we deduce

Corollary 2.2. *For all $\alpha>2$, for every positive function f and $t>0$,*

$$\Gamma(\ln P_tf)+\frac{t}{\alpha}(Z\ln P_tf)^2\leq\left(\frac{3\alpha-1}{\alpha-1}-\frac{2\rho t}{\alpha}\right)\frac{LP_tf}{P_tf}+\frac{\rho^2t}{\alpha}-\frac{\rho(3\alpha-1)}{\alpha-1}+\frac{(3\alpha-1)^2}{\alpha-2}\frac{1}{t}.$$

Observe that this takes a simpler form when $\rho\geq0$, since then one can use Proposition 2.1 with $\rho=0$ and get

Corollary 2.3. *When $\rho \geq 0$, there exist positive constants A, B and C such that, with $u = \ln(P_t f)$*

$$\partial_t u \geq A\Gamma(u) + Bt(Zu)^2 - \frac{C}{t}.$$

In particular, one gets $\partial_t u \geq -C/t$, which gives

$$P_t f \leq t^{-C} P_1 f.$$

On the Heisenberg group, one sees that the behavior of $P_t f$ when t goes to 0 is of order t^{-2} (a simple dilation argument shows that). Therefore, one sees that the optimal constant C in the previous inequality is $C = 2$. Unfortunately, it can be shown by some elementary considerations similar to those developed in the proof of corollary 2.5 that the best constant one may obtain from the previous proposition shall always produce a constant $C > 2$. This is a strong difference with the classical parabolic Li-Yau inequality where the inequality

$$\partial_t u \geq -\frac{n}{2t}$$

gives the right order of magnitude of the heat kernel near $t = 0$.

Now when $\rho > 0$, we easily get an exponential decay by using the function:

$$b(s) = \left(e^{-\frac{2\rho s}{3\alpha}} - e^{-\frac{2\rho t}{3\alpha}}\right)^{\alpha}, \quad \alpha > 2.$$

This writes:

Corollary 2.4. *For every $\alpha > 2$, for every positive function f, $x \in \mathbf{G}$ and $t > 0$,*

$$\Gamma(\ln P_t f)(x) + \frac{3}{2}\frac{1 - e^{-\frac{2\rho t}{3\alpha}}}{\rho}(Z \ln P_t f)^2(x) \leq \frac{3\alpha - 1}{\alpha - 1} e^{-\frac{2\rho t}{3\alpha}} \frac{LP_t f(x)}{P_t f(x)} + \frac{3}{2}\rho \frac{\left(1 - \frac{1}{3\alpha}\right)^2}{1 - \frac{2}{\alpha}} \frac{e^{-\frac{4\rho t}{3\alpha}}}{1 - e^{-\frac{2\rho t}{3\alpha}}}.$$

Moreover for $\rho > 0$ and t large, with more work we actually can do better.

Corollary 2.5. *Let us assume $\rho > 0$. There exist $t_0 > 0$ and $C > 0$, such that for any positive function f,*

$$|\partial_t \ln P_t f(x)| \leq C \exp\left(-\frac{\rho t}{3}\right), \quad t \geq t_0, x \in \mathbf{G}.$$

Proof. To prove this result we have to be more precise in the study of the differential inequality of Proposition 2.1. Start with this inequality and set $V(b) = -b^2 b'$ for b a positive decreasing function such that $b(t) = b'(t) = 0$. The constraints that the non negative function V on $[0, b_0]$ must satisfy are

$$t = \int_0^{b_0} \frac{x^2}{V(x)} dx$$

and

$$\left(\frac{V(x)}{x^2}\right)_{x=0} = 0.$$

We then get with $u_t = \ln P_t f$ and $a_0 = \frac{V(b_0)}{b_0^2}$

$$a_0 \Gamma(u_t) + b_0 (Z u_t)^2 \leq A \partial_t u_t + B,$$

where for any choice of such a function V, one has

$$A=\int_0^{b_0}\left(\frac{V'}{x^2}-2\rho t\right)dx,\ B=\frac{1}{4}\int_0^{b_0}\left(\frac{V'}{x^2}-2\rho t\right)^2 dx.$$

In this system, we see that changing $V(s)$ into $\frac{V(\lambda s)}{\lambda^3}$ and b_0 into $\frac{b_0}{\lambda}$ leaves t unchanged and multiply every constant a_0, A and B by $\frac{1}{\lambda}$. Therefore, we may assume that $b_0=1$ without any loss. Also, changing $V(s)$ into $cV(s)$ allows us to reduce to the case $t=1$. So finally we have rephrased the problem as follows. For any non negative function V on $[0,1]$ such that

$$\int_0^1\frac{x^2}{V}dx=1,\ \left(\frac{V(x)}{x^2}\right)_{x=0}=0,$$

and for any $u=\log P_t f$ with $f\geq 0$ one has

$$V(1)\Gamma(u)+t(Zu)^2\leq(\alpha(V)-2\rho t)\partial_t u+\frac{1}{4t}\left(\beta(V)-\alpha^2(V)+(\alpha(V)-2\rho t)^2\right),$$

where

$$\alpha(V)=\int_0^1\frac{V'}{x^2}dx,\ \beta(V)=\int_0^1\left(\frac{V'}{x^2}\right)^2dx.$$

The preceding calculus is valid for any ρ. An easy integration by parts shows us the term $\alpha(V)$ is non negative whatever V is. But now for $\rho>0$, observe that this time the term $\alpha(V)-2\rho t$ can be made negative, and therefore we may get as in the elliptic case with strictly positive Ricci bound a universal upper bound on $|\partial_t u|$.

One has the obvious inequalities

$$\alpha(V)>V(1)-\left(\frac{V(x)}{x^2}\right)_{x=0}+8,\ \beta(V)>\alpha(V)^2,$$

and in the previous, no equality may occur (in the first one because then $\beta=\infty$ and in the second one because of the constraint on V.) The first inequality comes from

$$\int_0^1\frac{V'}{x^2}dx=V(1)+2\int_0^1\frac{V}{x^3}dx,$$

and

$$\int_0^1\frac{V}{x^3}dx\int_0^1\frac{x^2}{V}dx\geq\left(\int_0^1\frac{dx}{\sqrt{x}}\right)^2=4.$$

To make the term $\beta(V)-\alpha^2(V)$ small we are led to choose $V=\lambda x^3$ on $[\epsilon,1]$ and $V=\lambda\epsilon^{3-\gamma}x^\gamma$ on $[0,\epsilon]$, for some fixed $\gamma\in(5/2,3)$. The constraint on V implies

$$\lambda=-\log\epsilon+\frac{1}{3-\gamma}.$$

Meanwhile, we have

$$\alpha=\lambda\left(3+2\epsilon\frac{3-\gamma}{\gamma-2}\right),$$

and

$$\beta=\lambda^2\left(9+\epsilon\frac{(15-\gamma)(3-\gamma)}{2\gamma-5}\right),$$

so that

$$\beta-\alpha^2=\lambda^2\epsilon\frac{(3-\gamma)^2}{\gamma-2}\left(\frac{\gamma+10}{2\gamma-5}+\epsilon\frac{4}{\gamma-2}\right).$$

By taking

$$\epsilon=\exp\left(-\frac{2\rho}{3}t+\frac{1}{3-\gamma}+R\right)$$

for t large enough to ensure $\varepsilon<1$, one obtains

$$\alpha-2\rho t\simeq -3R$$

and

$$\beta-\alpha^2\simeq Ct^2\varepsilon\simeq Ct^2\exp\left(-\frac{2\rho t}{3}\right).$$

With $R=ct\exp(-\frac{\rho}{3})$ the terms $(\alpha-2\rho t)^2$ and $\beta-\alpha^2$ are of the same order and playing now with the sign of c, one gets

$$|\partial_t u|\leq C\exp\left(-\frac{\rho t}{3}\right).$$

□

Interestingly, only from these estimates, we can deduce that for $\rho>0$ the Lie group $\mathbf{G}$ has to be compact. (This is of course not new since the Lie algebra is that of a compact semi-simple Lie group). But we also get an upper bound on the diameter similar to the classical upper bound of the Myers' theorem, together with some precise information on the Sobolev constants and the spectral gap. Those considerations in fact show that this parameter ρ may serve as a substitute for the Ricci lower bound on a Riemannian manifold. We proceed first by showing that in that case there is a spectral gap.

Proposition 2.6. *Let us assume* $\rho>0$. *The spectrum of* $-L$ *lies in* $\{0\}\cup[\frac{\rho}{3},+\infty]$.

Proof. We fix $x\in\mathbf{G}$ and denote by $p_t(x,\cdot)$ the heat kernel starting from x. We have for $t\geq t_0$,

$$|\,\partial_t\ln p_t(x,y)\,|\leq C\exp\left(-\frac{\rho t}{3}\right). \tag{2.2}$$

This shows that $\ln p_t$ converges when $t\to\infty$. Let us call $\ln p_\infty$ this limit. Moreover, from Corollary 2.4, $\Gamma(\ln p_t)$ is bounded above by a constant $C(t)$ which goes to 0 when t goes to ∞. Since the oscillation between $\ln p_t(x,y_1)$ and $\ln p_t(x,y_2)$ is bounded above by $\sqrt{C(t)}d(y_1,y_2)$, for the associated Carnot-Carathéodory distance, which may be defined (see [2]) as

$$d(x,y)=\sup_{\{f,\Gamma(f,f)\leq 1\}}f(x)-f(y), \tag{2.3}$$

such that if $\Gamma(f,f)\leq C$, then $f(x)-f(y)\leq\sqrt{C}d(x,y)$.

In the limit, $\ln p_\infty(x,\cdot)$ is a constant. We deduce from this that the invariant measure μ is finite. We may then as well suppose that this measure is a probability, in which case $p_\infty=1$. By integrating the inequality (2.2) from t to ∞ we therefore obtain for $t\geq t_0$:

$$|\ln p_t(x,y)\,|\leq C_2\exp\left(-\frac{\rho t}{3}\right)$$

and thus

$$\exp\left(-C_2\exp\left(-\frac{\rho t}{3}\right)\right)\le p_t(x,y)\le \exp\left(C_2\exp\left(-\frac{\rho t}{3}\right)\right),$$

and therefore

$$|p_t(x,y)-1|\le C_3\exp\left(-\frac{\rho t}{3}\right)$$

for t large enough. This implies by the Cauchy-Schwarz inequality that for $f \in L^2(\mu)$ such that $\int f d\mu = 0$,

$$(P_t f)^2 \le C_3\exp\left(-\frac{2\rho t}{3}\right)\int f^2 d\mu.$$

For a symmetric Markov semigroup P_t, this is a standard fact (see [2] for example) that this is equivalent to say that the spectrum of $-L$ lies in $\{0\}\cup[\rho/3,\infty)$, or equivalently that we have a spectral gap inequality: for any function f in L^2 such that ∇f is in L^2, one has

$$\int f^2 d\mu \le \left(\int f d\mu\right)^2 + \frac{3}{\rho}\int |\nabla f|^2 d\mu. \tag{2.4}$$

□

Remark 2.7. *It can be shown that the spectral gap is actually $\frac{\rho}{2}$ and not $\frac{\rho}{3}$.*

We can now conclude with a substitute of the Myers' theorem:

Proposition 2.8. *Assume that $\rho > 0$, then the diameter of L for the Carnot-Carathéodory distance is finite.*

Proof. We are now going to prove a Sobolev inequality for the invariant measure μ. Indeed, for $0 < t \le t_0$ we have

$$\partial_t \ln p_t \ge -C/t,$$

from which we get

$$\ln p_{t_0} - \ln p_t \ge -C\log(t_0/t),$$

and therefore

$$\ln p_t \le A - C\log t$$

where A is a constant. This gives the ultracontractivity of the semigroup P_t with a polynomial bound t^{-C} when $t\to 0$.

Now it is a well known fact (see [13, 2]) that this last property is equivalent to a Sobolev inequality

$$\left(\int f^{\frac{2C}{C-1}} d\mu\right)^{\frac{C-1}{C}} \le A\int f^2 d\mu + B\int \|\nabla f\|^2 d\mu. \tag{2.5}$$

When we have both Sobolev inequality (2.5) and spectral gap inequality (2.4) then (see [2]) we have a tight Sobolev inequality, that is the Sobolev inequality (2.5) with $A=1$.

In this situation, the diameter of G with respect to the distance defined in (2.3) is finite (see [5]), which concludes the proof.

□

Acknowledgments

The last author would like to express sincere thanks to China Scholarship council for financial support.

References

[1] D. Bakry, On Sobolev and logarithmic Sobolev inequalities for Markov semigroups. Taniguchi symposium. In: K. D. Elworthy, Shigeo Kusuoka, and Ichiro Shigekawa (Eds.), *New trends in stochastic analysis*, pp. 43-75. World Sci. Publ., River Edge, NJ, 1997.

[2] D. Bakry, L'hypercontractivité et son utilisation en théorie des semigroupes. *Lectures on probability theory de Saint Flour - 1992* (Lecture Notes in Math. **1581**), 1-114. Springer, Berlin, 1994.

[3] D. Bakry, F. Baudoin, M. Bonnefont and D. Chafai, *On gradient bounds for the heat kernel on the Heisenberg group*. Arxiv preprint 0710.3139, 2007.

[4] D. Bakry and M. Ledoux, A logarithmic Sobolev form of the Li-Yau parabolic inequality. *Revista Mat. Iberoamericana* **22** (2006), 683-702.

[5] D. Bakry and M. Ledoux, Myers' theorem and Sobolev inequalities. *Duke Math. J.* **85** (1996), 253-270.

[6] D. Bakry and Z. Qian, Harnack inequalities on a manifold with positive or negative Ricci curvature. *Rev. Mat. Iberoamericana* **15** (1999), 143-179.

[7] F. Baudoin and M. Bonnefont, *The subelliptic heat kernel on SU(2): Representations, Asymptotics and Gradient bounds*. Arxiv preprint, 2007.

[8] E.B. Davies, *Heat kernels and spectral theory* (Cambridge tracts in Math.). Cambridge Univ. Press **92**, 1989.

[9] H.-D. Cao and S.-T. Yau, Gradient estimates, Harnack inequalities and estimates for heat kernels of the sum of squares of vector fields. *Math. Z.* **121** (1992), 485-504.

[10] M. Ledoux, The geometry of Markov diffusion generators. Probability theory. *Ann. Fac. Sci. Toulouse Math.* **9** (2000), no. 2, 305-366.

[11] H.-Q. Li, Estimation optimale du gradient du semi-groupe de la chaleur sur le groupe de Heisenberg. *J. Funct. Anal.* **236** (2006), no. 2, 369-394.

[12] P. Li, S. T. Yau, On the parabolic kernel of the Schrödinger operator. *Acta. Math.* **56** (1986), 153-201.

[13] N. Varopoulos, Hardy-Littlewood theory for semigroups. *J. Funct. Anal.* **52** (1985), 240-260.

Institut de Mathématiques, Université de Toulouse et CNRS (UMR 5219), 118 route de Narbonne, 31062 Toulouse, FRANCE; and (D. Bakry)Institut Universitaire de France

E-mail address: bakry@math.univ-toulouse.fr, fbaudoin@math.univ-toulouse.fr, michel.bonnefont@math.univ-toulouse.fr, binqiancn@yahoo.com.cn

Potential Theory and Stochastics in Albac, 11–19
Aurel Cornea Memorial Volume

SELF-ORGANIZED CRITICALITY VIA STOCHASTIC PARTIAL DIFFERENTIAL EQUATIONS

VIOREL BARBU, PHILIPPE BLANCHARD, GIUSEPPE DA PRATO, AND MICHAEL RÖCKNER

ABSTRACT. Models of self-organized criticality which can be described as singular diffusions with or without (multiplicative) Wiener forcing term (as e.g. the Bak/Tang/Wiesenfeld- and Zhang-models) are analyzed. Existence and uniqueness of nonnegative strong solutions are proved. Previously numerically predicted transition to the critical state in 1-D is confirmed by a rigorous proof that this indeed happens in finite time with high probability.

1. INTRODUCTION

Within the past twenty years the notion of self-organized criticality (SOC) has become a new paradigm for the explanation of a huge variety of phenomena in nature and social sciences. Its origin lies in the attempt to explain the widespread appearance of power-law-statistics for characteristic events. In this paradigm an external perturbation may induce a chain reaction or avalanche in the system. Furthermore, a stationary state, the SOC-state, is reached where the average incoming flux is balanced by the average outgoing flux. This phenomenon was quite unexpected since attaining the critical state of a thermodynamic system usually requires a fine tuning of some control parameter, which is absent in the definition of the SOC models. Several models have been proposed to mimic this mechanism including the sand pile BTW-model [1, 2] and the Zhang-model [22]. The presence of thresholds in the definition of the dynamics implies that the energy can be accumulated locally, eventually generating a chain reaction which may transport energy on arbitrary large scales.

The literature on SOC is vast. We refer e.g. to [1, 2], [3], [7], [19], [21], [9, 10], [16], [18], [13], [8], [14], [17], [15], [22] for various studies. In [3] it was proposed to describe this phenomenon, e.g. in the case of the avalanche dynamics in the BTW- (see [1, 2]) and Zhang- (see [22]) models, by a singular diffusion. In the absense of noise the density $\varrho(t,\xi)$, $t \geq 0$, $\xi \in \mathbb{R}^d$, of this diffusion is formally described by the evolution equation

$$\frac{\partial}{\partial t}\varrho(t,\xi) = \Delta\Psi(\varrho(t,\xi)), \tag{1}$$

where $\Psi(\varrho) := f(\varrho)H(\varrho-\varrho_c)$, $\varrho_c \geq 0$, H is the Heaviside function and $f(\varrho) = \text{const.}$ in the BTW-model and $f(\varrho) = \varrho$ in the Zhang-model. Applying the chain rule informally,

2000 *Mathematics Subject Classification.* 60H15, 60H30.

Key words and phrases. Stochastic porous media equations, self-organized criticality, finite-time extinction.

the equation turns into

$$\begin{aligned}\frac{\partial}{\partial t}\varrho(t,\xi) &= \nabla\cdot[\Psi'(\varrho(t,\xi))\,\nabla\varrho(t,\xi)]\\ &= \Psi''(\varrho(t,\xi))\,|\nabla\varrho(t,\xi)|^2 + \Psi'(\varrho(t,\xi))\,\Delta\varrho(t,\xi),\end{aligned}\tag{2}$$

where

$$\Psi'(\varrho) = f'(\varrho)H(\varrho-\varrho_c) + f(\varrho)\delta(\varrho-\varrho_c).$$

To discuss the problem in a heuristic way, let us consider a smooth version of H, for example

$$H_\varepsilon(\varrho) = \frac{1}{2} + \frac{1}{\pi}\arctan(\tfrac{\varrho}{\varepsilon})$$

(or the mathematical more convenient one in (7) below). Since $H_\varepsilon(\varrho-\varrho_c)$ is convex left of ϱ_c and concave right of ϱ_c, we see that e.g. in the BTW-model $H_\varepsilon''(\varrho-\varrho_c)|\nabla\varrho|^2$ is positive if ϱ is below ϱ_c and negative if ϱ is above ϱ_c, i.e. according to (2), ϱ is "pushed" towards the critical value ϱ_c. This has been predicted numerically in 1-D by Bantay and Janosi in [3].

In [13] (see also [14]) Diaz–Guilera pointed out that it is more realistic to consider (1) perturbed by (an additive) noise to model a random amount of energy put into the system varying all over the underlying domain. So, (1) turns into a stochastic partial differential equation (SPDE). In [11] (see also the references therein) based on numerical tests, Carlson and Swindle observed that in the presence of such a noise the self-organized behaviour does not necesssarily occur, i.e. the system fails to converge to the critical value ϱ_c.

The purpose of this note is to provide rigorous mathematical proofs for the above phenomena. First, we sketch the proofs for existence and uniqueness of solutions to (1) perturbed by noise (more precisely for multiplicative noise, so that positivity of initial data is preserved). Second, we prove that at least in 1-D we have convergence to ϱ_c in finite time in the deterministic case (confirming the numerical results in [3]) and convergence to ϱ_c with high probability in the stochastic case. In regard to [11] one can probably not achieve more, but so far we failed to prove that this probability is really not equal to 1.

Let us introduce our framework, where we switch to common notation in SPDE, i.e. replace $\varrho(t,\xi)$ by $X(t,x)(\xi)$ with x being the density at $t=0$, $t\ge 0$, $\xi\in\mathcal{O}$, where $\mathcal{O}$ is an open bounded domain in $\mathbb{R}^d, d=1,2,3$, with smooth boundary $\partial\mathcal{O}$. The appropriate class of SPDE is then of the form

$$\begin{cases} dX(t) - \Delta\Psi(X(t))dt \ni \sigma(X(t))dW(t), & \text{in } (0,\infty)\times\mathcal{O},\\ \Psi(X(t)) \ni 0, \quad \text{on } (0,\infty)\times\partial\mathcal{O}, & \\ X(0,x) = x \quad \text{on } \mathcal{O}, & \end{cases}\tag{3}$$

where x is an initial datum, $\Psi:\mathbb{R}\to 2^{\mathbb{R}}$ a maximal monotone graph, i.e. Ψ is not strictly contained in another monotone graph, and

$$\sigma(X)dW = \sum_{k=1}^\infty \mu_k X d\beta_k\, e_k, \quad t\ge 0,$$

is a random forcing term, where $\{e_k\} \subset L^2(\mathscr{O})$ is the eigenbasis of the Laplacian $-\Delta$ on $\mathscr{O}$ with Dirichlet boundary conditions, μ_k are positive numbers and β_k independent standard Brownian motions on a filtered probability space $(\Omega, \mathscr{F}, \{\mathscr{F}_t\}_{t\geq 0}, \mathbb{P})$.

Throughout this note we make the following assumptions:

Hypothesis 1. (i) *$\Psi(r) = \rho$ sign $r + \widetilde{\Psi}(r)$, for $r \in \mathbb{R}$, where $0 \in \Psi(0)$, $\rho > 0$, $\widetilde{\Psi} : \mathbb{R} \to \mathbb{R}$ is Lipschitzian, $\widetilde{\Psi} \in C^1(\mathbb{R} \setminus \{0\})$ and for some $\delta > 0$ it satisfies $\widetilde{\Psi}'(r) \geq \delta$ for all $r \in \mathbb{R} \setminus \{0\}$.*

(ii) *The sequence $\{\mu_k\}$ is such that*

$$\sum_{k=1}^{\infty} \mu_k^2 \lambda_k^2 < +\infty,$$

where λ_k are the eigenvalues of $-\Delta$.

A typical example is given by

$$\Psi(r) = \psi_0(r) + c, \; r \in \mathbb{R},$$

where

$$\psi_0(r) := \begin{cases} \alpha_1 r, & r > 0 \\ [-\varrho, \varrho], & r = 0 \\ \alpha_2 r, & r < 0 \end{cases}$$

and $\alpha_1, \alpha_2 > 0$, $\varrho \geq 0$, $c \in [-\varrho, \varrho]$ are constants.

The following notations will be used. $L^p(\mathscr{O})$, $p \geq 1$, is the usual space of p-integrable functions on $\mathscr{O}$ with norm $|\cdot|_p$. The scalar product in $L^2(\mathscr{O})$ and the duality induced by the pivot space $L^2(\mathscr{O})$ will be denoted by $\langle \cdot, \cdot \rangle_2$. $H_0^1(\mathscr{O}) \subset L^2(\mathscr{O})$ is the first order Sobolev space on $\mathscr{O}$ with zero trace on the boundary. For a fixed measure space $(E, \mathscr{E}, m)$, a Banach space B and $p \in [1, \infty]$ we denote the space of all with respect to the measure m p-integrable maps from E to B by $L^p(E; B)$.

In the following by H we shall denote the distribution space

$$H = H^{-1}(\mathscr{O}) = (H_0^1(\mathscr{O}))'$$

endowed with the scalar product and norm defined by

$$\langle u, v \rangle = \int_{\mathscr{O}} (-\Delta)^{-1} u(\xi) v(\xi) d\xi, \quad |u|_{-1} = \langle u, u \rangle^{1/2}.$$

We recall that the operator $x \to -\Delta\Psi(x)$ with the domain

$$\{x \in L^1(\mathscr{O}) \cap H : \exists \eta \in H_0^1(\mathscr{O}) \text{ s.th. } \eta \in \Psi(x) \text{ a.e. on } \mathscr{O}\},$$

is essentially maximal monotone in H (see e.g. [4]) and so the distribution space H is the natural functional setting for equation (3). However, the general existence theory of infinite dimensional stochastic equations in Hilbert space with nonlinear maximal monotone operators (see e.g. [12], [20]) is not directly applicable and so a direct approach must be used.

2. Existence, uniqueness and positivity

Definition 2. *Let $x \in H$. An H-valued continuous $\mathscr{F}_t$-adapted process $X = X(t,x)$ is called a solution to* (3) *on* $[0,T]$ *if for some* $p \in [1,\infty[$

$$X \in L^p(\Omega \times (0,T) \times \mathscr{O}) \cap L^2(0,T;L^2(\Omega,H)),$$

and there exists $\eta \in L^p(\Omega \times (0,T) \times \mathscr{O})$ *such that* $\mathbb{P}$*-a.s.*

$$\begin{aligned} \langle X(t,x), e_j\rangle_2 &= \langle x, e_j\rangle_2 + \int_0^t \int_{\mathscr{O}} \eta(s,\xi)\Delta e_j(\xi)d\xi ds \\ &\quad + \sum_{k=1}^\infty \mu_k \int_0^t \langle X(s,x)e_k, e_j\rangle_2 d\beta_k(s), \\ &\qquad \forall\, j \in \mathbb{N},\ t \in [0,T], \end{aligned} \tag{4}$$

$$\eta \in \Psi(X) \quad a.e.\ in\ \Omega \times (0,T) \times \mathscr{O}. \tag{5}$$

Below for simplicity we often write $X(t)$ instead of $X(t,x)$.

Theorem 3. *Under Hypothesis* 1 *for each* $x \in L^4(\mathscr{O})$, *there is a unique solution* X *to* (3). *Moreover, if x is nonnegative a.e. in $\mathscr{O}$ then* $\mathbb{P}$*-a.s.*

$$X(t,x)(\xi) \geq 0, \quad for\ a.e.\ (t,\xi) \in (0,\infty) \times \mathscr{O}.$$

Sketch of proof. Consider the approximating equation

$$\begin{cases} dX_\lambda(t) - \Delta(\Psi_\lambda(X_\lambda(t))dt = \sigma(X_\lambda(t))dW(t), \\ X_\lambda(0,x) = x, \end{cases} \tag{6}$$

where $\lambda > 0$,

$$\Psi_\lambda(r) := \rho\,(\mathrm{sign})_\lambda(r) + \widetilde{\Psi}(r), \quad r \in \mathbb{R}, \tag{7}$$

$$(\mathrm{sign})_\lambda(r) := \begin{cases} 1 & \text{if } r > \lambda \\ \frac{r}{\lambda} & \text{if } r \in [-\lambda,\lambda] \\ -1 & \text{if } r < -\lambda. \end{cases}$$

By [6, Theorem 2.2] (applied with $m = 1$) equation (6) has a unique solution

$$X_\lambda \in L^2(\Omega \times (0,T) \times \mathscr{O}) \cap L^2(\Omega, C([0,T];H))$$

in the sense of Definition 2 which is nonnegative, if so is x. Here as usual the space of continuous H-valued paths $C([0,T];H)$ is equipped with the supremum norm.

By Itô's formula for $\alpha > 0$ large enough it follows that for all $\lambda, \mu \in (0,1)$ and $t \in [0,T]$

$$\begin{aligned} &\tfrac{1}{2}\,|X_\lambda(t) - X_\mu(t))|_{-1}^2 e^{-\alpha t} \\ &\leq C \max\{\lambda,\mu\} \int_0^t \int_{\mathscr{O}} \Big(|\Psi_\lambda(X_\lambda(s))|^2 + |X_\lambda(s)|^2 \\ &+ |\Psi_\mu(X_\mu(s))|^2 + |X_\mu(s)|^2 \Big) e^{-\alpha s} d\xi\, ds \\ &+ \int_0^t e^{-\alpha s} \langle X_\lambda(s) - X_\mu(s), \sigma(X_\lambda(s) - X_\mu(s))dW(s)\rangle, \end{aligned} \tag{8}$$

(where $|\cdot|$ denotes the absolute value in $\mathbb{R}$). Hence by the Burkholder-Davis-Gundy inequality $\{X_\lambda\}$ is a Cauchy net in $L^2(\Omega; C([0,T],H))$ and by a standard technique from stochatic partial differential equations one shows that the limit X is the desired solution to (3) (cf. [5]). □

Remark 4. One can also show (see [5, Prop. 3.4]) that $X, X_\lambda \in L^2(0,T;L^2(\Omega, H_0^1(\mathcal{O})))$, that

$$\lim_{\lambda\to 0} \mathbb{E}\int_0^T |X_\lambda - X|^2_{L^2(\mathcal{O})}\,dt = 0,$$

where $\mathbb{E}$ denotes expectation with respect to $\mathbb{P}$, and that both X and X_λ have continuous paths in $L^2(\mathcal{O})$. Theorem 3 is true for more general not linear growing Ψ (see [5]).

3. Extinction in finite time and self-organized criticality

In this section we assume $N<\infty$. Let τ be the stopping time

$$\tau = \inf\{t\ge 0 : |X(t,x)|_{-1} = 0\},$$

where $X(t,x)$, $t\ge 0$, is the solution from Theorem 3.

Proposition 5. *Assume there exists $N\in\mathbb{N}$ such that $\mu_k = 0 \quad \forall k\ge N+1$. Then*

$$X(t,x) = 0 \quad \textit{for } t\ge\tau,\ \mathbb{P}\textit{-a.s.}$$

Sketch of proof. For simplicity we consider the case with $\varrho\equiv 1$. Define

$$\mu(t) := -\sum_{k=1}^N \mu_k e_k \beta_k(t), \quad t\in[0,T], \quad \tilde{\mu} := \sum_{k=1}^N \mu_k^2 e_k^2$$

and

$$Y(t) := e^{\mu(t)} X(t), \quad t\ge 0.$$

Then by Itô's product rule Y satisfies $\mathbb{P}$-a.s. the following ordinary PDE

$$\frac{dY(t)}{dt} = e^{\mu(t)}\Delta\eta(t) - \frac{1}{2}\tilde{\mu}Y(t), \quad t\ge 0,$$

with $\eta(t)\in\Psi(X(t))$. Setting $Y_\lambda := e^{\mu}X_\lambda$ we consider the approximating equation

$$\frac{dY_\lambda(t)}{dt} = e^{\mu(t)}\Delta\eta_\lambda(t) - \frac{1}{2}\,\tilde{\mu}Y_\lambda(t), \quad t\ge 0, \tag{9}$$

where

$$\eta_\lambda(t) = \Psi_\lambda(X_\lambda(t)) \in H_0^1(\mathcal{O}). \tag{10}$$

Hence

$$\begin{aligned} &\left\langle \frac{dY_\lambda(t)}{dt}, Y_\lambda(t)\right\rangle_2 \\ &= \langle \eta_\lambda(t), \Delta(e^{\mu(t)}Y_\lambda(t))\rangle_2 - \frac{1}{2}\langle\tilde{\mu}Y_\lambda(t), Y_\lambda(t)\rangle_2, \end{aligned} \tag{11}$$

where by (10), (7) and integrating by parts we have

$$\begin{aligned} &\langle \eta_\lambda(t), \Delta(e^{\mu(t)}Y_\lambda(t))\rangle_2 \\ &= -\frac{1}{\lambda}\int_{\{e^{-\mu(t)}|Y_\lambda(t)|<\lambda\}} (|\nabla Y_\lambda(t)|^2 \\ &\qquad - |Y_\lambda(t)|^2\,|\nabla\mu(t)|^2)\,d\xi \\ &\quad - \int_{\mathcal{O}} \widetilde{\Psi}'(e^{-\mu(t)}Y_\lambda(t)) \\ &\qquad (|\nabla Y_\lambda(t)|^2 - |Y_\lambda(t)|^2\,|\nabla\mu(t)|^2)\,d\xi. \end{aligned}$$

This yields

$$\langle \eta_\lambda(t), \Delta(e^{\mu(t)} Y_\lambda(t))\rangle_2 \le C\big(|Y_\lambda(t)|_2^2 + \lambda\big). \tag{12}$$

Hence (11) and Gronwall's lemma imply

$$|Y_\lambda(t)|_2^2 \le e^{C(t-s)}\big(|Y_\lambda(s)|_2^2 + C\lambda T\big), \quad t \ge s.$$

Now letting $\lambda \to 0$ we get

$$|Y(t)|_2^2 \le e^{C(t-s)}|Y(s)|_2^2, \quad t \ge s. \tag{13}$$

Taking in (13) $s = \tau$ we get $Y(t) = 0$ for all $t \ge \tau$ as claimed. □

For proving the extinction result we need $\mathscr{O} \subset \mathbb{R}$, i.e. $d = 1$. In this case we can easily prove the assertion of Proposition 5 by a supermartingale argument without having to assume that $\mu_k = 0\ \forall k \ge N+1$ for some $N \in \mathbb{N}$ (see the proof of Theorem 6 below). To be more specific let $\mathscr{O} = (0,\pi)$. Then $e_k(\xi) = \sqrt{\frac{2}{\pi}}\ \sin k\xi, \quad \xi \in [0,\pi],\ \lambda_k = k^2$ and $L^1(0,\pi) \subset H$ continuously, so

$$\gamma = \inf\left\{\frac{|x|_{L^1}}{|x|_{-1}} : x \in L^1(0,\pi)\right\} > 0. \tag{14}$$

Theorem 6. *Assume $0 \in \operatorname{int}\Psi(0)$ and that $x \in L^4(\mathscr{O})$. Consider the equation*

$$\begin{cases} dX(t) - \Delta(\rho\, \text{sign}\,(X(t)-x_c) + \widetilde{\Psi}(X(t)-x_c))dt \\ \qquad\qquad \ni \sigma(X(t)-x_c)\displaystyle\sum_{k=1}^{\infty}\mu_k e_k d\beta_k, \quad t \ge 0, \\ \rho\,\text{sign}\,(X(t)-x_c) + \widetilde{\Psi}(X(t)-x_c) \ni 0,\ on\,\partial(0,\pi), \\ X(0,x) = x. \end{cases} \tag{15}$$

where $x_c \in \mathbb{R}$.

Define $\tilde\rho := \rho - |\tilde\Psi(0)|$ (>0 since $0 \in \operatorname{int}\Psi(0)$) and assume that

$$|x - x_c|_{-1} < \tilde\rho\gamma C_\infty^{-1},$$

where $C_\infty := \frac{\pi}{4}\sum_{k=1}^{\infty}(1+k)^2\mu_k^2$ and γ is as in (14). Then for all $t > 0$

$$\mathbb{P}(\tau_c \le t) \ge 1 - \frac{|x - x_c|_{-1}}{\tilde\rho\gamma\int_0^t e^{-C_\infty s}ds}, \tag{16}$$

where

$$\begin{aligned}\tau_c &= \inf\{t \ge 0 : |X(t) - x_c|_{-1} = 0\} \\ &= \sup\{t \ge 0 : |X(t) - x_c|_{-1} > 0\}\end{aligned}$$

and $X - x_c$ is the solution from Theorem 3.

Sketch of proof. For simplicity we assume $x_c = 0$. An application of Itô's formula for $\varphi_\varepsilon(|X|_{-1}^2) = (|X|_{-1}^2 + \varepsilon^2)^{1/2}$ yields

$$
\begin{aligned}
&\varphi_\varepsilon(|X(t)|_{-1}^2)+\tilde\rho\int_r^t \frac{|X(s)|_{L^1((0,\pi))}}{(|X(s)|_{-1}^2+\varepsilon^2)^{1/2}}ds\\
\le&\varphi_\varepsilon(|X(r)|_{-1}^2)+C_\infty\int_r^t \frac{|X(s)|_{-1}^2}{(|X(s)|_{-1}^2+\varepsilon^2)^{1/2}}ds\\
&+2\int_r^t \langle\sigma(X(s))dW(s),\varphi_\varepsilon'(|X(t)|_{-1}^2)X(s)\rangle \quad \mathbb{P}\text{-a.s., } r<t.
\end{aligned}
$$

Hence by (14)

$$
\begin{aligned}
&\varphi_\varepsilon(|X(t)|_{-1}^2)+\gamma\tilde\rho\int_r^t \frac{|X(s)|_{-1}}{(|X(s)|_{-1}^2+\varepsilon^2)^{1/2}}ds\\
\le&\ \varphi_\varepsilon(|X(r)|_{-1}^2)+C_\infty\int_r^t \frac{|X(s)|_{-1}^2}{(|X(s)|_{-1}^2+\varepsilon^2)^{1/2}}ds\\
&+2\int_r^t \langle\sigma(X(s))dW(s),\varphi_\varepsilon'(|X(s)|_{-1}^2)X(s)\rangle. \quad \mathbb{P}\text{-a.s., } r<t.
\end{aligned}
$$

Now, letting ε tend to zero we get

$$
\begin{aligned}
&|X(t)|_{-1}+\gamma\tilde\rho\int_r^t 1_{\{|X(s)|_{-1}>0\}}ds\\
\le&|X(r)|_{-1}+C_\infty\int_r^t |X(s)|_{-1}ds\\
&+\int_r^t 1_{\{|X(s)|_{-1}>0\}}\langle\sigma(X(s))dW(s),X(s)|X(s)|_{-1}^{-1}\rangle \quad \mathbb{P}\text{-a.s., } r<t.
\end{aligned}
$$

Hence by Itô's product rule

$$
\begin{aligned}
&e^{-C_\infty t}|X(t)|_{-1}+\gamma\tilde\rho\int_r^t e^{-C_\infty s}1_{\{|X(s)|_{-1}>0\}}ds\\
\le&e^{-C_\infty r}|X(r)|_{-1} \qquad (17)\\
&+\int_r^t e^{-C_\infty s}1_{\{|X(s)|_{-1}>0\}}\langle\sigma(X(s))dW(s),X(s)|X(s)|_{-1}^{-1}\rangle \quad \mathbb{P}\text{-a.s., } r<t.
\end{aligned}
$$

In particular, $e^{-C_\infty t}|X(t)|_{-1}$, $t\ge 0$, is an $(\mathscr{F}_t)$-supermartingale, hence $|X(t)|_{-1}=0\ \forall t\ge\tau$. Since $\tilde\rho>0$, taking expectation in (17) with $r=0$ yields

$$\int_0^t e^{-C_\infty s}\mathbb{P}(\tau>s)ds\le\frac{|x|_{-1}}{\gamma\tilde\rho}.$$

and (16) follows.

□

Corollary 7. *If in the situation of the above Theorem 6, the noise is zero, i.e.* $C_\infty=0$, *then*

$$\tau_c\le\frac{|x-x_c|_{-1}}{\tilde\varrho\gamma}.$$

Remark 8. We note that there is a small error in the corresponding result in [5], namely [5, Theorem 4.2, Corollary 4.3 and Remark 4.4]. There we forgot to replace ρ by $\tilde\rho$ and to assume that $0\in\mathrm{int}\Psi(0)$, rather than just $0\in\Psi(0)$.

Acknowledgments

This work has been supported in part by the CEEX Project 05 of Romanian Minister of Research, the DFG-International Graduate School "Stochastics and Real World Models", the SFB-701, NSF-Grants 0603742, 0606615 as well as the BiBoS-Research Center, the research programme "Equazioni di Kolmogorov" from the Italian "Ministero della Ricerca Scientifica e Tecnologica" and "FCT, POCTI-219, FEDER".

References

[1] P. Bak, C. Tang, and K. Wiesenfeld, Self-organized criticality: An explanation of the $1/f$ noise. *Phys. Rev. Lett.* **59** (1987), 381-384.

[2] P. Bak, C. Tang, and K. Wiesenfeld, Self-organized criticality. *Phys. Rev. A* **38** (1988), 364-374.

[3] P. Bantay and M. Janosi, Self-organization and anomalous diffusion. *Physica A* **185** (1992), 11-18.

[4] V. Barbu, *Nonlinear semigroups and differential equations in Banach spaces.* International Publishing, Leyden, 1976.

[5] V. Barbu, G. Da Prato, and M. Röckner, Stochastic porous media equation and self-organized criticality. *Commun. Math. Phys.* **285** (2009), 901-923.

[6] V. Barbu, G. Da Prato, and M. Röckner, Existence and uniqueness of nonnegative solutions to the stochastic porous media equation. *Indiana Univ. Math. J.* **57** (2008), 187-211.

[7] Ph. Blanchard, B. Cessac, and T. Krüger, What can one learn about self-Organized criticality from dynamical systems theory. *J. Stat. Phys.* **98** (2000), 375-404.

[8] R. Cafiero, V. Loreto, A. Pietronero, A. Vespignani, and S. Zapperi, Local rigidity and self-organized criticality for avalanches. *Europhys. Lett. EPL* **29** (1995), 111-116.

[9] J.M. Carlson, J.T. Chayes, E.R. Grannan, and G.H. Swindle, Self-organized criticality in sandpiles: Nature of the critical phenomenon. *Phys. Rev. A* **42** (1990), 2467-2470.

[10] J.M. Carlson, J.T. Chayes, E.R. Grannan, and G.H. Swindle, Self-organized criticality and singular diffusion. *Phys. Rev. Lett.* **65** (1990), 2547-2550.

[11] J.M. Carlson and G.H. Swindle, Self-organized criticality: Sandpiles, singularities and scaling. *Proc. National Acad. Sci. USA* **92** (1995), 6712-6719.

[12] G. Da Prato and J. Zabczyk, *Ergodicity for infinite dimensional systems* (London Math. Soc. Lect. Notes vol. **259**). Cambridge Univ., 1996.

[13] A. Díaz-Guilera, Dynamic renormalization group approach to self-organized critical phenomena. *Europhys. Lett. EPL* **26** (1994), 177-182.

[14] A. Giacometti and A. Díaz-Guilera, Dynamical properties of the Zhang model of self-organized criticality. *Phys. Rev. E* **58** (1998), 247-253.

[15] G. Grinstein, D.H. Lee, and S. Sachdev, Conservation laws, anisotropy, and "self-organized criticality" in noisy nonequilibrium systems. *Phys. Rev. Lett.* **64** (1990), 1927-1930.

[16] H.G.E. Hentschel and F. Family, Scaling in open dissipative systems. *Phys. Rev. Lett.* **66** (1991), 1982-1985.

[17] T. Hwa and M. Kardar, Dissipative transport in open systems: An investigation of self-organized criticality. *Phys. Rev. Lett.* **62** (1989), 1813-1816.

[18] I.M. Janosi and J. Kertesz, Self-organized criticality with and without conservation. *Physica A* **200** (1993), 179-188.

[19] H.J. Jensen, *Self-organized criticality.* Cambridge Univ. Press, 1988.

[20] C. Prévôt and M. Röckner, *A concise course on stochastic partial differential equations* (Lect. Notes Math. **1905**). Springer, 2007.
[21] D.L. Turcotte, Self-organized criticality. *Reports on Progress in Physics* **621** (1999), 1377-1429.
[22] Y.C. Zhang, Scaling theory of self-organized criticality. *Phys. Rev. Lett.* **63** (1989), 470-473.

"OCTAV MAYER" INSTITUTE OF MATHEMATICS, IASI, ROMANIA

FACULTY OF PHYSICS, UNIVERSITY OF BIELEFELD, GERMANY

SCUOLA NORMALE SUPERIORE DI PISA, ITALY

FACULTY OF MATHEMATICS, UNIVERSITY OF BIELEFELD, GERMANY AND DEPARTMENTS OF MATHEMATICS AND STATISTICS, PURDUE UNIVERSITY, U. S. A.

Potential Theory and Stochastics in Albac, 21–34
Aurel Cornea Memorial Volume

ON A NONHOMOGENOUS QUASILINEAR PROBLEM IN SOBOLEV SPACES WITH VARIABLE EXPONENT

KHALED BEN ALI AND MOUNIR BEZZARGA

ABSTRACT. We consider the nonlinear eigenvalue problem $-\operatorname{div}\left(|\nabla u(x)|^{p(x)-2}\nabla u(x)\right) = \lambda|u|^{q(x)-2}u + f(u)$ in Ω, $u = 0$ on $\partial\Omega$, where Ω is a bounded open set in $\mathbb{R}^N$ with smooth boundary and p, q are continuous functions on $\overline{\Omega}$ such that $2 \le \inf_\Omega q < \inf_\Omega p < \sup_\Omega q$, $\sup_\Omega p < N$, and $q(x) < Np(x)/(N - p(x))$ for all $x \in \overline{\Omega}$. The main result of this paper establishes that any $\lambda > 0$ sufficiently small the above nonhomogeneous quasilinear problem have a non-trivial weak solution. And, we show that the absolute value of this solution is also a solution of the problem. The proof relies on simple variational arguments based on Ekeland's variational principle.

1. INTRODUCTION

The study of differential equations and variational problems with nonstandard $p(x)$-growth conditions have received more and more interest in recent years. The specific attention accorded to such kind of problems is due to their applications in mathematical physics. More precisely, such equations are used to model phenomena which arise in elastic mechanics or electrorheological fluids (sometimes referred to as "smart fluids").

In this paper we discuss the existence of weak solutions for the following problem

$$\begin{cases} -\Delta_{p(x)} u(x) = \lambda|u(x)|^{q(x)-2}u + f(u(x)), & \text{for } x \in \Omega \\ u \not\equiv 0, & \text{for } x \in \Omega \\ u = 0, & \text{for } x \in \partial\Omega \end{cases} \tag{1}$$

where $\Omega \subset \mathbb{R}^N (N \ge 3)$ is a bounded domain with smooth boundary, $\lambda > 0$ is a real number, and $p, q : \overline{\Omega} \to \mathbb{R}$ are continuous functions.

In this paper we study problem (1) under the basic assumption

(A) $2 \le \min_{\overline{\Omega}} q(x) < \min_{\overline{\Omega}} p(x) < \max_{\overline{\Omega}} q(x)$
and $f : \mathbb{R} \to \mathbb{R}_+$ satisfies the condition :

2000 *Mathematics Subject Classification.* 35D05, 35J60, 35J70, 58E05, 76A02.
Key words and phrases. $p(x)$-Laplace operator, generalized Lebesgue–Sobolev spaces, Ekeland's variational principle, weak solution.

(B) $f(u) \leq C_0|u|^{\alpha(x)-1}$ for $x \in \Omega$ with $C_0 > 0$, $\alpha : \overline{\Omega} \to \mathbb{R}$ a continuous functions, f a hölderian function with exponent $\beta > \max_{\overline{\Omega}} \alpha(x)$ and

$$2 \leq \min_{\overline{\Omega}} q(x) < \min_{\overline{\Omega}} \alpha(x) < \max_{\overline{\Omega}} p(x).$$

We denote by $\Delta_{p(x)}$ the $p(x)$-Laplace operator, i.e.

$$\Delta_{p(x)} u := \operatorname{div}\left(|\nabla u|^{p(x)-2} \nabla u\right),$$

where p is a continuous non-constant function. This differential operator is a natural generalization of the p-Laplace operator $\Delta_p u = \operatorname{div}\left(|\nabla u|^{p-2}\nabla u\right)$, where $p > 1$ is a real constant. However, the $p(x)$-Laplace operator possesses more complicated nonlinearities than the p-Laplace operator, due to the fact that $\Delta_{p(x)}$ is not homogeneous. Recent qualitative propreties of solutions to quasilinear problems in Sobolev spaces with variable exponent have been obtained by Alves and Souto [1], Chabrowski and Fu [3], and Mihăilescu and Rădulescu [5]. Equation (1) will be study in the framework of the variable Lebesgue and Sobolev spaces $L^{p(x)}$ and $W_0^{1,p(x)}$ which will be briefly described in the next section.

2. Preliminary results

We recall in what follows some definitions and basic properties of the generalized Lebesgue spaces $L^{p(x)}(\Omega)$ and generalized Sobolev spaces $W^{1,p(x)}(\Omega)$, where Ω is an open subset of $\mathbb{R}^N$. In that context we refer to the book of Musielak [4] and the papers of Kovacik and Rakosnik [8] and Fan et al. [11, 10].

Set

$$C_+(\overline{\Omega}) = \{h;\ h \in C(\overline{\Omega}),\ h(x) > 1 \text{ for all } x \in \overline{\Omega}\}.$$

For any $h \in C_+(\overline{\Omega})$ we define

$$h^+ = \sup_{x\in\Omega} h(x) \quad \text{and} \quad h^- = \inf_{x\in\Omega} h(x).$$

For any $p(x) \in C_+(\overline{\Omega})$, we define the variable exponent Lebesgue space

$$L^{p(x)}(\Omega) = \{u : \text{is a Borel real-valued function on } \Omega : \int_\Omega |u(x)|^{p(x)}\, dx < \infty\}.$$

We define on $L^{p(x)}$, the so-called *Luxemburg norm*, by the formula

$$|u|_{p(x)} := \inf\Big\{\mu > 0;\ \int_\Omega \Big|\frac{u(x)}{\mu}\Big|^{p(x)}\, dx \leq 1\Big\}.$$

Variable exponent Lebesgue spaces resemble classical Lebesgue spaces in many respects: they are separable and Banach spaces [8, Theorem 2.5; Corollary 2.7] and the Hölder inequality holds [8, Theorem 2.1]. The inclusion between Lebesgue spaces are also naturally generalized [8, Theorem 2.8]: if $0 < |\Omega| < \infty$ and r_1, r_2 are variable exponents so that $r_1(x) \leq r_2(x)$ almost everywhere in Ω then there exists the continuous embedding $L^{r_2(x)}(\Omega) \hookrightarrow L^{r_1(x)}(\Omega)$.

We denote by $L^{p'(x)}(\Omega)$ the conjugate space of $L^{p(x)}(\Omega)$, where $1/p(x) + 1/p'(x) = 1$. For any $u \in L^{p(x)}(\Omega)$ and $v \in L^{p'(x)}(\Omega)$ the Hölder type inequality

$$\Big|\int_\Omega uv\, dx\Big| \leq \Big(\frac{1}{p^-} + \frac{1}{p'^-}\Big)|u|_{p(x)}|v|_{p'(x)} \tag{2}$$

holds.

An important role in manipulating the generalized Lebesgue–Sobolev spaces is played by the *modular* of the $L^{p(x)}(\Omega)$ space, which is the mapping $\rho_{p(x)} : L^{p(x)}(\Omega) \to \mathbb{R}$ defined by

$$\rho_{p(x)}(u) = \int_\Omega |u|^{p(x)}\,dx.$$

The space $W^{1,p(x)}(\Omega)$ is equipped with the following norm

$$\|u\| = |u|_{p(x)} + |\nabla u|_{p(x)}.$$

If (u_n), u, $\in W^{1,p(x)}(\Omega)$ and $p^+ < \infty$ then the following relations hold

$$\min(|u|_{p(x)}^{p^-}, |u|_{p(x)}^{p^+}) \le \rho_{p(x)}(u) \le \max(|u|_{p(x)}^{p^-}, |u|_{p(x)}^{p^+}) \tag{3}$$

$$\min(|\nabla u|_{p(x)}^{p^-}, |\nabla u|_{p(x)}^{p^+}) \le \rho_{p(x)}(|\nabla u|) \le \max(|\nabla u|_{p(x)}^{p^-}, |\nabla u|_{p(x)}^{p^+}) \tag{4}$$

$$\begin{aligned} |u|_{p(x)} \to 0 &\Leftrightarrow \rho_{p(x)}(u) \to 0, \\ \lim_{n\to\infty} |u_n - u|_{p(x)} = 0 &\Leftrightarrow \lim_{n\to\infty} \rho_{p(x)}(u_n - u) = 0, \\ |u|_{p(x)} \to \infty &\Leftrightarrow \rho_{p(x)}(u) \to \infty. \end{aligned} \tag{5}$$

We define also $W_0^{1,p(x)}(\Omega)$ as the closure of $C_0^\infty(\Omega)$ under the norm

$$\|u\| = |\nabla u|_{p(x)}.$$

Using the result of Fan and Zhao [9], we remark that the norm $\|u\| = |u|_{p(x)} + |\nabla u|_{p(x)}$ is equivalent to $\|u\| = |\nabla u|_{p(x)}$ in $W_0^{1,p(x)}(\Omega)$. The space $(W_0^{1,p(x)}(\Omega), \|.\|)$ is a separable and reflexive Banach space.

Next, we remember some embedding results regarding variable exponent Lebesgue–Sobolev spaces. We note that if $s(x) \in C_+(\overline{\Omega})$ and $s(x) < p^*(x)$ for all $x \in \overline{\Omega}$ then the embedding $W_0^{1,p(x)}(\Omega) \hookrightarrow L^{s(x)}(\Omega)$ is compact and continuous, where $p^*(x) = Np(x)/(N - p(x))$ if $p(x) < N$ or $p^*(x) = +\infty$ if $p(x) \ge N$. We refer to [8] for more properties of Lebesgue and Sobolev spaces with variable exponent. We also refer to the recent papers [11, 12, 10, 5, 6, 7] for the treatment of nonlinear boundary value problems in Lebesgue–Sobolev spaces with variable exponent.

3. The main result and an auxiliary result

Definition 1. *We say that $u \in W_0^{1,p(x)}(\Omega)$ is a weak solution of* (1) *if*

$$\int_\Omega \left(|\nabla u|^{p(x)-2}\nabla u \nabla v - \lambda |u|^{q(x)-2} uv\right) dx = \int_\Omega f(u) v\,dx,$$

for any $v \in W_0^{1,p(x)}(\Omega)$.

The main result of this paper is given by the following theorem

Theorem 1. *Assume that condition (A) is fulfilled,* $\max_{\overline{\Omega}} p(x) < N$, *$q(x) < p^*(x)$ for all $x \in \overline{\Omega}$, $f : \mathbb{R} \to \mathbb{R}_+$ satisfies the condition (B), and $\alpha(x) < p^*(x)$ for all $x \in \overline{\Omega}$. Then there exists $\lambda^* > 0$ such that any $\lambda \in (0, \lambda^*)$ the problem* (1) *has a non-trivial weak solution.*

Let E denote the generalized Sobolev space $W_0^{1,p(x)}(\Omega)$. The energy functional corresponding to problem (1) is defined as $J_\lambda : E \to \mathbb{R}$,

$$J_\lambda(u) := \int_\Omega \frac{1}{p(x)} |\nabla u|^{p(x)} dx - \lambda \int_\Omega \frac{1}{q(x)} |u|^{q(x)} dx - \int_\Omega F(u) dx,$$

where $F(u) = \int_0^u f(s) ds$.

Remark 1. *According the following convention, for any $\alpha(x) \in C_+(\overline{\Omega})$*

$$u^{\alpha(x)} = \begin{cases} u^{\alpha(x)} & \text{for } u \geq 0 \\ -(-u)^{\alpha(x)} & \text{for } u < 0. \end{cases}$$

We have

$$If \quad u \geq 0 \quad then \quad F(u) \leq C_0 \int_0^u s^{\alpha(x)-1} ds = C_0 \frac{u^{\alpha(x)}}{\alpha(x)},$$

and

$$if \quad u < 0 \quad then \quad F(u) \leq C_0 \int_u^0 (-s)^{\alpha(x)-1} ds = -C_0 \left(-\frac{(-u)^{\alpha(x)}}{\alpha(x)}\right) = C_0 \frac{(-u)^{\alpha(x)}}{\alpha(x)}.$$

It follows that

$$|F(u)| \leq C_0 \frac{|u|^{\alpha(x)}}{\alpha(x)}, \quad \text{for all } u \in E. \tag{6}$$

Proposition 1. *The functional J_λ is well-defined on E.*

Proof. Using the above hypothesis and the condition (A) and (B) we get:

$$J_\lambda(u) \leq \frac{1}{p^-} \int_\Omega |\nabla u|^{p(x)} dx - \frac{\lambda}{q^+} \int_\Omega |u|^{q(x)} dx + \frac{C_0}{\alpha^-} \int_\Omega |u|^{\alpha(x)} dx.$$

The proof, then holds by the Sobolev embeddings $E \hookrightarrow L^{q(x)}$ and $E \hookrightarrow L^{\alpha(x)}$. □

We set

$$J_{1\lambda}(u) := \int_\Omega \frac{1}{p(x)} |\nabla u|^{p(x)} dx - \lambda \int_\Omega \frac{1}{q(x)} |u|^{q(x)} dx$$

and

$$J_{2\lambda}(u) := \int_\Omega F(u) dx.$$

We first prove an auxiliary result

Theorem 2. *Suppose we are under hypotheses of Theorem 1. Then $J_\lambda \in C^1(E, \mathbb{R})$ and $u \in E$ is a critical point of J_λ if and only if u is a weak solution for the problem* (1).

Proof. To show that $J_\lambda \in C^1(E, \mathbb{R})$, we show that for all $\varphi \in E$,

$$\lim_{t \to 0^+} \frac{J_\lambda(u + t\varphi) - J_\lambda(u)}{t} = \langle dJ_\lambda(u), \varphi \rangle,$$

and $dJ_\lambda : E \to E^*$ continuous, where we denote by E^* the dual space of E.

For all $\varphi \in E$ we have

$$\begin{aligned}
\lim_{t\to 0^+} \frac{J_{2\lambda}(u+t\varphi) - J_{2\lambda}(u)}{t} &= \frac{d}{dt} J_{2\lambda}(u+t\varphi)|_{t=0} = \frac{d}{dt}\int_\Omega F(u+t\varphi)dx|_{t=0} \\
&= \int_\Omega \frac{\partial}{\partial t}\left(F(u+t\varphi)\right)|_{t=0} dx = \int_\Omega f(u+t\varphi)\varphi|_{t=0} dx \\
&= \int_\Omega f(u)\varphi dx \\
&= \langle dJ_{2\lambda}(u), \varphi\rangle.
\end{aligned}$$

The differentiation under the integral is allowed if for t close to zero, $f(u+t\varphi)\varphi$ can be dominated by one fixed function $g \in L^1(\Omega)$ which does not depend on t. Indeed we have, say if $|t| < 1$ that

$$|f(u+t\varphi)\varphi| \le C_0|u+t\varphi|^{\alpha(x)-1}|\varphi| \le C_0(|u|+|t||\varphi|)^{\alpha(x)-1}|\varphi| \le C_0(|u|+|\varphi|)^{\alpha(x)-1}|\varphi|.$$

Set

$$g(x) := C_0(|u|+|\varphi|)^{\alpha(x)-1}|\varphi|.$$

We observe that $g \in L^1(\Omega)$ because $u, \varphi \in E$ imply

$$|u|, |\varphi| \in E \hookrightarrow L^{\alpha(x)}(\Omega),$$

and then

$$(|u|+|\varphi|)^{\alpha(x)-1} \in L^{\frac{\alpha(x)}{\alpha(x)-1}}(\Omega), |\varphi| \in L^{\alpha(x)}(\Omega) = \left(L^{\frac{\alpha(x)}{\alpha(x)-1}}(\Omega)\right)^*.$$

For $u \in E$ chosen we show that $dJ_{2\lambda}(u) \in W^{-1,p'(x)}(\Omega) = E^*$, where $1/p(x) + 1/p'(x) = 1$. It is easy to see that $dJ_2(u)$ is linear. We know $\alpha : \Omega \to \mathbb{R}$ is continuous and $\alpha(x) < p^*(x)$, there is a continuous embedding $E \hookrightarrow L^{\alpha(x)}(\Omega)$. Then there exists a constant $M > 0$ such that, for all $u \in E$

$$|u|_{\alpha(x)} \le M\|u\|. \tag{7}$$

Using (2), (7) and (B) we obtain

$$\begin{aligned}
|\langle dJ_{2\lambda}(u), \varphi\rangle| &= |\int_\Omega f(u)\varphi dx| \le \int_\Omega |f(u)||\varphi| dx \\
&\le C_0 \int_\Omega |u|^{\alpha(x)-1}|\varphi| dx \le C_0 ||u|^{\alpha(x)-1}|_{\frac{\alpha(x)}{\alpha(x)-1}} |\varphi|_{\alpha(x)} \\
&\le MC_0 ||u|^{\alpha(x)-1}|_{\frac{\alpha(x)}{\alpha(x)-1}} \|\varphi\|.
\end{aligned}$$

Hence there exists $M_1 = MC_0||u|^{\alpha(x)-1}|_{\frac{\alpha(x)}{\alpha(x)-1}} > 0$ such that

$$|\langle dJ_{2\lambda}(u), \varphi\rangle| \le M_1\|\varphi\|, \quad \text{for all} \quad \varphi \in E.$$

Using the linearity of $dJ_{2\lambda}(u)$ and the above formula we deduce that $dJ_{2\lambda}(u) \in E^* = W^{-1,p'(x)}(\Omega)$. Now we show $dJ_{2\lambda} : E \to E^*$ is continuous. Let sequence $\{u_n\} \subset E$ such that $\|u_n - u\| \longrightarrow 0$.

Using (2) and the condition (B) we obtain

$$\begin{aligned}|\langle dJ_{2\lambda}(u_n)-dJ_{2\lambda}(u),\varphi\rangle| &= |\langle f(u_n)-f(u),\varphi\rangle| \\ &\leq \int_\Omega M_0|u_n-u|^\beta\varphi dx \\ &\leq M_0||u_n-u|^\beta|_{\frac{\alpha(x)}{\alpha(x)-1}}|\varphi|_{\alpha(x)}.\end{aligned}$$

Now, using the continuous embedding $E\hookrightarrow L^{\alpha(x)}(\Omega)$, $L^{\alpha(x)}(\Omega)\hookrightarrow L^{\frac{\beta\alpha(x)}{\alpha(x)-1}}(\Omega)$ and relation (5) we have $dJ_{2\lambda}$ is a continuous function. We conclude that $J_{2\lambda}$ is Fréchet differentiable.

Now we show that $J_{1\lambda}$ is Fréchet differentiable. For all $\varphi\in E$ we have

$$\lim_{t\to0^+}\frac{J_{1\lambda}(u+t\varphi)-J_{1\lambda}(u)}{t}=\frac{d}{dt}J_{1\lambda}(u+t\varphi)|_{t=0}$$

$$=\frac{d}{dt}\int_\Omega\left[\frac{1}{p(x)}|\nabla u+t\nabla\varphi|^{p(x)}-\frac{\lambda}{q(x)}|u+t\varphi|^{q(x)}\right]dx|_{t=0}$$

$$=\int_\Omega\frac{\partial}{\partial t}\left[\frac{1}{p(x)}|\nabla u+t\nabla\varphi|^{p(x)}-\frac{\lambda}{q(x)}|u+t\varphi|^{q(x)}\right]|_{t=0}dx.$$

Since

$$\frac{\partial}{\partial t}|u+t\varphi|^{q(x)}=q(x)|u+t\varphi|^{q(x)-1}sgn(u+t\varphi)\varphi=q(x)|u+t\varphi|^{q(x)-2}(u+t\varphi)\varphi,$$

$$\frac{\partial}{\partial t}|\nabla u+t\nabla\varphi|^{p(x)}=p(x)|\nabla(u+t\varphi)|^{p(x)-2}\nabla(u+t\varphi)\nabla\varphi.$$

So

$$\lim_{t\to0^+}\frac{J_{1\lambda}(u+t\varphi)-J_{1\lambda}(u)}{t}$$

$$=\int_\Omega\left(|\nabla(u+t\varphi)|^{p(x)-2}\nabla(u+t\varphi)\nabla\varphi-\lambda|u+t\varphi|^{q(x)-2}(u+t\varphi)\varphi\right)|_{t=0}dx$$

$$=\int_\Omega\left(|\nabla u|^{p(x)-2}\nabla u\nabla\varphi-\lambda|u|^{q(x)-2}u\varphi\right)dx.$$

The differentiation under the integral is allowed if for t close to zero

$$|\nabla(u+t\varphi)|^{p(x)-2}\nabla(u+t\varphi)\nabla\varphi-\lambda|u+t\varphi|^{q(x)-2}(u+t\varphi)\varphi$$

can be dominated by one fixed function $h\in L^1(\Omega)$ which does not depend on t. Indeed we have, say if $|t|<1$ that

$$\begin{aligned}&||\nabla(u+t\varphi)|^{p(x)-2}\nabla(u+t\varphi)\nabla\varphi-\lambda|u+t\varphi|^{q(x)-2}(u+t\varphi)\varphi| \\ \leq\;& |\nabla u+t\nabla\varphi|^{p(x)-1}|\nabla\varphi|+\lambda|u+t\varphi|^{q(x)-1}|\varphi| \\ \leq\;& (|\nabla u|+|t||\nabla\varphi|)^{p(x)-1}|\nabla\varphi|+\lambda(|u|+|t||\varphi|)^{q(x)-1}|\varphi| \\ \leq\;& (|\nabla u|+|\nabla\varphi|)^{p(x)-1}|\nabla\varphi|+\lambda(|u|+|\varphi|)^{q(x)-1}|\varphi|.\end{aligned}$$

Set

$$h(x):=(|\nabla u|+|\nabla\varphi|)^{p(x)-1}|\nabla\varphi|+\lambda(|u|+|\varphi|)^{q(x)-1}|\varphi|.$$

We observe that $h\in L^1(\Omega)$ because $u,\varphi\in E$ imply

$$|u|,\ |\varphi|\in E\hookrightarrow L^{q(x)}(\Omega),$$

so

$$|u|,\ |\varphi| \in L^{q(x)},\ \text{yield } (|u|+|\varphi|)^{q(x)-1} \in L^{\frac{q(x)}{q(x)-1}} = \left(L^{q(x)}\right)^*$$

and

$$|\nabla u|,\ |\nabla\varphi| \in L^{p(x)},\ \text{yield } (|\nabla u|+|\nabla\varphi|)^{p(x)-1} \in L^{\frac{p(x)}{p(x)-1}} = \left(L^{p(x)}\right)^*.$$

For all $u \in E$ we have

$$\langle dJ_{1\lambda}(u),\varphi\rangle = \int_\Omega \left(|\nabla u|^{p(x)-2}\nabla u \nabla\varphi - \lambda|u|^{q(x)-2}u\varphi\right)dx.$$

It is easy to see that $dJ_{1\lambda}(u)$ is a linear function. We also have the continuous embedding $E \hookrightarrow L^{q(x)}(\Omega)$, which implies that there exists $M' > 0$ such that, for all $u \in E$,

$$|u|_{q(x)} \le M'\|u\|. \tag{8}$$

Using (2) and (8) we obtain

$$\begin{aligned}
|\langle dJ_{1\lambda}(u),\varphi\rangle| &= |\int_\Omega \left(|\nabla u|^{p(x)-2}\nabla u\nabla\varphi - \lambda|u|^{q(x)-2}u\varphi\right)dx| \\
&\le \int_\Omega |\nabla u|^{p(x)-1}|\nabla\varphi|dx + \lambda\int_\Omega |u|^{q(x)-1}|\varphi|dx \\
&\le ||\nabla u|^{p(x)-1}|_{\frac{p(x)}{p(x)-1}}|\nabla\varphi|_{p(x)} + \lambda||u|^{q(x)-1}|_{\frac{q(x)}{q(x)-1}}|\varphi|_{q(x)} \\
&\le ||\nabla u|^{p(x)-1}|_{\frac{p(x)}{p(x)-1}}\|\varphi\| + M'\lambda||u|^{q(x)-1}|_{\frac{q(x)}{q(x)-1}}\|\varphi\|.
\end{aligned}$$

Thus there exists $M_2 = ||\nabla u|^{p(x)-1}|_{\frac{p(x)}{p(x)-1}} + M'\lambda||u|^{q(x)-1}|_{\frac{q(x)}{q(x)-1}} > 0$ such that

$$|\langle dJ_{1\lambda}(u),\varphi\rangle| \le M_2\|\varphi\|,\ \text{for all } \varphi \in E.$$

The above relation and the linearity of $dJ_{1\lambda}(u)$ imply that $dJ_{1\lambda}(u) \in E^*$ and

$$\langle dJ_{1\lambda}(u),\varphi\rangle = \int_\Omega \left(|\nabla u|^{p(x)-2}\nabla u\nabla\varphi - \lambda|u|^{q(x)-2}u\varphi\right)dx.$$

For the continuity of $dJ_{1\lambda} : E \longrightarrow E^*$ we need the following results

Proposition 2. $\forall u, v \in L^{q(x)}(\Omega), |u-v|_{q(x)} \to 0 \Rightarrow ||u-v|^{\frac{q(x)}{q(x)-1}}|_{q(x)-1} \to 0.$

Proof. Let $\varepsilon > 0$, let us find $\eta > 0$ such that

$$|u-v|_{q(x)} < \eta \Rightarrow ||u-v|^{\frac{q(x)}{q(x)-1}}|_{q(x)-1} < \varepsilon.$$

If $||u-v|^{\frac{q(x)}{q(x)-1}}|_{q(x)-1} < \varepsilon$, then, $\inf\{\mu > 0 : \int_\Omega \frac{|u-v|^{q(x)}}{\mu^{q(x)-1}}\ dx \le 1\} < \varepsilon$ and consequently, there exists $\eta > 0$ such that $\displaystyle\int_\Omega \frac{|u-v|^{q(x)}}{\eta^{q(x)-1}}\ dx \le 1$ and $\eta < \varepsilon < 1$.

Let $u,\ v \in L^{q(x)}(\Omega)$ such that $|u-v|_{q(x)} < \eta$, then

$$|u-v|_{q(x)} = \inf\{\mu \in]0,\eta[: \int_\Omega \frac{|u-v|^{q(x)}}{\mu^{q(x)}}\ dx \le 1\}.$$

Since $0<\mu<\eta<1$, then $\int_\Omega \frac{|u-v|^{q(x)}}{\mu^{q(x)-1}}\,dx \le \int_\Omega \frac{|u-v|^{q(x)}}{\mu^{q(x)}}\,dx$ and consequently

$$\inf\{\mu\in]0,\eta[: \int_\Omega \frac{|u-v|^{q(x)}}{\mu^{q(x)-1}}\,dx\le 1\} \le \inf\{\mu\in]0,\eta[: \int_\Omega \frac{|u-v|^{q(x)}}{\mu^{q(x)}}\,dx\le 1\}$$
$$= |u-v|_{q(x)} < \eta < \varepsilon.$$

We end the proof using the fact $|| u-v|^{\frac{q(x)}{q(x)-1}}|_{q(x)-1} \le \inf\{\mu\in]0,\eta[: \int_\Omega \frac{|u-v|^{q(x)}}{\mu^{q(x)-1}}\,dx\le 1\}$.
□

In the sequel, for any $q\in C_+(\overline{\Omega})$ and for simplicity of calculate, we introduce the convention that

$$u^{q(x)} = \begin{cases} u^{q(x)} & \text{for } u\ge 0\\ -(-u)^{q(x)} & \text{for } u<0. \end{cases}$$

So, we have the following lemma

Lemma 1. *The map* $u\in L^{q(x)}(\Omega)\to |u|^{q(x)-2}u\in L^{\frac{q(x)}{q(x)-1}}(\Omega)$ *is continuous.*

Proof. Our intention is to prove

$$|u-v|_{q(x)}\to 0 \Rightarrow |u^{q(x)-1}-v^{q(x)-1}|_{\frac{q(x)}{q(x)-1}}\to 0.$$

For the case $q(x)=2$ the result is trivial. We claim to show the result for the case $q(x)>2$, since

$$\rho_{\frac{q(x)}{q(x)-1}}(u^{q(x)-1}-v^{q(x)-1}) = \int_\Omega |u^{q(x)-1}-v^{q(x)-1}|^{\frac{q(x)}{q(x)-1}}\,dx,$$

then, for fixed $x\in\Omega$, by Lagrange theorem applied to the function $f_0(t)=t^{q(x)-1}$, there exists $C_0(x)$ somewhere between $u(x)$ and $v(x)$ such that

$$\frac{f_0(u(x))-f_0(v(x))}{u(x)-v(x)} = f_0'(C_0(x)).$$

Thus

$$\rho_{\frac{q(x)}{q(x)-1}}(u^{q(x)-1}-v^{q(x)-1}) \le {q^+}^{\frac{q^+}{q^--1}} \int_\Omega |u-v|^{\frac{q(x)}{q(x)-1}} \max(|u|,|v|)^{\frac{q(x)(q(x)-2)}{q(x)-1}}\,dx.$$

Because $|u-v|\in L^{q(x)}$, then $|u-v|^{\frac{q(x)}{q(x)-1}}\in L^{q(x)-1}(\Omega) = (L^{\frac{q(x)-1}{q(x)-2}}(\Omega))^*$, moreover, $|u|,|v|\in L^{q(x)}(\Omega)$, imply $|u|^{\frac{q(x)(q(x)-2)}{q(x)-1}}, |v|^{\frac{q(x)(q(x)-2)}{q(x)-1}} \in L^{\frac{q(x)-1}{q(x)-2}}(\Omega)$. Now, using (2), (5) and Proposition 2, we end the proof. □

Using Lemma 1, we conclude that $J_{1\lambda}$ is Frechet differentiable.
We remark that $J_\lambda\in C^1(E,\mathbb{R})$ because $J_{1\lambda}$, $J_{2\lambda}\in C^1(E,\mathbb{R})$. Moreover

$$\begin{aligned} \langle dJ_\lambda(u),v\rangle &= \langle dJ_{1\lambda}(u)-dJ_{2\lambda}(u),v\rangle = \langle dJ_{1\lambda}(u),v\rangle - \langle dJ_{2\lambda}(u),v\rangle \\ &= \int_\Omega \left(|\nabla u|^{p(x)-2}\nabla u\nabla v - \lambda|u|^{q(x)-2}uv\right)dx - \int_\Omega f(u)v\,dx,\ \forall\, v\in E. \end{aligned}$$

Let u be a critical point of J_λ. Then we have $dJ_\lambda(u)=0_{E^*}$ that is

$$\langle dJ_\lambda(u),v\rangle = 0, \text{ for all } v\in E$$

which yields

$$\int_\Omega \left(|\nabla u|^{p(x)-2}\nabla u \nabla v - \lambda |u|^{q(x)-2} uv\right) dx = \int_\Omega f(u) v dx, \ \forall v \in E.$$

It follows that u is a weak solution for the problem (1).

Now we assume that u is a weak solution of the problem (1). By Definition 1 it results that

$$\int_\Omega \left(|\nabla u|^{p(x)-2}\nabla u \nabla v - \lambda |u|^{q(x)-2} uv\right) dx = \int_\Omega f(u) v dx, \ \forall v \in E,$$

that is $\langle dJ_\lambda(u), v\rangle = 0, \forall v \in E$. We obtain $dJ_\lambda(u) = 0_{E^*}$. Hence u is a critical point of J_λ. This completes the proof of Theorem 2. □

4. Proof of Main Theorem

Our idea is to prove Theorem 1 applying Ekeland's variational principle (see e.g. [2]).

Lemma 2. *Suppose we are under hypotheses of Theorem 1. Then there exist $C_2 > 0$ such that for all $\rho \in (0,1)$, for all $C_0 < \dfrac{q^-}{p^+ C_2^{\alpha^-}} \rho^{p^+ - q^-}$, there exist $\lambda^* > 0$ and $a > 0$ such that for all $u \in E$ with $\|u\| = \rho$*

$$J_\lambda(u) \geq a > 0 \qquad \textit{for all} \qquad \lambda \in (0, \lambda^*).$$

Proof. Since $q(x) < p^*(x)$ for all $x \in \overline{\Omega}$ it follows that E is continuously embedded in $L^{q(x)}(\Omega)$. So, there exists a positive constant C_1 such that, for all $u \in E$,

$$|u|_{q(x)} \leq C_1 \|u\|. \tag{9}$$

Moreover, $\alpha(x) < p^*(x)$ for all $x \in \overline{\Omega}$. We have then the continuous embedding $E \hookrightarrow L^{\alpha(x)}(\Omega)$, which implies that there exists a positive constant C_2 such that, for all $u \in E$,

$$|u|_{\alpha(x)} \leq C_2 \|u\|. \tag{10}$$

We fix $\rho \in (0,1)$ such that $\rho < \min(1, 1/C_1, 1/C_2)$. Then relations (9) and (10) implies that for all $u \in E$ with $\|u\| = \rho$

$$|u|_{q(x)} < 1 \quad \text{and} \quad |u|_{\alpha(x)} < 1. \tag{11}$$

Furthermore, relations (3) yields for all $u \in E$ with $\|u\| = \rho$

$$\int_\Omega |u|^{q(x)} dx \leq |u|_{q(x)}^{q^-} \quad \text{and} \quad \int_\Omega |u|^{\alpha(x)} dx \leq |u|_{\alpha(x)}^{\alpha^-}. \tag{12}$$

Relations (9), (10) and (12) imply for all $u \in E$ with $\|u\| = \rho$

$$\int_\Omega |u|^{q(x)} dx \leq C_1^{q^-} \|u\|^{q^-} \quad \text{and} \quad \int_\Omega |u|^{\alpha(x)} dx \leq C_2^{\alpha^-} \|u\|^{\alpha^-}. \tag{13}$$

Using relations (4), (6) and (13) we deduce that for any $u \in E$ with $\|u\| = \rho$ the following inequalities hold true

$$
\begin{aligned}
J_\lambda(u) &\geq \frac{1}{p^+}\int_\Omega |\nabla u|^{p(x)}dx - \frac{\lambda}{q^-}\int_\Omega |u|^{q(x)}dx - \frac{C_0}{\alpha^-}\int_\Omega |u|^{\alpha(x)}dx \\
&\geq \frac{1}{p^+}\|u\|^{p^+} - \frac{\lambda}{q^-}C_1^{q^-}\|u\|^{q^-} - \frac{C_0}{\alpha^-}C_2^{\alpha^-}\|u\|^{\alpha^-} \\
&\geq \frac{1}{p^+}\rho^{p^+} - \frac{\rho^{q^-}}{q^-}\left(\lambda C_1^{q^-} + C_0 C_2^{\alpha^-}\right) \\
&= \rho^{q^-}\left(\frac{1}{p^+}\rho^{p^+-q^-} - \frac{1}{q^-}\left(\lambda C_1^{q^-} + C_0 C_2^{\alpha^-}\right)\right).
\end{aligned}
$$

By the above inequality we remark that if we define for all $C_0 < \dfrac{q^-}{p^+ C_2^{\alpha^-}}\rho^{p^+-q^-}$,

$$
\lambda^* = \frac{1}{C_1^{q^-}}\left(\frac{q^-}{p^+}\rho^{p^+-q^-} - C_0 C_2^{\alpha^-}\right) \tag{14}
$$

then for any $\lambda \in (0, \lambda^*)$ and $u \in E$ with $\|u\| = \rho$ there exists $a > 0$ such that

$$
J_\lambda(u) \geq a > 0.
$$

The proof of Lemma 2 is complete. □

Lemma 3. *There exists $\varphi \in E$ such that $\varphi \geq 0$, $\varphi \neq 0$ and $J_\lambda(t\varphi) < 0$, for $t > 0$ small enough.*

Proof. Assumption (A) implies that $q^- < p^-$. Let $\epsilon_0 > 0$ be such that $q^- + \epsilon_0 < p^-$. On the other hand, since $q \in C(\overline{\Omega})$ it follows that there exists an open set $\Omega_0 \subset\subset \Omega$ such that $|q(x) - q^-| < \epsilon_0$ for all $x \in \Omega_0$. Thus, we conclude that $q(x) \leq q^- + \epsilon_0 < p^-$ for all $x \in \overline{\Omega}_0$.
Let $\varphi \in C_0^\infty(\Omega)$ be such that $\text{supp}(\varphi) \supset \overline{\Omega}_0$, $\varphi(x) = 1$ for all $x \in \overline{\Omega}_0$ and $0 \leq \varphi \leq 1$ in Ω. Then using the above information for any $t \in (0,1)$ we have

$$
\begin{aligned}
J_\lambda(t\varphi) &= \int_\Omega \frac{t^{p(x)}}{p(x)}|\nabla\varphi|^{p(x)}dx - \lambda\int_\Omega \frac{t^{q(x)}}{q(x)}|\varphi|^{q(x)}dx - \int_\Omega F(t\varphi)dx \\
&\leq \frac{t^{p^-}}{p^-}\int_\Omega |\nabla\varphi|^{p(x)}dx - \frac{\lambda}{q^+}\int_\Omega t^{q(x)}|\varphi|^{q(x)}dx \\
&\leq \frac{t^{p^-}}{p^-}\int_\Omega |\nabla\varphi|^{p(x)}dx - \frac{\lambda t^{q^-+\epsilon_0}}{q^+}\int_{\Omega_0} |\varphi|^{q(x)}dx \\
&= \frac{t^{p^-}}{p^-}\int_\Omega |\nabla\varphi|^{p(x)}dx - \frac{\lambda t^{q^-+\epsilon_0}}{q^+}|\Omega_0|.
\end{aligned}
$$

Therefore

$$
J_\lambda(t\varphi) < 0
$$

for $t<\delta^{1/(p^+-q^--\epsilon_0)}$ with

$$0<\delta<\min\left\{1,\frac{\frac{\lambda p^-}{q^+}|\Omega_0|}{\int_\Omega|\nabla\varphi|^{p(x)}dx}\right\}.$$

Finally, we point out that $\int_\Omega|\nabla\varphi|^{p(x)}dx>0$. We suppose that $|\nabla\varphi|=0$ on $\Omega\backslash\overline{\Omega}_0$ then φ is null that's an absurdity. That's why $|\nabla\varphi|>0$ on an open subset of $\Omega\backslash\overline{\Omega}_0$ inside's no empty. □

Proof of Theorem 1. Let $\lambda^*>0$ be defined as in (14) and $\lambda\in(0,\lambda^*)$. By Lemma 2 it follows that on the boundary of the ball centered at the origin and of radius ρ in E, denoted by $B_\rho(0)$, we have

$$\inf_{\partial B_\rho(0)}J_\lambda>0. \tag{15}$$

On the other hand, by Lemma 3, there exists $\varphi\in E$ such that $J_\lambda(t\varphi)<0$ for all $t>0$ small enough. Moreover, relations (4), (6) and (13) imply that for any $u\in B_\rho(0)$ we have

$$J_\lambda(u)\geq\frac{1}{p^+}\|u\|^{p^+}-\frac{\lambda}{q^-}C_1^{q^-}\|u\|^{q^-}-\frac{C_0}{\alpha^-}C_2^{\alpha^-}\|u\|^{\alpha^-}$$

it follows that

$$-\infty<\underline{c}:=\inf_{\overline{B_\rho(0)}}J_\lambda<0.$$

We let now $0<\epsilon<\inf_{\partial B_\rho(0)}J_\lambda-\inf_{B_\rho(0)}J_\lambda$. Using the above information, the functional $J_\lambda:\overline{B_\rho(0)}\longrightarrow\mathbb{R}$, is lower bounded on $\overline{B_\rho(0)}$ and $J_\lambda\in C^1(\overline{B_\rho(0)},\mathbb{R})$. Then by Ekeland's variational principle there exists $u_\epsilon\in\overline{B_\rho(0)}$ such that

$$\begin{cases}\underline{c}\leq J_\lambda(u_\epsilon)\leq\underline{c}+\epsilon\\[2mm] 0<J_\lambda(u)-J_\lambda(u_\epsilon)+\epsilon\cdot\|u-u_\epsilon\|,\quad u\neq u_\epsilon.\end{cases}$$

Since

$$J_\lambda(u_\epsilon)\leq\inf_{\overline{B_\rho(0)}}J_\lambda+\epsilon\leq\inf_{B_\rho(0)}J_\lambda+\epsilon<\inf_{\partial B_\rho(0)}J_\lambda,$$

we deduce that $u_\epsilon\in B_\rho(0)$. Now, we define $I_\lambda:\overline{B_\rho(0)}\longrightarrow\mathbb{R}$ by $I_\lambda(u)=J_\lambda(u)+\epsilon\cdot\|u-u_\epsilon\|$. It is clair that u_ϵ is a minimum point of I_λ and thus

$$\frac{I_\lambda(u_\epsilon+t\cdot v)-I_\lambda(u_\epsilon)}{t}\geq0$$

for small $t>0$ and any $v\in B_1(0)$. The above relation yields

$$\frac{J_\lambda(u_\epsilon+t\cdot v)-J_\lambda(u_\epsilon)}{t}+\epsilon\cdot\|v\|\geq0$$

Letting $t\to0$ it follows that $<dJ_\lambda(u_\epsilon),v>+\epsilon\cdot\|v\|\geq0$ and we infer that $\|dJ_\lambda(u_\epsilon)\|\leq\epsilon$. We deduce that there exists a sequence $\{w_n\}\subset B_\rho(0)$ such that

$$J_\lambda(w_n)\longrightarrow\underline{c}\quad\text{and}\quad dJ_\lambda(w_n)\longrightarrow0. \tag{16}$$

It is clair that $\{w_n\}$ is bounded in E. Thus, there exists w in E such that, up to a subsequence, $\{w_n\}$ converges weakly to w in E. Since $q(x)<p^*(x)$ for all $x\in\overline{\Omega}$ we deduce

that there exists a compact embedding $E \hookrightarrow L^{q(x)}(\Omega)$. Then $\{w_n\}$ converges strongly in $L^{q(x)}(\Omega)$. So we have:

Proposition 3. $\lim_{n\to\infty}\int_\Omega |w_n|^{q(x)-2}w_n(w_n-w)dx=0.$

Proof. Using (2) we have $\int_\Omega |w_n|^{q(x)-2}w_n(w_n-u)dx \le ||w_n|^{q(x)-2}w_n|_{\frac{q(x)}{q(x)-1}}|w_n-w|_{q(x)}$, then if $||w_n|^{q(x)-2}w_n|_{\frac{q(x)}{q(x)-1}}>1$, by (3), there exists $C>0$ such that $||w_n|^{q(x)-2}w_n|_{\frac{q(x)}{q(x)-1}} \le |w_n|^C_{q(x)}$ and this ends the proof. □

On the other hand, Since $\alpha(x)<p^*(x)$ for all $x\in\overline{\Omega}$ we deduce that there exists a compact embedding $E\hookrightarrow L^{\alpha(x)}(\Omega)$. Then $\{w_n\}$ converges strongly in $L^{\alpha(x)}(\Omega)$. So we have:

Proposition 4. $\lim_{n\to\infty}\int_\Omega f(w_n)(w_n-w)dx=0.$

Proof. Using (2) and (B) we have

$$\int_\Omega f(w_n)(w_n-w)dx \le C_0||w_n|^{\alpha(x)-1}|_{\frac{\alpha(x)}{\alpha(x)-1}}|w_n-w|_{\alpha(x)},$$

then if $||w_n|^{\alpha(x)-1}|_{\frac{\alpha(x)}{\alpha(x)-1}}>1$, by (3), there exists $C'>0$ such that $||w_n|^{\alpha(x)-1}|_{\frac{\alpha(x)}{\alpha(x)-1}} \le |w_n|^{C'}_{\alpha(x)}$ and this ends the proof. □

Using the above information we find

$$\begin{aligned}|\langle dJ_\lambda(w_n),w_n-w\rangle| &\le |\langle dJ_\lambda(w_n),w_n\rangle|+|\langle dJ_\lambda(w_n),w\rangle|\\ &\le \|dJ_\lambda(w_n)\|\|w_n\|+\|dJ_\lambda(w_n)\|\|w\|.\end{aligned}$$

So

$$\lim_{n\to\infty}\langle dJ_\lambda(w_n),w_n-w\rangle=0.$$

Combining Proposition 3, Proposition 4 and the last relation we deduce that

$$\lim_{n\to\infty}\int_\Omega |\nabla w_n|^{p(x)-2}\nabla w_n\nabla(w_n-w)dx=0. \tag{17}$$

Now, if we note $L:E\to E^*$, such that

$$\langle L(u),v\rangle=\int_\Omega |\nabla u|^{p(x)-2}\nabla u\nabla v dx \ \forall u,v\in E,$$

then

$$\begin{aligned}|\langle L(w_n)-L(w),w_n-w\rangle| &\le |\langle L(w_n),w_n-w\rangle|+|\langle L(w),w_n-w\rangle|\\ &\le |\langle L(w_n),w_n-w\rangle|+|\langle L(w),w_n\rangle-\langle L(w),w\rangle|.\end{aligned}$$

From (17) and the fact that $\{w_n\}$ converges weakly to w in E it follows that

$$\lim_{n\to\infty}\langle L(w_n)-L(w),w_n-w\rangle=0$$

and by Theorem 3.1 in Fan and Zhang [12] we deduce that $\{w_n\}$ converges strongly to w in E. Then since $J_\lambda \in C^1(E,\mathbb{R})$ we conclude

$$dJ_\lambda(w_n) \to dJ_\lambda(w), \tag{18}$$

as $n \to \infty$.

Relations (16) and (18) show that $dJ_\lambda(w) = 0$ and thus w is a weak solution for problem (1). Moreover, by relation (16) it follows that $J_\lambda(w) < 0$ and thus, w is a nontrivial weak solution for (1). The proof of Theorem 1 is complete. □

Now, we show that the absolute value of the solution given in Theorem 1 is a nontrivial weak solution for our problem.

Corollary 1. *Suppose we are under hypotheses of Theorem 1. Then the absolute value of w is a non-trivial weak solution of problem* (1).

Proof. Let $w \in E$ is a non-trivial weak solution for our problem. Then

$$J_\lambda(w) = \inf_{B_\rho(0)} J_\lambda \quad \text{and} \quad dJ_\lambda(w) = 0.$$

So,

$$\begin{aligned} J_\lambda(|w|) &= \int_\Omega \frac{1}{p(x)}|\nabla(|w|)|^{p(x)}dx - \lambda\int_\Omega \frac{1}{q(x)}|w|^{q(x)}dx - \int_\Omega F(|w|)dx \\ &\le \int_\Omega \frac{1}{p(x)}|\nabla w|^{p(x)}dx - \int_\Omega \frac{1}{q(x)}|w|^{q(x)} - \int_\Omega F(w)dx \\ &= J_\lambda(w). \end{aligned}$$

It follows that $J_\lambda(|w|) = \inf_{B_\rho(0)} J_\lambda$ and $dJ_\lambda(|w|) = 0$. We conclude that $|w|$ is a nontrivial weak solution of problem (1). □

REFERENCES

[1] C. O. Alves and M. A. Suoto, Existance of solutions for a class of problems in $\mathbb{R}^N$ involving the $p(x)$-Laplacian. In: T. Cazenave, D. Costa, O. Lopes, R. Manásevich, P. Rabinowitz, B. Ruf, C. Tomei (Eds.), *Contributions to Nonlinear Analysis, A Tribute to D. G. de Figueiredo on the Occasion of this* 70*th Birthday* (Progress in Nonlinear Differential Equations and Their Applications, Vol. **66**), pp. 17-32. Birkhäuser, Basel, 2006. MR2187792 (2006g35050).

[2] H. Le Dret, M2-Équations aux dérivées partielles elliptiques, *Notes de cours*. Université Pierre Marie Curie, 2005.

[3] J. Chabrowski and Y. Fu, Existence of solutions for $p(x)$-Laplacian problems on a bounded domain. *J. Math. Anal. Appl.* **306** (2005), 604-618. MR2136336 (2006e:35087).

[4] J. Musielak, *Orlicz Spaces and Modular Spaces* (Lecture Notes in Math. **1034**). Springer, Berlin, 1983.

[5] M. Mihăilescu and V. Rădulescu, A multiplicity result for a nonlinear degenerate problem arising in the theory of electrorheological fluids. *Proc. Roy. Soc. London Ser. A* **462** (2006), 2625-2641.

[6] M. Mihăilescu and V. Rădulescu, On a nonhomogeneous quasilinear eigenvalue problem in Sobolev spaces with variable exponent. *Proc. Amer. Math. Soc.* **135** (2007), 2929-2937.

[7] M. Mihăilescu and V. Rădulescu, Existence and multiplicity of solutions for quasilinear nonhomogeneous problems: an Orlicz-Sobolev space setting. *J. Math. Anal. Appl.*, in press.

[8] O. Kováčik and J. Rákosník, On spaces $L^{p(x)}$ and $W^{1,p(x)}$. *Czechoslovak Math. J.* **41** (1991), 592-618.

[9] X. Fan and D. Zhao, On the generalized Orlicz-Sobelev space $W^{k,p(x)}(\Omega)$. *J. Gancu Educ. College 12* **1** (1998), 1-6.

[10] X. L. Fan and D. Zhao, On the Spaces $L^{p(x)}(\Omega)$ and $W^{m,p(x)}(\Omega)$. *J. Math. Anal. Appl.*, **263** (2001), 424-446.

[11] X. Fan, J. Shen and D. Zhao, Sobolev Embedding Theorems for Spaces $W^{k,p(x)}(\Omega)$. *J. Math. Anal. Appl.* **262** (2001), 749-760.

[12] X. L. Fan and Q. H. Zhang, Existence of solutions for $p(x)$-Laplacian Dirichlet problem. *Nonlinear Anal.* **52** (2003), 1843-1852.

INSTITUT PRÉPARATOIRE AUX ETUDES D'INGÉNIEURS DE TUNIS, TUNISIA
E-mail address: benali.khaled@yahoo.fr

2 RUE JAWAHER LEL NEHRU, 1008 MONTFLEURY-TUNIS, TUNISIA
E-mail address: mounir.bezzarga@yahoo.fr

Potential Theory and Stochastics in Albac, 35–47
Aurel Cornea Memorial Volume

Γ-CONVERGENCE FOR STRONGLY LOCAL DIRICHLET FORMS IN OPEN SETS WITH HOLES

MARCO BIROLI

ABSTRACT. We consider the Γ-convergence of problems with holes relative to strongly local Dirichlet forms in the cases in which we have Dirichlet or Neumann homogeneous conditions on the boundaries of holes. In the first case the results we give here are mainly contained in the papers [26], [27] (see also [6], [7], [2]); in the second case the main difficulties arrive from the absence of a group structure on the underlying space and then from the non-periodic distribution of the holes. The complete proofs of the results are contained in the paper [8].

1. INTRODUCTION

In this paper we will give some results on the Γ-convergence of problems with holes relative to a Riemannian or a Poincaré type Dirichlet form. We consider both the cases of homogeneous Dirichlet or Neumann boundary conditions on the boundary of holes. In this second case one of the main difficulties is that we operate on a locally compact metrizable Hausdorff space, then no structure of group is available so there is no notion of periodicity on our space. We divide the introduction of the paper into two parts: in the first part we recall previous results given in elliptic or subelliptic case and in the second one we give the notion of Poincaré type and Riemannian Dirichlet form with some properties. We observe that some of the results hold also for more general type of Dirichlet forms and you can find the results in all their generality in the references cited; for sake of simplicity we prefer to use a framework in which all the results hold.

We consider firstly the case of Dirichlet homogeneous conditions. In this case the Euclidean elliptic case was studied firstly in [22], [25], but our reference papers for the Euclidean case will be [13], [14], where a result of Γ-convergence is given for a general distribution of the holes. The papers [13],[14] point out that in the limit problems a term of the type μu appears where μ is a positive Borel measure (u is the unknown function). This result suggest to embed our problem in a more general class of problems i.e. the relaxed Dirichlet problems as defined in [18], [19] where the Γ-convergence for this type of problems is studied (see also [16]). For the case of Neumann boundary condition the first paper in which for a particular Euclidean elliptic problem with periodically distributed holes the convergence to a limit problem is rigorously proved [15], more general Euclidean cases are considered in [23]. L. Tartar

2000 *Mathematics Subject Classification.* 31C25, 35B27, 49J45, 35H20.

Key words and phrases. Dirichlet forms, homogenization problems, variational convergence, subelliptic operators.

made the remark that all the proofs depend on the existence of suitable extension operators (in the holes); the problem in Euclidean setting without periodicity but with the assumption of the existence of suitable extension operators (in the holes) has been solved in [12]. We remark that the question of existence of extension operators with the required properties is solved in Euclidean setting with some assumption on the regularity of the boundary of the holes but becomes very delicate in the subelliptic setting so it is important to investigate methods, which can operate in absence of classical extension theory; for paper in this direction see [23], [1], [11], [20], [29], [30], [31] for the Euclidean case and [9] for the periodic case in the Heisenberg (subelliptic) group.

We will now introduce the notion of Poincaré type or Riemannian Dirichlet forms. Let X be a metrizable locally compact Hausdorff space with a positive Radon measure m such that $supp[m] = X$; we assume that we are given a *strongly local of diffusion type, regular Dirichlet form* in the Hilbert space $L^2(X,m)$, in the sense of M. Fukushima, [21], whose domain is denoted by $D[a]$. Such a form a admits the following integral representation $a(u,v) = \int_X \alpha(u,v)(dx)$ for every $u, v \in D[a]$ where $\alpha(u,v)$ is a signed Radon measure on X, uniquely associated with the functions u, v with $\alpha(u.v) = \alpha(v,u)$. Moreover for any open subset Ω of X the restriction of $\alpha(u,v)$ to Ω depends only on the restrictions of u and v to Ω and $\alpha(u,v) = 0$ whenever $u = cst$ on a neighbourhood of the support of v. Let Ω be an open set; by $D_0[a,\Omega]$ we denote the closure of $C_0(\Omega)\cap D[a]$ in $D[a]$. By $D_{loc}[a,\Omega]$ we denote the space of all m-measurable functions u, v in X, that coincide m-a.e. on every compact subset of Ω with some function of $D[a]$. The measure $\alpha(u,v)$ is defined unambigously in Ω for all $u, v \in D_{loc}[a,\Omega]$. We refer to [4], [21], [26], [27] for the properties of $\alpha(u,v)$ with respect to Leibnitz, chain and troncation rules. The above assumptions allow us to define a capacity relative to the form a (and to the measure space (X,m)). The capacity of an open set O is defined as

$$cap(O) = \inf\{a(v,v);\ v \in D[a],\ v \geq 1 \text{ a.e. on } O\}$$

if the set $\{v \in D_0\ , v \geq 1 \text{ a.e. on } O\}$ is not empty and

$$cap(O) = +\infty$$

otherwise. Let E be a subset of X, we define

$$cap(E) = \inf\{cap(O); O \text{ open set with } E \subset O\}.$$

We recall that the above defined capacity is a Choquet capacity. The strong locality property allows us to define , given an open set O in X, a Choquet capacity $cap(E;O)$ of a set E with respect to O, where $E \subset \overline{E} \subset O$. Moreover the set $B(x,r) = \{y; d(x,y) < r\}$, $B(r)$ will be balls $B(x,r)$ with a fixed center x. We say that the Dirichlet form is of *Poincaré type* if the following two assumptions hold:

(H_1) For every fixed compact set K there exist positive constants c_0, ν and r_0 such that

$$0 < c_0(\frac{r}{R})^\nu m(B(x,R)) \leq m(B(x,r))$$

where $x \in K$, $0 < r < R < r_0$ and c_0 is independent of $x \in K, 0 < r < r_0$.

(H_2) For every fixed compact set K there exist positive constants c_0 and r_1 such that for every $f \in D_{loc}[\Omega]$ the following *scaled Poincaré inequality* holds

$$\int_{B(x,r)} |f - f_{x,r}|^2 m(dx) \leq c_1 r^2 \int_{B(x,kr)} \alpha(f,f)(dx)$$

where $0 < r < r_1$, $x \in K$ and $c_1, k \geq 1$, are independent of $x \in K, 0 < r < r_1$.

We can assume without loss of generality $r_0 = r_1$.

Remark 1.1 We observe that (H_1) holds if *a doubling property* holds for every compact set K, that is for every fixed compact set K there exist positive constants c_0 and r_1 such that

$$m(B(x,2r)) \leq c_0^\star \; m(B(x,r)),$$

where $x \in K$, $0 < r < R < r_0$ and $c_0^\star$ is independent of $x \in K, 0 < r < r_0$.

If d is a distance on X and $d \in D_{loc}[a]$ with $\alpha(d,d) \leq m$ we say that our Dirichlet form is a *Riemannian* Dirichlet form

Remark 1.2 We observe that if

$$d_a(x,y) = sup\{\phi(x) - \phi(y); \forall \phi \in D[a] \cap C_0(X) \; with \; \alpha(\phi,\phi) \leq m\}$$

and $d_a(x,y)$ is finite and separating (i.e. if $x \neq y$ then $d_a(x,y) \neq 0$) then d_a is a distance and $d_a \in D_{loc}[a]$ with $\alpha(d_a, d_a) \leq m$.

Remark 1.3 If a is Riemannian there exists cut-off functions between balls (i.e. given two balls $B(x,r)$ and $B(x,R)$, $r < R < r_0$, there exists a function ϕ with $\phi = 1$ on $B(x,r)$, $\phi = 0$ out of $B(x,R)$ and $\alpha(\phi,\phi) \leq \frac{1}{(R-r)^2}$

We recall that for a Dirichlet form satisfying the above assumptions a Green function $G_\Omega(x,y) = G_\Omega(y,x)$ relative to a relatively compact open set Ω can be defined and that, in the case $\Omega = B(x_0,R)$, the following estimate holds:

$$G_{B(x_0,R)}(x_0,y) \approx \int_{d(x_0,y)}^{R} \frac{s^2}{m(B(x,s))} \frac{ds}{s} \tag{1.1}$$

for every y such that $d(x_0,y) \leq \frac{R}{2}$. As a consequence of the above estimate we have also an estimate on the capacity of a ball with respect to a concentric ball

$$cap(B(x,r);B(x,R)) \approx (\int_{r}^{R} \frac{s^2}{m(B(x,s))} \frac{ds}{s})^{-1} \tag{1.1}$$

for every y such that $r \leq \frac{R}{q}$, $q > 1$, where the constants in the equivalence depend on q. Finally we observe that all our assumptions hold for Dirichlet forms associated to wide classes of

(a) weighted uniformly elliptic operators;

(b) weighted degenerate elliptic operators generated by vector fields satisfying an Hörmander condition;

(c) subelliptic operators;

see [3], [4], [24] for more details. Moreover also the Laplacian on p.c.f. fractals define a Poincarè type Dirichlet form, [28].

2. THE CASE OF HOMOGENOUS DIRICHLET BOUNDARY CONDITION

In this section we will consider the case of Dirichlet homogeneous boundary conditions on the boundary of the holes. Let Ω be a relatively compact open proper subset of X. We consider a sequence of open sets Ω_ϵ contained in Ω, in this case the hole is given by the set $\Omega_\epsilon^c = \Omega - \Omega_\epsilon$.
We consider the problems

$$\int_{\Omega_\epsilon} \alpha(u,v)(dx) = < f, v >_{(D_0[a,\Omega])', D_0[a,\Omega]}$$

$$(2.1_\epsilon) \qquad u \in D_0[a,\Omega_\epsilon]$$

$$\text{for all } v \in D_0[a,\Omega_\epsilon],$$

where $\alpha(u,v)$ is the energy density of a Riemannian Dirichlet form on X. The equation (2.1_ϵ) makes sense why every function $v \in D_0[a,\Omega_\epsilon]$ can be extended by 0 to a function (denoted again by v) in $D_0[a,\Omega]$. We have that:

Theorem 2.1 *Let u_ϵ be the solution of problem (2.1_ϵ). The sequence u_ϵ weakly converges (at least after extraction of subsequences) in $D_0[a,\Omega]$ to u_0 which is the unique solution of the problem*

$$\int_{\Omega} \alpha(u,v)(dx) + \int_{\Omega} uv\, \mu(dx) = < f, v >_{(D_0[a,\Omega])', D_0[a,\Omega]}$$

$$(2.1_0) \qquad u \in D_0[a,\Omega]$$

$$\textit{for all } v \in D_0[a,\Omega],$$

where μ is a positive Borel measure that does not charge sets of zero capacity (relative to a).

In the following we denote by $\mathcal{M}^+$ the set of positive Borel measures on Ω, which does not charge sets of zero capacity.
We recall that the proof of Theorem 2.1 is obtained by the embedding of the class of problems (2.1_ϵ) into the more general class of relaxed Dirichlet problems. Consider the problems

$$\int_{\Omega} \alpha(u,v)(dx) + \int_{\Omega} uv\, \mu_\epsilon(dx) = < f, v >_{(D_0[a,\Omega])', D_0[a,\Omega]}$$

$$(2.1'_\epsilon) \qquad u \in D[a,\Omega] \cap L^2(\Omega, \mu_\epsilon)$$

$$\text{for all } v \in D[a,\Omega] \cap L^2(\Omega, \mu_\epsilon),$$

where μ_ϵ are positive Borel measures that does not charge sets of zero capacity (relative to a). We observe that $(2.1'_\epsilon)$ reduces to (2.1_ϵ) if we choose

$$(2.2) \qquad \mu_\epsilon(B) = 0$$

if B is a Borel set such that $B \cap \Omega_\epsilon^c$ has zero capacity and

$$(2.3) \qquad \mu_\epsilon(B) = +\infty$$

if B is a Borel set such that $B \cap \Omega_\epsilon^c$ has positive capacity.

Definition 2.1 We say that the sequence μ_ϵ Γ*-converges to* μ if the sequence u_ϵ of solutions of problems ($2.1'_\epsilon$) weakly converges in $D_0[a,\Omega]$ to u_0 which is the unique solution of the problem

$$\int_\Omega \alpha(u,v)(dx) + \int_\Omega uv\ \mu(dx) =< f, v >_{(D_0[a,\Omega])', D_0[a,\Omega]}$$

$$(2.1'_0) \qquad u \in D_0[a,\Omega] \cap L^2(\Omega,\mu)$$

$$\text{for all } v \in D_0[a,\Omega] \cap L^2(\Omega,\mu).$$

Remark 2.1 The limit measure μ does not change if we replace Ω by a relatively compact set Ω' containing Ω and we define as zero the density of μ_ϵ and μ out of Ω.

The proof of Theorem 2.1 is done if we prove the following result:

Theorem 2.2 *The set $\mathcal{M}^+$ (of positive Borel measures on Ω that does not charge sets of zero capacity) is compact under Γ-convergence.*

For the proof of this result we refer to [16], [19] for the Euclidean case and the proof in the Euclidean case can be carried on for the general case, see [26]. We recall that the method of the proof stands on the equivalence between the Γ-convergence of the measures μ_ϵ and the Γ-convergence of the functionals $F_{\mu_\epsilon} : L^2(\Omega) \to [0,+\infty]$

$$F_{\mu_\epsilon} = a(u,u) + \int_\Omega u^2 \mu_\epsilon(dx); \quad u \in D_0[a,\Omega] \cap L^2(\Omega,\mu_h),$$

$F_{\mu_\epsilon} = +\infty$ otherwise, to the functional

$$F_\mu = a(u,u) + \int_\Omega u\mu(dx); \quad u \in D_0[a,\Omega] \cap L^2(\Omega,\mu),$$

$F_\mu = +\infty$ otherwise, following the definition of Γ-convergence of functionals in [26]. We observe also that Theorem 2.2 gives also that the Borel measure μ in Theorem 2.1 does not change if we enlarge Ω conserving the same holes. We also observe that the result in Theorem 2.2 (and then the one in Theorem 2.1) holds again in the framework of Poincaré type Dirichlet forms. Moreover the following result has been proved in [7]:

Theorem 2.3 *Assume that there is no point of positive capacity in Ω. The Borel measures defined by holes, i.e. defined as in* (2.2), (2.3), *are dense for Γ-convergence in $\mathcal{M}^+$.*

The proof uses the following lemma:

Lemma 2.1 *Let $\mu \in (D_0[a,\Omega])' \cap \mathcal{M}^+$. If there exist two positive measures ν_ϵ, $\lambda_\epsilon \in (D_0[a,\Omega])'$ and a sequence of functions $w_\epsilon \in D_{loc}[a,\Omega]$ with $\int_\Omega \alpha(w_\epsilon,w_\epsilon)(dx)$ bounded such that*

$$\lim_{\epsilon\to 0} \int_\Omega \alpha(w_\epsilon, v)(dx) + \int_\Omega w_\epsilon v\ m(dx) \int_\Omega v\ m(dx)$$

for any $v \in D_{loc}[a,\Omega]$ with $\int_\Omega \alpha(v,v)(dx)$ finite,

$$w_\epsilon = 0 \text{ q.e. in } \Omega_\epsilon,$$

$$a(w_\epsilon, v) =< \nu_\epsilon - \lambda_\epsilon, v >_{(D_0[a,\Omega])', D_0[a,\Omega]}$$

for any $v \in D_0[a,\Omega]$,

$$\lim_{\epsilon\to 0} \nu_h = \mu$$

strongly in $(D_0[a,\Omega])'$,

$$\lim_{\epsilon\to 0}\lambda_h=\mu$$

weakly in $(D_0[a,\Omega])'$,

$$<\lambda_\epsilon, v>_{(D_0[a,\Omega])',D_0[a,\Omega]}=0$$

for $v\in D_0[a,\Omega]$ *with* $v=0$ *in* Ω_ϵ. *Then the measures corresponding to the holes* Ω_ϵ *(by (2.2)–(2.3))* Γ*-converge to the measure* μ.

The lemma is proved in [7], where the functions w_ϵ and the measures ν_ϵ, μ_ϵ corresponding to a measure μ which is a Kato measure (for the definition see [5]), then we use the representation result for measures in $\mathcal{M}^+$ in [17] to prove the result in the general case.

Remark 2.2 It is possible to prove by similar methods the result of Theorem 2.3 avoiding the assumption of the absence of points of positive capacity in Ω if we assume that the measure μ is diffuse (i.e. does not charge the points). This last assumption can not be suppressed, since also in the Euclidean case with dimension one (where all the points have positive capacity) the result is false for non diffuse measures.

As in the Euclidean case, [19], in the Riemannian case it is possible to characterize the points, where a measure in $\mathcal{M}^+$ represents holes [2].

Definition 2.2 Let μ be in $\mathcal{M}^+$. A point $x_0\in\Omega$ is called a *regular point* for μ if any solution of the problem

$$\int_\Omega a(u,v)(dx)+\int_\Omega (u-g)v\,\mu(dx)=0$$

$$u\in D_{loc}[a,\Omega]\cap L^2_{loc}(\Omega,\mu) \tag{2.4}$$

for all $v\in D_0[a,\Omega]\cap L^2(\Omega,\mu)$, *with compact support in* Ω,

where Ω is any relatively compact neighborhood of x_0 and g is an arbitrary function $D_0[a,\Omega]\cap C(\Omega)$, assumes the value $g(x_0)$ with continuity at x_0.

It is possible to characterize the regular point for μ by a Wiener type criterion using a suitable notion of capacity:

Definition 2.3 Let μ be in $\mathcal{M}^+$, E be a Borel set contained in the open set Ω and μ_E be the restriction of μ to E. The set E is μ*-admissible* (in Ω) if there exists $w\in D_0[a,\Omega]$ such that $(w-1)\in L^2(X,\mu_E)$; if E is μ-admissible we define

$$cap_\mu(E,\Omega)=\min\{a(v,v)+\int (v-1)^2\mu_E(dx);\ v\in D[a,\Omega]\}.$$

Remark 2.3 Let B be an open set with closure contained in Ω and assume that μ is the Borel measure relative to the hole $B^c=\Omega-B$, then if $x_0\in\partial B$

$$cap_\mu(B(x_0,r),B(x_0,k^\star r))=cap(B(x_0,r)\cap B^c,B(x_0,k^\star r)),$$

$k^\star>1$.

Let
$$\delta(r,x_0)=\frac{cap_\mu(B(x_0,r),B(x_0,k^\star r))}{cap_(B(x_0,r),B(x_0,k^\star r))}$$
where $k^\star>1$ depends on k in (H_2). The following result is proved in [2]:

Theorem 2.3 *A point x_0 is a regular point for $\mu\in\mathcal{M}^+$ if and only if x_0 is a Wiener point for μ, i.e.*
$$\lim_{r\to 0}\int_r^1 \delta(\rho,x_0)\frac{d\rho}{\rho}=+\infty.$$
The proof of Theorem 2.3 has been given in [2] using a refinement of methods used in Euclidean case in [19] due to the presence of points of positive μ-capacity. The refinement is inspired to the methods used in [4] to prove the Harnack inequality for the harmonic functions relative to a Riemannian Dirichlet form and is founded on the estimate on the Green function relative to balls obtained in [4].

Remark 2.4 We remark that in[2] is also proved an estimate of the rate of convergence to 0 in terms of the Wiener integral defined in Theorem 2.3. Moreover in the case of Remark 2.3 the Theorem 2.3 is equivalent to the Wiener criterion for the regularity of a point x_0 in the boundary of B.

Remark 2.5 If we are in the Riemannian case and the energy density $\alpha(u,u)(dx)$ is absolutely continuous with respect to m, i.e. $\alpha(u,u)(dx)=\alpha(u,u)(x)m(dx)$ where $\alpha(u,u)(.)\in L^1(X,m)$, for every $u\in D[a]$ and there exist n bounded linear operators $L_i:D[a]\to L^2(X,m)$, $i=1,...,n$, such that
$$\lambda\sum_{i=1}^n|L_iu(x)|^2\le\alpha(u,u)(x)\le\Lambda\sum_{i=1}^n|L_iu(x)|^2 \quad \text{a.e. in } X$$
we can define $D_0^p[a,\Omega]$, $p>1$, as the closure of $D[a]\cap C_0(\Omega)$ for the norm
$$||u||_p=(\int_\Omega\sum_{i=1}^n|L_iu(x)|^p m(dx)+\int_\Omega|u|^p m(dx))^{\frac{1}{p}}$$
where Ω is an open relatively compact set. The space $D_0^p[a,\Omega]$ is an uniformly convex Banach space and it can be proved that if Ω is relatively compact the solutions of (2.1_ϵ) converges strongly (at least after extraction of subsequences) in $D_0^p[a,\Omega]$ for $1<p<2$, see [6].

3. THE CASE OF HOMOGENEOUS NEUMANN BOUNDARY CONDITION

In this section we outline the proofs of the results, [8]. We consider a countable covering $\mathcal{F}_\epsilon$ of the space obtained by balls of radius $(1+\delta)\epsilon$ ($\epsilon>0$, $\delta>0$ fixed), such that the homotetic balls of radius ϵ do not intersect.We assume also that $\mathcal{B}_\epsilon$ has a property of finite intersection uniform with respect to ϵ (i.e. every point of X belongs to at most Q balls in the covering, where Q does not depend on ϵ).

We denote by $B_{i,(1+\delta)\epsilon}$, $i=1,2,\dots$, the balls in $\mathcal{B}_\epsilon$ and by $B_{i,\epsilon}$, $i=1,2,\dots$, the homotetic balls of radius ϵ. Consider now a relatively compact open set Ω strictly contained in X with boundary Γ and denote by $B_{j,\epsilon}$, $j=1,2,\dots,q$, a subfamily of the balls $B_{i,\epsilon}$, such that $B_{i,\epsilon(1+\delta)}$ is contained in Ω, where Ω is a relatively compact open set with boundary Γ (the number of the balls $B_{j,\epsilon}$ is finite due to the homogeneous structure of X). In every ball $B_{j,\epsilon(1-\delta)}$ we consider a compact set $T_{j,\epsilon}$. We denote $\Omega_\epsilon = \Omega - \cup_j T_{j,\epsilon}$. Let θ_ϵ be the characteristic function of Ω_ϵ We assume that θ_ϵ converges in the weak* topology of $L^\infty(\Omega, m)$ to a function θ this property always hold at least after extraction of subsequences) with $0 < \sigma < \theta \le 1$. We consider on X a Riemannian type form $a(u, v) = \int \alpha(u, v)(dx)$ with domain $D[a]$; we denote by $D_0[a, \Omega]$ the domain of the restriction of the form to Ω and we denote by V_ϵ the closure of the space of the functions v in $C(\Omega_\epsilon) \cap D_{loc}[a, \Omega_\epsilon]$ with $supp(v) \cap \Gamma = \emptyset$ for the norm

$$||v||_\epsilon = [\int_{\Omega_\epsilon} \alpha(v, v)(dx) + \int_{\Omega_\epsilon} |v|^2 m(dx)]^{\frac{1}{2}}.$$

We assume that the following scaled Poincaré inequalities hold

$$(P_1) \qquad \int_{B_{j,\epsilon} - T_{j,\epsilon}} |v - \bar{v}_{j,\epsilon}|^2 m(dx) \le C_2 \epsilon^2 \int_{B_{j,\epsilon(1+\delta)} - T_{j,\epsilon}} \alpha(v, v)(dx)$$

for every $v \in V_\epsilon$ where C_2 is a constant independent of j, ϵ and $\bar{v}_{j,\epsilon}$ is the average of v for the measure $m(dx)$ on $B_{j,\epsilon} - B_{j,\epsilon(1-\delta)}$.

Remark 3.1 We can replace in (P_1) $\bar{v}_{j,\epsilon}$ by the average of v on $B_{j,\epsilon} - T_{j,\epsilon}$.

Using (P_1) we can prove the following coercivity inequality:

Proposition 3.1 *There exists $\epsilon_0 > 0$ such that for $\epsilon \le \epsilon_0$ we have*

$$\int_{\Omega_\epsilon} \alpha(v, v)(dx) \ge \lambda ||v||^2_{L^2(\Omega_\epsilon)}$$

where $v \in V_\epsilon$ and λ is a positive constant.

Let now $\eta_{j,\epsilon}$ be the cut-off function of $B_{j,\epsilon(1-\delta)}$ with respect to $B_{j,\epsilon}$ and define the linear operator $P^\epsilon : V_\epsilon \to D_0[a, \Omega]$ as

$$P^\epsilon v = (1 - \sum_j \eta_{j,\epsilon}) v + \sum_j \eta_{j,\epsilon} \bar{v}_{j,\epsilon}, \text{ on } \Omega_\epsilon,$$

$$P^\epsilon v = \sum_j \eta_{j,\epsilon} \bar{v}_{j,\epsilon}, \text{ on } \Omega - \Omega_\epsilon.$$

We have

Proposition 3.2 *The linear operator P^ϵ is bounded. Moreover denoted by $\tilde{v}_\epsilon$ the extension of v by $\bar{v}_{j,\epsilon}$ to $T_{j,\epsilon}$ ($\tilde{v}_\epsilon$ is so defined on Ω) we have that $\lim_{\epsilon\to 0}(P^\epsilon v - \tilde{v}_\epsilon) = 0$ in $L^2(\Omega, m)$. Consider now a sequence $v_\epsilon \in V_\epsilon$ such that $||v_\epsilon||_\epsilon$ is bounded; then there exists a subsequence, denoted again by v_ϵ, such that $\lim_{\epsilon\to 0} P^\epsilon v_\epsilon = w$ weakly in $D_0[a, \Omega]$, strongly in $L^2(\Omega, m)$ and $\lim_{\epsilon\to 0}(P^\epsilon v_\epsilon - \widetilde{(v_\epsilon)}_\epsilon = 0$ in $L^2(\Omega, m)$ (equivalently $||P^\epsilon v_\epsilon - v_\epsilon||_{L^2(\Omega_\epsilon, m)}$ converges to 0), more exactly*

$$||(P^\epsilon v_\epsilon - \widetilde{(v_\epsilon)}_\epsilon||^2_{L^2(\Omega)} \le C\epsilon^2 \int_{\Omega_\epsilon} \alpha(v_\epsilon, v_\epsilon)(dx),$$

$$||(P^\epsilon v_\epsilon - v_\epsilon)||^2_{L^2(\Omega_\epsilon)} \le C\epsilon^2 \int_{\Omega_\epsilon} \alpha(v_\epsilon, v_\epsilon)(dx),$$

$\widetilde{(v_\epsilon)}_\epsilon$ converges to w in $L^2(\Omega, m)$, and $||v_\epsilon - w||_{L^2(\Omega_\epsilon, m)}$ converges to 0. Moreover if v is a function in $D_0[a, \Omega]$ then $(P^\epsilon v - v)$ converges to 0 in $L^2(\Omega, m)$.

The proof is based on the Poincaré inequality (P_1) and the compact embedding of $D_0[a,\Omega]$ into $L^2(X,m)$, [10].

In the following we will use P^ϵ as an approximate extension operator.

We consider now the following problems

$$\int_{\Omega_\epsilon} \alpha(u_\epsilon, v)(dx) = < f, P^\epsilon v >_{(D_0[a,\Omega])', D_0[a,\Omega]}, \tag{3.1$_\epsilon$}$$

$$u_\epsilon \in V_\epsilon, \quad \forall v \in V_\epsilon.$$

The above problems have a unique solution u_ϵ. Choosing $v = u_\epsilon$ we obtain that

$$\int_{\Omega_\epsilon} \alpha(u_\epsilon, u_\epsilon)(dx) \le C$$

where C is a constant independent of ϵ. We define the linear operator $B_\epsilon : (D_0[a,\Omega])' \to D_0[a,\Omega]$ as

$$B_\epsilon f = P^\epsilon u_\epsilon.$$

It is easy to prove that the operators B_ϵ are uniformly bounded and, at least after extraction of a subsequence denoted again by B_ϵ , weakly converge at every $f \in (D_0[a,\Omega])'$ to a linear operator $B_0 : (D_0[a,\Omega])' \to D_0[a,\Omega]$ (i.e. $\lim_{\epsilon\to 0} B_\epsilon u = B_0 f$ weakly in $D_0[a,\Omega]$ for every fixed $f \in (D_0[a,\Omega])'$); it means that

$$\lim_{\epsilon\to 0} P^\epsilon u_\epsilon = u_0$$

weakly in $D_0[a,\Omega]$ and strongly in $L^2(\Omega,m)$. Then the operator $B_0 : (D_0[a,\Omega])' \to D_0[a,\Omega]$ is defined as

$$B_0 f = u_0.$$

At first we observe that for $f, g \in (D_0[a,\Omega])'$

$$< g, B_0 f >_{(D_0[a,\Omega])', D_0[a,\Omega]} = < f, B_0 g >_{(D_0[a,\Omega])', D_0[a,\Omega]}$$

$$\begin{aligned} &< f, B_0 f >_{(D_0[a,\Omega])', D_0[a,\Omega]} = < f, u_0 >_{(D_0[a,\Omega])', D_0[a,\Omega]} \\ &= \lim_{\epsilon\to 0} < f, P^\epsilon u_\epsilon >_{(D_0[a,\Omega])', D_0[a,\Omega]} = \lim_{\epsilon\to 0} \int_{\Omega_\epsilon} \alpha(u_\epsilon, u_\epsilon)(dx) \\ &\ge C \limsup_{\epsilon\to 0} ||P^\epsilon u_\epsilon||^2_{D_0[a,\Omega]} \ge C||u_0||^2_{D_0[a,\Omega]} \ge C||B_0 f||^2_{D_0[a,\Omega]}. \end{aligned}$$

We prove now that B_0 is coercive

$$\begin{aligned} &< f, B_0 f >_{(D_0[a,\Omega])', D_0[a,\Omega]} = < f, u_0 >_{(D_0[a,\Omega])', D_0[a,\Omega]} \\ &= \lim_{\epsilon\to 0} < f, P^\epsilon u_\epsilon >_{(D_0[a,\Omega])', D_0[a,\Omega]} \lim_{\epsilon\to 0} \int_{\Omega_\epsilon} \alpha(u_\epsilon, u_\epsilon)(dx). \end{aligned} \tag{3.2}$$

We have also

$$\int_{\Omega_\epsilon} \alpha(u_\epsilon, v)(dx) = \int_\Omega \alpha(A^{-1} f, P^\epsilon v)(dx)$$

where A is the operator defined by the form a. Choose now v as the restriction to Ω_ϵ of $A^{-1} f$, we obtain

$$\int_{\Omega_\epsilon} \alpha(u_\epsilon, A^{-1} f)(dx) = \int_\Omega \alpha(A^{-1} f, P^\epsilon A^{-1} f)(dx),$$

then

$$\frac{1}{2}\int_{\Omega_\epsilon}\alpha(u_\epsilon,u_\epsilon)(dx)+\frac{1}{2}\int_{\Omega_\epsilon}\alpha(A^{-1}f,A^{-1}f)(dx)$$
$$\geq\int_{\Omega}\alpha(A^{-1}f,P^\epsilon A^{-1}f)(dx).$$

We obtain in the limit as $\epsilon\to 0$

$$\lim_{\epsilon\to 0}\int_{\Omega_\epsilon}\alpha(u_\epsilon,u_\epsilon)(dx)\geq\int_{\Omega}\alpha(A^{-1}f,A^{-1}f)(dx)=||f||_{(D_0[a,\Omega])'},$$

then using (3.2)

$$<f,B_0f>_{(D_0[a,\Omega])',D_0[a,\Omega]}\geq||f||_{(D_0[a,\Omega])'}.$$

Taking into account the previous properties we have that the operator B_0 is invertible and we denote by $A_0:D_0[a,\Omega]\to(D_0[a,\Omega])'$ the operator B_0^{-1}. We observe that we can prove that A_0 is defined on all $D_0[a,\Omega]$ and is bounded and coercive from $D_0[a,\Omega]$ into $(D_0[a,\Omega])'$ i.e.

$$\lambda<Au,u>_{(D_0[a,\Omega])',D_0[a,\Omega]}\leq<A_0u,u>_{(D_0[a,\Omega])',D_0[a,\Omega]}$$
$$\leq\Lambda<Au,u>_{(D_0[a,\Omega])',D_0[a,\Omega]}.$$

Then the operator A_0 defines a bilinear form $a_0(u,v)$ on $D_0[a,\Omega]\times D_0[a,\Omega]$, which is positive and such that

$$\lambda a(u,u)\leq a_0(u,u)\leq\Lambda a(u,u)$$

for every $u\in D_0[a,\Omega]$. The above domination inequality implies that the bilinear form a_0 is regular ($D_0[a_0,\Omega]=D_0[a,\Omega]$ and the norms of the graph are equivalent in the two cases). Finally we observe that $u_0=B_0f$ is the unique solution of the problem

$$a_0(u_0,v)=<f,v>_{(D_0[a,\Omega])',D_0[a,\Omega]},\quad\text{for every } v\in D_0[a,\Omega]D_0[a_0,\Omega].$$

We will now prove that a_0 is a Dirichlet form. To prove that a_0 is a Dirichlet form it is enough to prove that for a_0 a Markov type property holds. The proof of Markov property uses essentially the following semicontinuity property:

Proposition 3.3 *Let $v_\epsilon\in V_\epsilon$ be a sequence such that*

$$\int_{\Omega_\epsilon}\alpha(v_\epsilon,v_\epsilon)(dx)\leq C$$

and $P^\epsilon v_\epsilon$ converges to v_0 as $\epsilon\to 0$ weakly in $D_0[a,\Omega]$, then

$$\liminf_{\epsilon\to 0}\int_{\Omega_\epsilon}\alpha(v_\epsilon,v_\epsilon)(dx)\geq a_0(v_0,v_0).$$

Now we assume that a has an *energy density in* $L^1(\Omega,m)$, i.e. we have $\alpha(u,u)(dx)=\alpha(u,u)(x)m(dx)$ where $\alpha(u,u)(.)\in L^1(\Omega,m)$ for all $u\in D_0[a,\Omega]$.

As in [27] we can prove also that a_0 is strongly local and we denote by $\alpha_0(u,u)(dx)$ its energy density. As a consequence of the boundness and coercivity of a_0 with respect to a and we obtain that

$$\lambda\alpha(u,u)(dx)\leq\alpha_0(u,u)(dx)\leq\Lambda\alpha(u,u)(dx)$$

(see [27]), then $\alpha_0(u,u)(dx)=\alpha_0(u,u)(x)m(dx)$ where $\alpha_0(u,u)(.)\in L^1(\Omega,m)$ for all $u\in D_0[a,\Omega]$. Moreover if a is of a Riemannian form also a_0 is a Riemannian form with the same domain as a with respect to a distance which is of the form $\Lambda^{-1}d$ where d is the distance associated with a. We have so the following result:

Theorem 3.4 *Let u_ϵ be a sequence of solutions of the problems (3.1_ϵ) such that $P^\epsilon u_\epsilon$ strongly converges in $L^2(\Omega, m)$ to u_0; then u_0 is the solution of the problem*

$$(3.1_0) \qquad \int_\Omega \alpha_0(u,v)(x)dx = < f, u_0 >_{(D_0[a,\Omega])', D_0[a,\Omega]}$$

$$u_0 \in D_0[a,\Omega], \quad \forall v \in D_0[a,\Omega]$$

where $a_0(u,v) = \int_\Omega \alpha(u,v)(x)dx$ is a Riemannian type Dirichlet form.

We can also prove the weak* convergence in the measures for the energy densities:

Theorem 3.5 *Let u_ϵ be a sequence of solutions of the problems (3.1_ϵ) such that $P^\epsilon u_\epsilon$ strongly converges in $L^2(\Omega, m)$ to u_0; then*

$$\int_\Omega \theta_\epsilon \alpha(u_\epsilon, u_\epsilon)\phi m(dx) \int_\Omega \alpha_0(u_0,u_0)\phi m(dx)$$

for all $\phi \in C(\Omega)$ with compact support in Ω.

We observe that the convergence result in Theorems 3.4, 3.5 hold again if in the problems (3.1_ϵ) we replace f by a sequence f_ϵ converging to f in $(D_0[a,\Omega])'$; then we obtain:

Theorem 3.5 *Let u_ϵ be solutions of the problems*

$$(3.1'_\epsilon) \qquad \int_{\Omega_\epsilon} \alpha(u_\epsilon, v)(dx) = \int_{\Omega_\epsilon} f v \;\; m(dx)$$

$$u_\epsilon \in V_\epsilon \;\; \forall v \in V_\epsilon$$

where $f \in L^2(\Omega, m)$ and assume that $P^\epsilon u_\epsilon$ strongly converges in $L^2(\Omega, m)$ to u_0; then u_0 is the solution of the problem

$$(3.1'_0) \qquad \int_\Omega \alpha_0(u,v)(x)dx = \int_\Omega \theta f u_0 \;\; m(dx)$$

$$u_0 \in D_0[a,\Omega] \;\; \forall v \in D_0[a,\Omega]$$

where α_0 is as in Theorem 3.4. Moreover we have

$$\int_\Omega \theta_\epsilon \alpha(u_\epsilon, u_\epsilon)\phi m(dx) \int_\Omega \alpha_0(u_0,u_0)\phi m(dx)$$

for all $\phi \in C(\Omega)$ with compact support in Ω.

ACKNOWLEDGMENTS

Research supported by MIUR Research Project 20050173.

References

[1] G. Allaire and F. Murat, Homogenization of the Neumann problem with non-isolated holes. *Asymp. An.* **7** (1993), 81-95.

[2] M. Biroli, The Wiener test for Poincaré-Dirichlet forms. In: *Classical and Modern Potential Theory and Applications*, pp. 93-104, Kluwer Publishers, Amsterdam, 1994.

[3] M. Biroli and U. Mosco, Formes de Dirichlet et estimations structurelles dans les milieux discontinus. *C.R.A.S. Paris* **315** (1991), 193-198.

[4] M. Biroli and U. Mosco, A Saint-Venant principle for Dirichlet forms on discontinuous media. *Ann. Mat. Pura Appl.* **169**, pp. 125-181.

[5] M. Biroli and U. Mosco, Kato space for Dirichlet forms. *Potential Analysis* **10** (1999), 327-345.

[6] M. Biroli and N.A. Tchou, Asymptotic behavior of relaxed Dirichlet problems involving a Dirichlet-Poincarè inequality. *Z.A.A.* **16** (1997), 281-309.

[7] M. Biroli and N.A. Tchou, Relaxation for Dirichlet problems involving a Dirichlet form. *Z.A.A.* **19** (1997), 203-225.

[8] M. Biroli and N.A. Tchou, Γ-convergence for strongly local Dirichlet forms in perforated domains with homogeneous Neumann boundary conditions. *Adv. Math. Sci. Appl.* **17** (2007), 149-179.

[9] M. Biroli, N.A. Tchou and V.V. Zhikov, Homogenization for Heisenberg operator with Neumann boundary conditions. *Ric. Mat.* **48** (1999) Supplemento, 45-59.

[10] M. Biroli and S. Tersian, On the existence of nontrivial solutions to a semilinear equation relative to a Dirichlet form. *Rend. Istituto Lombardo Acc. di Scienze e Lettere, A* **131** (1997), 151-168.

[11] M. Briane, Poincarè-Writinger inequality for homogenization in perforated domains. *Boll. UMI* **11-B** (1987), 53-82.

[12] M. Briane, A. Damlamian and P. Donato, H-convergence in perforated domains. *Nonlinear Partial Differential Equations and Applications, Collège de France Seminar* **13** (1998) Research Notes in Math. Pitman London, 62-100.

[13] D. Cioranescu and F. Murat, Un terme étrange venu d' ailleurs, I. *Nonlinear Partial Differential Equations and Applications, Collège de France Seminar* **2** (1998) Research Notes in Math. Pitman London, 93-138.

[14] D. Cioranescu and F. Murat, Un terme étrange venu d' ailleurs, II. *Nonlinear Partial Differential Equations and Applications, Collège de France Seminar* **3** (1983) Research Notes in Math. Pitman London, 154-178.

[15] D. Cioranescu and J. Saint Jean Paulin, Homogenization in open sets with holes. *J. Math. An. Appl.* **71** (1979), 590-607.

[16] G. Dal Maso: Γ-convergence and μ capacities. *Ann. Scuola Norm. Sup. Pisa Cl. Scienze* **14** (1987), 423-464.

[17] G. Dal Maso and V. De Cicco, Evans-Vasilescu theorem in Dirichlet spaces. *Rendiconti di Mat.* **19** (1999), 1-15.

[18] G. Dal Maso and U. Mosco, Wiener criteria and energy decay for relaxed Dirichlet problems. *Arch. Rat. Mech. An.* **95** (1986), 345–387.

[19] G. Dal Maso and U. Mosco, Wiener's criterion and Γ-convergence. *Appl. Math. Optim.* **15** (1987), 15-63.

[20] A. Damlamian and P. Donato, Periodic homogenization in perforated domains with Neumann conditions. *Homogenization 2001* **18** (2003) (Gakuto international Series in Mathematical Sciences), 169-182.

[21] M. Fukushima: *Dirichlet Forms and Markov Processes.* North Holland Math. Lib., North Holland Amsterdam, 1980.

[22] E.Ja. Hruslov, The method of hortogonal projections and the Dirichlet problem in domains with a fine-grained boundary. *Mat. Sb. USSR* **17** (1972), 37-59.

[23] E.Ja. Hruslov, The asymptotic behavior of solutions of the second boundary value problems under fragmentation of the boundary of the domain. *Mat. Sb. USSR* **35** (1979), 266-282.

[24] G. Lu, *On Harnack's inequality for a class of degenerate Schródinger operators formed by vector fields*, preprint.

[25] V.A. Marčenko and E.Ja. Hruslov, *Boundary value problems in domains with fine-grained boundaries*, Nauk. Dumka, Kiev, 1974.

[26] U. Mosco, Compact families of Dirichlet forms. *Cours 3ème cycle*, Université Paris VI, 1992.
[27] U. Mosco, Composite media and asymptotic Dirichlet forms. *J. Func. An.* **123**(1994), 368–421.
[28] U. Mosco, Dirichlet forms and self-similarity. *New Directions in Dirichlet forms* **8** (1998) (AMS/IP Studies in Advanced Mathematics), International Press, Cambridge (MA), 117-155.
[29] V.V. Zhikov, Averaging in perforated random domains of general type. *Math. Notes* **53** (1993), 30-42.
[30] V.V. Zhikov, On the homogenization of nonlinear variational problems in perforated domains. *Russian J. of Math. Phys.* **2** (1994), 393-408
[31] V.V. Zhikov, Connectedness and homogenization. Examples of fractal conductivity. *Mat. Sb. USSR* **187** (1996), 1109-1147.

DIPARTIMENTO DI MATEMATICA "F. BRIOSCHI", POLITECNICO DI MILANO E ACCADEMIA NAZIONALE DELLE SCIENZE DETTA DEI XL, ITALIA
E-mail address: marco.biroli@polimi.it

Potential Theory and Stochastics in Albac, 49–64
Aurel Cornea Memorial Volume

DIVERGENCE OF RATIONAL APPROXIMANTS TO NON-ANALYTIC FUNCTIONS

HANS-PETER BLATT

Dedicated to the memory of Aurel Cornea

ABSTRACT. The convergence behaviour of best uniform rational approximations $r^*_{n,m}$ with numerator degree n and denominator degree m on $[-1,1]$ is investigated for non-analytic functions if ray sequences in the lower half of the Walsh table are considered, i. e. for sequences $\{(n,m(n)\}_{n=1}^{\infty}$ with

$$\frac{n}{m(n)} \to c \in (1,\infty] \text{ as } n \to \infty.$$

For $f(x) = |x|^{\alpha}$, $\alpha \in \mathbb{R}_+ \setminus 2\mathbb{N}$, it is known that the best rational approximants diverge everywhere outside $[-1,1]$. We investigate the situation for more general functions f that are not analytic on $[-1,1]$.

1. INTRODUCTION AND MAIN RESULTS

Our aim is to investigate the convergence of ray sequences in the Walsh table to a function f that is not analytic on $[-1,1]$.

Let $\mathscr{P}_n$ denote the collection of all real polynomials of degree at most n and

$$\mathscr{R}_{n,m} := \{p/q : p \in \mathscr{P}_n,\ q \in \mathscr{P}_m,\ q \not\equiv 0\},$$

where $n,m \in \mathbb{N}$. Let $C[-1,1]$ denote the set of all real continuous functions on $[-1,1]$ and $r^*_{n,m} = r^*_{n,m}(f) \in \mathscr{R}_{n,m}$ the best uniform rational approximation to fixed $f \in C[-1,1]$.

An infinite sequence $N_c \subset \mathbb{N}^2$ of indices (n,m) as well as the corresponding sequence $\{r^*_{n,m}\}_{(n,m)\in N_c}$ is called a *ray sequence* with associated asymptotic *numerator-denominator ratio c* if

$$\frac{n}{m} \to c \in [0,\infty] \text{ as } n+m \to \infty,\ (n,m) \in N_c.$$

The sequence $\{(n,n)\}_{n=0}^{\infty}$ is called *diagonal sequence*, sequences $N_c = \{(n,m)\}$ with $c \in (1,\infty]$ are called *subdiagonal sequences.*

If $f \in C[-1,1]$ is not analytic on $[-1,1]$, then Blatt and Saff [2] have shown in the case $m = 0$:

There exists a subsequence $\Lambda \subset \mathbb{N}$ and a countable set $S \subset \mathbb{C} \setminus [-1,1]$ such that

$$\lim_{n\in\Lambda,\, n\to\infty} |r^*_{n,0}(z)|^{1/n} = e^{G(z,\infty)} \text{ for all } z \notin [-1,1] \cup S,$$

where $G(z,\infty)$ is Green's function of $\Omega = \overline{\mathbb{C}} \setminus [-1,1]$ with pole at ∞.

2000 *Mathematics Subject Classification.* 41A05, 41A20, 41A25, 41A50.

Key words and phrases. Rational approximation, rational interpolation, weighted Fekete points, harmonic majorants, Green's function.

In contrast to this, Blatt, Iserles and Saff [3] have shown that for $f(x) = |x|$ the sequence $\{r^*_{n,n}\}$ of best approximations converge in the right half-plane H_+, resp. in the left half-plane H_-.

Precisely,

$$\lim_{n\to\infty} r^*_{n,n}(z) = \begin{cases} z & \text{for } z \in H_+ \\ -z & \text{for } z \in H_- \end{cases} .$$

Moreover, all zeros and poles of $r^*_{n,n}$ are lying on the imaginary axis.

The divergence phenomenon of the polynomial case occurs again in the subdiagonal case. We cite a result of E. B. Saff and H. Stahl [6] :

Let $\alpha \in \mathbb{R}_+ \setminus 2\mathbb{N}$ and $f(x) = |x|^\alpha$. Then for any ray sequence N_c with $c > 1$ we have

$$\lim_{\substack{n+m\to\infty \\ (n,m)\in N_c}} |r^*_{n,m}(z)|^{1/(n+m)} = \exp\left[\frac{c-1}{c+1} G(z,\infty)\right]$$

uniformly on compact subsets of $\overline{\mathbb{C}} \setminus [-1,1]$.

In the following let $c > 1$, $\{(n, m(n))\}_{n=1}^{\infty} \in N_c$ and

$$m(n) \le m(n+1) \le m(n) + 1, \tag{1.1}$$

and we denote by

$$E_{n,m(n)}(f) = \|f - r^*_{n,m(n)}\| = \inf_{r\in\mathscr{R}_{n,m(n)}} \|f - r\| \tag{1.2}$$

the minimal deviation of f with respect to $\mathscr{R}_{n,m(n)}$, where $\|\cdot\|$ denotes the sup norm on $[-1,1]$.

In the proof of our main result, we need a condition on the poles of $r^*_{n,m(n)}$ to a given function f.

Definition. The sequence $\{r^*_{n,m(n)}\}_{n=1}^{\infty}$ satisfies the *pole condition* (P_τ), $\tau > 1$, if for any $(n, m(n))$ with $r^*_{n,m(n)} \notin \mathscr{R}_{n,m(n-1)}$ there exists a pole ξ of $r^*_{n,m(n)}$ with

$$G(\xi,\infty) > \log\tau.$$

Theorem 1. *Let $c > 1$, $N_c = \{(n, m(n))\}_{n=1}^{\infty}$, $\limsup_{n\to\infty} (E_{n,m(n)}(f))^{1/n} = 1$ and let the pole condition (P_τ) hold for some fixed $\tau > 1$. Then for any continuum $S \subset \mathbb{C} \setminus [1,1]$ we have*

$$\limsup_{n\to\infty} \|r^*_{(n,m(n))}\|_S^{1/(n+m(n))} \ge \exp\left[\frac{c-1}{c+1} \|G(\cdot,\infty)\|_S\right],$$

where $\|\cdot\|_S$ is the sup norm on S.

We remark that $\limsup_{n\to\infty} (E_{n,m(n)}(f))^{1/n} = 1$ holds for example for piecewise analytic functions (cf. Gončar [4]).

2. Green functions

Let $\Omega = \overline{\mathbb{C}} \setminus [-1,1]$ and let $G(z,a)$ denote Green´s function of Ω with pole at $a \in \Omega$. Then

$$G(z,\infty) = \log|z + \sqrt{z^2-1}| = \log|\varphi(z)|, \tag{2.1}$$

where φ is the conformal mapping of Ω onto $\Delta = \{t \in \overline{\mathbb{C}} : |t| > 1\}$,

$$\varphi(z) = z + \sqrt{z^2 - 1} \tag{2.2}$$

normalized by $\sqrt{z^2-1} > 0$ for $z > 1$, $z \in \mathbb{R}$. For $b \in \Delta$,

$$\psi(t,b) = \frac{1 - t\overline{b}}{t - b}$$

is the Möbius transformation that maps Δ onto Δ, the unit circle onto itself and $\psi(b,b) = \infty$. Hence,

$$G(z,a) = \log\left|\frac{1 - \varphi(z)\overline{\varphi(a)}}{\varphi(z) - \varphi(a)}\right|. \tag{2.3}$$

Let us consider the level lines of Green´s function $G(z,\infty)$,

$$\Gamma_\sigma = \{z \in \Omega : G(z,\infty) = \log\sigma\} \tag{2.4}$$

for $\sigma \geq 1$, and the exterior domain

$$R_\sigma = \{z \in \Omega : G(z,\infty) > \log\sigma\}. \tag{2.5}$$

We denote for $a \in R_\sigma$ by $G_\sigma(z,a)$ Green´s function of R_σ with pole at a. Then

$$G_\sigma(z,a) = \log\left|\frac{\sigma^2 - \varphi(z)\overline{\varphi(a)}}{\sigma(\varphi(z) - \varphi(a))}\right|. \tag{2.6}$$

3. Bernstein´s lemma for rational functions

Let $r = p/q$ be a rational function with

$$\|r\| = \max_{-1\leq x\leq 1} |r(x)| < \infty. \tag{3.1}$$

Lemma 1. *Let $r = p/q$ with degree $p \leq n$ and degree $q \leq m$ and*

$$q(z) = \prod_{i=1}^{m}(z - z_i)$$

with $z_i \notin [-1,1]$ for $1 \leq i \leq m$. Then for $z \in \Omega = \overline{\mathbb{C}} \setminus [-1,1]$

$$\log|r(z)| \leq \sum_{i=1}^{m} G(z,z_i) + (n-m)\,G(z,\infty) + \log\|r\|. \tag{3.2}$$

Proof. Consider the subharmonic function

$$h(z) := \log|r(z)| - \sum_{i=1}^{m} G(z,z_i) - (n-m)\,G(z,\infty)$$

which is continuous in $\overline{\mathbb{C}}$. Hence the maximum principle implies for $z \in \Omega$

$$\begin{aligned} h(z) \leq \max_{z\in[-1,1]} h(z) &= \max_{z\in[-1,1]} \log|r(z)| \\ &= \log\|r\|, \end{aligned}$$

equivalent to inequality (3.2). □

Lemma 1 is well-known and generalizes Bernstein's lemma for polynomials (cf. Walsh [8], Gončar [5]).

4. Rational interpolation

Let us consider n points

$$-1 \le x_1 < x_2 < \ldots < x_n \le 1 \tag{4.1}$$

and

$$q_m(z) = \prod_{i=1}^{m} (z - z_i) \tag{4.2}$$

with $z_i \notin [-1,1]$. Let f be analytic on $[-1,1]$, we consider the interpolation of f at the points x_i, $1 \le i \le n$, with respect to rational functions of type

$$r = \frac{p_{n-1}}{q_m} \in \mathcal{R}_{n-1,m}. \tag{4.3}$$

Let $r_{n-1}(z)$ be the interpolating rational of type (4.3) at the points (4.1) then

$$r_{n-1}(z) = \sum_{i=1}^{n} f(x_i)\, L_i(z), \tag{4.4}$$

where

$$L_i(z) = \frac{q_m(x_i)}{w_n'(x_i)} \frac{w_n(z)}{q_m(z)} \frac{1}{z - x_i} \tag{4.5}$$

and

$$w_n(z) = \prod_{i=1}^{n} (z - x_i). \tag{4.6}$$

Concerning the error of interpolation the following representation generalizes some statements of Walsh [8], pp. 184 ff.:

Let Γ be a Jordan curve with $[-1,1]$ in the interior, let f be analytic on Γ and in the interior of Γ except for points $z_1, z_2, \ldots, z_s$ that are poles of f of order $m_1, m_2, \ldots, m_s$. Let

$$q(z) = \prod_{i=1}^{s} (z - z_i)^{m_i} \text{ and } q | q_m. \tag{4.7}$$

If we consider the interpolation problem at the points (4.1) with respect to rational functions of type (4.3), the error of interpolation at any point $z \in \operatorname{int}(\Gamma)$, $z \neq z_i$ for $i = 1, 2, \ldots, s$, is represented by

$$f(z) - r_{n-1}(z) = \frac{w_n(z)}{q_m(z)} \frac{1}{2\pi i} \int_{\Gamma} \frac{q_m(t) f(t)}{w_n(t)(t - z)}\, dt. \tag{4.8}$$

Lemma 2. *Let $m \le n$ and let $r_{n-1}(t)$ be the Lagrange interpolation of type (4.3) to the function $f(z) = 1/(t - z)$, then*

$$\frac{1}{t - z} - r_{n-1}(z) = \frac{1}{t - z} \frac{w_n(z)}{q_m(z)} \frac{q_m(t)}{w_n(t)} \tag{4.9}$$

for $z \in \mathbb{C}$, $z \neq z_i$ $(1 \le i \le s)$ and $z \neq t$.

The proof can be based on the residues of both sides of (4.9).

Lemma 2 provides a representation of the kernel in the error formula (4.8).

5. Weighted Fekete points

For obtaining good estimates for the error of interpolation we introduce *weighted Fekete points in* $[-1,1]$ *to the weight* $w(z)$ *with* $w(z) > 0$ *in* $[-1,1]$.

Definition. The set of n points

$$-1 \le x_1^{(n)} < x_2^{(n)} < \ldots < x_n^{(n)} \le 1$$

is called *weighted n-th Fekete point set* if in the Vandermonde expression

$$V_{w,n}(z_1,\ldots,z_n) := \prod_{1\le i<j=n} |z_i - z_j|\, w(z_i)\, w(z_j) \tag{5.1}$$

the points $x_1^{(n)},\ldots,x_n^{(n)}$ are maximizing the value $V_{w,n}(z_1,\ldots,z_n)$ for $z_1,\ldots,z_n$ in $[-1,1]$.

For $w \equiv 1$ we obtain the classical Fekete points.

Let $z \in [-1,1]$, then

$$V_{w,n}(x_1^{(n)},\ldots,x_{i-1}^{(n)},z,x_{i+1}^{(n)},\ldots,x_n^{(n)}) \le V_{w,n}(x_1^{(n)},\ldots,x_n^{(n)})$$

and

$$\begin{aligned} 1 \;&\ge\; \frac{V_{w,n}(x_1^{(n)},\ldots,x_{i-1}^{(n)},z,x_{i+1}^{(n)},\ldots,x_n^{(n)})}{V_{w,n}(x_1^{(n)},\ldots,x_n^{(n)})} \\ &=\; \frac{w(z)^{n-1}}{w(x_i^{(n)})^{n-1}}\,\frac{V_{1,n}(x_1^{(n)},\ldots,x_{i-1}^{(n)},z,x_{i+1}^{(n)},\ldots,x_n^{(n)})}{V_{1,n}(x_1^{(n)},\ldots,x_n^{(n)})} \\ &=\; \frac{w(z)^{n-1}}{w(x_i^{(n)})^{n-1}}\,\frac{|w_n(z)|}{|w_n'(x_i^{(n)})(z-x_i^{(n)})|}, \end{aligned} \tag{5.2}$$

where

$$w_n(z) = \prod_{i=1}^{n}(z - x_i^{(n)}). \tag{5.3}$$

As in (4.5), the analogue rational functions to the fundamental polynomials of Lagrange are

$$L_{w,i}(z) = \frac{q_m(x_i^{(n)})}{q_m(z)}\,\frac{w_n(z)}{w_n'(x_i^{(n)})}\,\frac{1}{(z-x_i^{(n)})},\quad 1 \le i \le n \tag{5.4}$$

and we get for the interpolating rational function $r_{n-1}(z)$ at the points $x_i^{(n)}$, $1 \le i \le n$,

$$r_{n-1}(z) = \sum_{i=1}^{n} f(x_i^{(n)}) L_{w,i}(z). \tag{5.5}$$

Lemma 3. *Let*

$$w(z) = \sqrt[n-1]{\frac{1}{|q_m(z)|}} \tag{5.6}$$

be the weight function with $q_m(z)$ of (4.2) and $\{x_i^{(n)}\}_{i=1}^n$ the n-th weighted Fekete point set with respect to $w(z)$, then

$$\left|\frac{1}{t-z}\frac{w_n(z)}{q_m(z)}\frac{q_m(t)}{w_n(t)}\right| \le \frac{n+1}{d_1(t)},\ z\in[-1,1], \tag{5.7}$$

where

$$d_1(t) = \min_{z\in[-1,1]} |t-z|.$$

Proof. We know that

$$\widetilde{r}_{n-1}(z) = \sum_{i=1}^{n} \frac{1}{t-x_i^{(n)}} L_{w,i}(z)$$

is the interpolating rational to $f(z) = 1/(t-z)$ at the point set $\{x_i^{(n)}\}_{i=1}^n$. Hence, by (4.9)

$$\left|\frac{1}{t-z}\frac{w_n(z)}{q_m(z)}\frac{q_m(t)}{w_n(t)}\right| \le \frac{1}{|t-z|} + \sum_{i=1}^{n}\frac{1}{|t-x_i^{(n)}|}|L_{w,i}(z)|.$$

Now, by the choice of $w(z)$ and (5.2)

$$|L_{w,i}(z)| = \left|\frac{w(z)^{n-1}}{w(x_i^{(n)})^{n-1}}\frac{w_n(z)}{w_n'(x_i^{(n)})}\frac{1}{z-x_i^{(n)}}\right| \le 1$$

and we get

$$\left|\frac{1}{t-z}\frac{w_n(z)}{q_m(z)}\frac{q_m(t)}{w_n(t)}\right| \le \frac{n+1}{d_1(t)}.$$ □

6. Harmonic majorants

Let

$$r_n^* = r_{n,m(n)}^* = \frac{p_n^*}{q_{m(n)}^*} \in \mathscr{R}_{n,m(n)}$$

be the best rational approximation to f where p_n^* and $q_{m(n)}^*$ have no common factor. Since

$$E_{n,m(n)}(f) = \|f-r_n^*\| = \inf_{r_n\in\mathscr{R}_{n,m(n)}} \|f-r_n\| \le \|f\|$$

we get

$$\|r_n^*\| \le 2\|f\|$$

and by Lemma 1

$$\log|r_n^*(z)| \le \sum_{i=1}^{\mu_n} G(z,z_i^{(n)}) + (n-\mu_n)G(z,\infty) + \log(2\|f\|),$$

where

$$q_{m(n)}^*(z) = \prod_{i=1}^{\mu_n}(z-z_i^{(n)}),\ \mu_n = \text{degree}\, q_{m(n)}^*.$$

Hence, the Green function $G(z,\infty)$ is a harmonic majorant for the subharmonic functions

$$H_n(z) := \frac{1}{n-\mu_n}\left[\log|r_n^*(z)| - \sum_{i=1}^{\mu_n} G(z, z_i^{(n)}) - \log(2\|f\|)\right]. \tag{6.1}$$

Here, we have used the notion of harmonic majorant due to Walsh [7]. For completeness, we will give below the precise definition.

Definition. Let $\{F_n\}_{n=1}^{\infty}$ be a sequence of subharmonic functions in a domain D of $\overline{\mathbb{C}}$. Then the harmonic function V in D is called *harmonic majorant* of $\{F_n\}_{n=1}^{\infty}$ if for any continuum $S \subset D$

$$\limsup_{n\to\infty} \max_{z\in S} F_n(z) \le \max_{z\in S} V(z). \tag{6.2}$$

The harmonic majorant V of $\{F_n\}_{n=1}^{\infty}$ is called *exact* if there exists a compact set $S' \subset D$ with

$$\limsup_{n\to\infty} \max_{z\in S'} F_n(z) = \max_{z\in S'} V(z). \tag{6.3}$$

We remark that V is an exact harmonic majorant if and only if (6.3) holds for any continuum $S' \subset D$ (cf. Walsh [7]).

Let us return to the functions $H_n(z)$ of (6.1). Then the main tool for proving Theorem 1 is the following lemma.

Lemma 4. *Let $c > 1$, $N_c = \{(n, m(n))\}_{n=1}^{\infty}$ and the pole condition (P_τ) is satisfied for some $\tau > 1$. If Green´s function $G(z,\infty)$ is not an exact harmonic majorant to the sequence $\{H_n\}_{n=1}^{\infty}$, defined in (6.1), then there exists $r_{n-1} \in \mathcal{R}_{n-1,m(n-1)}$ such that*

$$\limsup_{n\to\infty} \|r_n^* - r_{n-1}\|^{1/n} < 1.$$

Proof. Let us consider the best approximation

$$r_n^* = \frac{p_n^*}{q_{m(n)}^*} \in \mathcal{R}_{n,m(n)}$$

of f where p_n^* and $q_{m(n)}^*$ have no common factor and $q_{m(n)}^*(z) > 0$ for all $z \in [-1,1]$.

Then we want to approximate r_n^* with respect to $\mathcal{R}_{n-1,m(n-1)}$ and we will do this by interpolation. By condition (P_τ), there exists a pole z_0 of r_n^* (resp. $\overline{z}_0$) in R_τ if $\mu_n =$ degree $q_{m(n)}^* > m(n-1)$. We define $q_{n-1} \in \mathcal{P}_{m(n-1)}$ by the representation

$$q_{m(n)}^*(z) = \begin{cases} \operatorname{sgn}(z_0)(z_0 - z)q_{n-1}(z), & \text{if } z_0 \in \mathbb{R} \text{ and } \mu_n > m(n-1) \\ (z-z_0)(z-\overline{z}_0)q_{n-1}(z), & \text{if } z_0 \notin \mathbb{R} \text{ and } \mu_n > m(n-1)\ . \\ q_{n-1}(z), & \text{if } \mu_n \le m(n-1) \end{cases} \tag{6.4}$$

Then q_{n-1} is a real polynomial and $q_{n-1}(z) > 0$ for $z \in [-1,1]$.

Moreover,

$$\operatorname{grad} q_{n-1} = \nu_n = \begin{cases} \mu_n - 1, & \text{if } z_0 \in \mathbb{R} \text{ and } \mu_n > m(n-1) \\ \mu_n - 2, & \text{if } z_0 \notin \mathbb{R} \text{ and } \mu_n > m(n-1) \\ \mu_n, & \text{if } \mu_n \le m(n-1) \end{cases} . \tag{6.5}$$

We consider now the interpolation of r_n^* at the weighted n-th Fekete points of $[-1,1]$ to the positive weight

$$w(z) = \sqrt[n-1]{\frac{1}{q_{n-1}(z)}} \tag{6.6}$$

with respect to rational functions of type

$$r(z) = \frac{p_{n-1}(z)}{q_{n-1}(z)} \in \mathscr{R}_{n-1,m(n-1)}. \tag{6.7}$$

By (4.8), we obtain for the error

$$r_n^*(z) - r_{n-1}(z)$$

of the interpolating rational function r_{n-1} to $r_n^*(z)$:

$$r_n^*(z) - r_{n-1}(z) = \frac{w_n(z)}{q_{n-1}(z)} \frac{1}{2\pi i} \int\limits_{\Gamma_\rho} \frac{q_{n-1}(t)}{w_n(t)} \frac{1}{t-z} r_n^*(t)\, dt, \tag{6.8}$$

where

$$-1 \le x_1^{(n)} < x_2^{(n)} < \ldots < x_n^{(n)} \le 1$$

are the n-th weighted Fekete points to the weight $w(z)$ of (6.6),

$$w_n(z) = \prod_{i=1}^{n} (z - x_i^{(n)})$$

and Γ_ρ is the level line of Green´s function $G(z,\infty)$ to the parameter ρ as in (2.4) and $\rho < \tau$. Because of condition (P_τ), z_0 resp. $\overline{z}_0$ lies in R_τ and therefore the representation of the error according to (4.8) holds.

Now, let us choose $1 < \sigma < \rho < \tau$. We consider the function

$$\begin{aligned} g_{n,\sigma}(t) \quad := \quad & \log \left| \frac{1}{t-z} \frac{w_n(z)}{q_{n-1}(z)} \frac{q_{n-1}(t)}{w_n(t)} \right| \\ & + \sum_{i=1}^{m_\sigma} G_\sigma(t, z_i^{(n)}) + (n+1-\nu_n) G(t,\infty) \end{aligned} \tag{6.9}$$

for fixed $z \in [-1,1]$, where m_σ is the number of poles of r_n^*, different from z_0 and $\overline{z_0}$, in R_σ, and we have enumerated all poles

$$z_1^{(n)}, \ldots, z_{\mu_n}^{(n)}$$

in such a way that

$$\{z_1^{(n)}, \ldots, z_{m_\sigma}^{(n)}\} \tag{6.10}$$

is the set of poles of r_n^* in R_σ. $g_{n,\sigma}$ is harmonic in R_σ and subharmonic in Ω. Moreover,

$$g_{n,\sigma}(\infty) = \sum_{i=1}^{m_\sigma} G_\sigma(\infty, z_i^{(n)}) - (n+1-\nu_n)\log\frac{1}{2},$$

since

$$\lim_{t\to\infty}(G(t,\infty) - \log|t|) = -\log\frac{1}{2}.$$

By the maximum principle,

$$\max_{t\in\Gamma_\sigma} g_{n,\sigma}(t) \geq \max_{t\in\Gamma_\rho} g_{n,\sigma}(t) \tag{6.11}$$

and with Lemma 3 we get

$$\max_{t\in\Gamma_\sigma} g_{n,\sigma}(t) \leq \log\frac{n+1}{d_1} + (n+1-\nu_n)\log\sigma, \tag{6.12}$$

where

$$d_1 = \min_{t\in\Gamma_\sigma} d_1(t) = \min_{t\in\Gamma_\sigma}\min_{z\in[-1,1]}|t-z|. \tag{6.13}$$

Hence, (6.9), (6.11) and (6.12) imply for $t \in \Gamma_\rho$

$$\left|\frac{1}{t-z}\,\frac{w_n(z)}{q_{n-1}(t)}\,\frac{q_{n-1}(z)}{w_n(t)}\right| \exp\left\{\sum_{i=1}^{m_\sigma} G_\sigma(t, z_i^{(n)}) + (n+1-\nu_n)\log\rho\right\}$$
$$\leq \exp\left(\log\frac{n+1}{d_1} + (n+1-\nu_n)\log\sigma\right)$$

or

$$\left|\frac{1}{t-z}\,\frac{w_n(z)}{q_{n-1}(t)}\,\frac{q_{n-1}(z)}{w_n(t)}\right| \leq \frac{n+1}{d_1}\left(\frac{\sigma}{\rho}\right)^{n+1-\nu_n} \exp\left\{-\sum_{i=1}^{m_\sigma} G_\sigma(t, z_i^{(n)})\right\} \tag{6.14}$$

for all $t \in \Gamma_\rho$.

Now, $G(z,\infty)$ is not an exact harmonic majorant for $\{H_n\}_{n=1}^{\infty}$. That implies

$$\limsup_{n\to\infty}\max_{z\in\Gamma_\rho} H_n(z) < \max_{z\in\Gamma_\rho} G(z,\infty) = \log\rho.$$

Hence, there exists $0 < \varepsilon < \rho - 1$ and $n_0 = n_0(\varepsilon)$ such that

$$H_n(t) \leq \log(\rho - \varepsilon) \text{ for } t \in \Gamma_\rho,\ n \geq n_0. \tag{6.15}$$

The structure of $H_n(z)$ in (6.1) yields, for $t \in \Gamma_\rho$,

$$H_n(t) = \frac{1}{n-\mu_n}\left[\log|r_n^*(t)| - \sum_{i=1}^{\mu_n} G(t, z_i^{(n)}) - \log(2\|f\|)\right] \leq \log(\rho - \varepsilon)$$

or

$$|r_n^*(t)| \leq 2\|f\|(\rho-\varepsilon)^{n-\mu_n}\exp\left(\sum_{i=1}^{\mu_n} G(z, z_i^{(n)})\right).$$

Using this bound and (6.14), we obtain for $t \in \Gamma_\rho$ that

$$\begin{aligned}
&\left| \frac{1}{t-z} \frac{w_n(z)}{q_{n-1}(t)} \frac{q_{n-1}(z)}{w_n(t)} r_n^*(t) \right| \\
&\le 2\|f\| \frac{n+1}{d_1} \left(\frac{\sigma}{\rho}\right)^{n+1-\nu_n} (\rho-\varepsilon)^{n-\mu_n} \cdot \exp(S_{\sigma,n}(t)), \qquad (6.16)
\end{aligned}$$

where

$$\begin{aligned}
S_{\sigma,n}(t) &= \sum_{i=1}^{\mu_n} G(t, z_i^{(n)}) - \sum_{i=1}^{m_\sigma} G_\sigma(t, z_i^{(n)}) \\
&= S_{\sigma,n}^1(t) + S_{\sigma,n}^2(t)
\end{aligned}$$

and

$$S_{\sigma,n}^1(t) = \sum_{i=1}^{m_\sigma} [G(t, z_i^{(n)}) - G_\sigma(t, z_i^{(n)})], \qquad (6.17)$$

$$S_{\sigma,n}^2(t) = \sum_{i=m_\sigma+1}^{\mu_n} G(t, z_i^{(n)}). \qquad (6.18)$$

Now, we have to estimate $S_{\sigma,n}^1(t)$ and $S_{\sigma,n}^2(t)$ for $t \in \Gamma_\rho$.

(a) Bound for $S_{\sigma,n}^1(t)$ and $t \in \Gamma_\rho$:

According to (2.6) we have for $a \in R_\sigma$, $\sigma \ge 1$,

$$G_\sigma(t,a) = \log \left| \frac{\sigma^2 - \varphi(t)\overline{\varphi(a)}}{\sigma(\varphi(t) - \varphi(a))} \right|$$

and $G(z,a) = G_1(z,a)$. Hence,

$$\begin{aligned}
G(t,a) - G_\sigma(t,a) &= \log \left| \frac{1 - \varphi(t)\overline{\varphi(a)}}{\varphi(t) - \varphi(a)} \frac{\sigma(\varphi(t) - \varphi(a))}{\sigma^2 - \varphi(t)\overline{\varphi(a)}} \right| \\
&= \log \left| \sigma \frac{1 - \varphi(t)\overline{\varphi(a)}}{\sigma^2 - \varphi(t)\overline{\varphi(a)}} \right|. \qquad (6.19)
\end{aligned}$$

In the following, we will choose

$$\sigma = \sigma_n = 1 + \frac{\kappa}{n} < \rho, \qquad (6.20)$$

where

$$0 < \kappa < 1$$

and κ will be a constant independent of n, specified later. Then we note that

$$\sigma_n^2 = \left(1 + \frac{\kappa}{n}\right)^2 = 1 + \frac{2\kappa}{n} + \left(\frac{\kappa}{n}\right)^2 \le 1 + \frac{3\kappa}{n}$$

and

$$\begin{aligned}
\left|\frac{1-\varphi(t)\overline{\varphi(a)}}{\sigma_n^2-\varphi(t)\overline{\varphi(a)}}\right| &= \left|\frac{\sigma_n^2-\varphi(t)\overline{\varphi(a)}+(1-\sigma_n^2)}{\sigma_n^2-\varphi(t)\overline{\varphi(a)}}\right| \\
&= \left|1+\frac{1-\sigma_n^2}{\sigma_n^2-\varphi(t)\overline{\varphi(a)}}\right| \\
&\le 1+(\sigma_n^2-1)\frac{1}{|\sigma_n^2-\varphi(t)\overline{\varphi(a)}|} \\
&\le 1+\frac{3\kappa}{n}\,\frac{1}{|\sigma_n^2-\varphi(t)\overline{\varphi(a)}|}.
\end{aligned}$$

Since $a \in R_{\sigma_n}$, we have

$$|\varphi(a)| \ge 1+\frac{\kappa}{n}$$

and

$$|\varphi(t)| = \rho \quad \text{for } t \in \Gamma_\rho.$$

Therefore

$$\begin{aligned}
|\sigma_n^2-\varphi(t)\overline{\varphi(a)}| &\ge |\varphi(t)\varphi(a)|-\sigma_n^2 \ge \rho\left(1+\frac{\kappa}{n}\right)-\left(1+\frac{\kappa}{n}\right)^2 \\
&= (\rho-1)+\frac{\kappa}{n}\left(\rho-2-\frac{\kappa}{n}\right)
\end{aligned}$$

and we can choose

$$n_1 \ge n_0(\varepsilon),$$

independent of $0<\kappa<1$, such that

$$|\sigma_n^2-\varphi(t)\overline{\varphi(a)}| \ge \frac{\rho-1}{2} \quad \text{for } n \ge n_1.$$

Therefore, for $n \ge n_1$ and $t \in \Gamma_\rho$

$$\begin{aligned}
\log \quad &\left|\sigma_n\frac{1-\varphi(t)\overline{\varphi(a)}}{\sigma_n^2-\varphi(t)\overline{\varphi(a)}}\right| \\
&\le \log\sigma_n+\log\left(1+\frac{3\kappa}{n}\,\frac{2}{\rho-1}\right) \\
&\le \frac{\kappa}{n}+\frac{\kappa}{n}\,\frac{6}{\rho-1} \\
&= \frac{\kappa}{n}\left(1+\frac{6}{\rho-1}\right) \le \frac{1}{n},
\end{aligned} \tag{6.21}$$

if we choose

$$\kappa = \frac{\rho-1}{\rho+5}. \tag{6.22}$$

If κ satisfies (6.22), then $\sigma_n < \rho$ for sufficiently big n, $n \ge n_2 \ge n_1$, and with $z_i^{(n)} \in R_\sigma$, $1 \le i \le m_\sigma$, we obtain by (6.19) and (6.21) for $t \in \Gamma_\rho$ and $n \ge n_2$

$$S^1_{\sigma_n,n}(t) = \sum_{i=1}^{m_\sigma} [G(t, z_i^{(n)}) - G_{\sigma_n}(t, z_i^{(n)})] \le \frac{m_\sigma}{n} \le 1,$$

and therefore

$$\exp(S^1_{\sigma_n,n}(t)) \le e \text{ for all } t \in \Gamma_\rho,\ n \ge n_2 \ge n_0(\varepsilon). \tag{6.23}$$

(b) Bound for $S^2_{\sigma,n}(t)$ and $t \in \Gamma_\rho$, where $\sigma = \sigma_n = 1 + \frac{\kappa}{n}$:

Since $\rho < \tau$ there exists a constant $\alpha > 0$, such that

$$G(t, z_0) \le \alpha \text{ and } G(t, \overline{z_0}) \le \alpha \tag{6.24}$$

for all $t \in \Gamma_\rho$. α is independent of $z_0, \overline{z}_0 \in R_\tau$.

Let $\xi \in \mathbb{C}$ be in the complement of R_{σ_n}, hence

$$G(\xi, \infty) \le \log \sigma_n.$$

For fixed $t \in \Gamma_\rho$ we consider

$$G(t, \xi) = G(\xi, t) = \log \left| \frac{1 - \varphi(\xi)\overline{\varphi(t)}}{\varphi(\xi) - \varphi(t)} \right|,$$

where $\varphi : \Omega \to \Delta$ is the conformal mapping as in section 2. Then

$$\exp G(t, \xi) = \left| \frac{1 - \varphi(\xi)\overline{\varphi(t)}}{\varphi(\xi) - \varphi(t)} \right| = F(t; \varphi(\xi)),$$

where

$$F(t; y) = \frac{1 - y\overline{\varphi(t)}}{y - \varphi(t)}.$$

For fixed $t \in \Gamma_\rho$ the function $F(t; y)$ is holomorphic with respect to y in the domain

$$\Delta_{\widetilde{\varepsilon}} := \{y : 1 - \widetilde{\varepsilon} < |y| < \rho\},$$

where $\widetilde{\varepsilon} > 0$ is independent of $t \in \Gamma_\rho$.

Let

$$y = |y| e^{i\theta} \text{ with } \theta \in [0, 2\pi],$$

then

$$\begin{aligned} |F(t; y)| &\le |F(t; e^{i\theta})| + |F(t; y) - F(t; e^{i\theta})| \\ &= 1 + |F(t; y) - F(t; e^{i\theta})| \\ &\le 1 + |y - e^{i\theta}| \max_{\eta \in \Gamma(y)} |F'(t; \eta)|, \end{aligned} \tag{6.25}$$

where $F'(t; y)$ is the derivative of $F(t; y)$ with respect to y and

$$\Gamma(y) = \{\eta = r\, e^{i\theta} : 1 \le r \le |y|\}.$$

If we define

$$\beta_n = \max_{\theta \in [0, 2\pi]} \max_{\eta \in \Gamma(\sigma_n e^{i\theta})} |F'(t; \eta)|$$

then β_n exists and is bounded for all $n \geq n_2$, namely

$$\beta_n \leq \beta_{n_2} = \beta.$$

Hence, we get with (6.25) for all ξ in the complement of $\mathscr{R}_{\sigma_n}$

$$\exp G(t,\xi) \leq |F(t;\varphi(\xi))| \leq 1 + (\sigma_n - 1)\beta \leq 1 + \frac{\beta\kappa}{n}. \tag{6.26}$$

Since each $z_i^{(n)} \neq z_0, \overline{z_0}$ lies in the complement of $\mathscr{R}_{\sigma_n}$, we deduce for $t \in \Gamma_\rho$ together with (6.24) that

$$\exp\left(\sum_{i=m_\sigma+1}^{\mu_n} G(t, z_i^n)\right) \leq c_1 \left(1 + \frac{\beta\kappa}{n}\right)^m, \; n \geq n_2$$

and

$$\limsup_{n\to\infty} \max_{t\in\Gamma_\rho} \exp S^2_{\sigma_n,n}(t) \leq c_1 e^{\beta\kappa/c}. \tag{6.27}$$

Combining (6.16), (6.23) and (6.27) we get for $n \geq n_2$

$$\left|\frac{1}{t-z}\,\frac{w_n(z)}{q_{n-1}(t)}\,\frac{q_{n-1}(z)}{w_n(t)}\,r_n^*(t)\right| \leq c_2 \frac{n+1}{d_1}\left(\frac{\sigma_n}{\rho}\right)^{n+1-\nu_n} (\rho-\varepsilon)^{n-\mu_n}$$

with some constant c_2.

d_1 is defined by (6.13), therefore exists a constant $c_3 > 0$ such that

$$d_1 \geq \frac{c_3}{n^2} \text{ for all } n,$$

taking into account that $\sigma = \sigma_n = 1 + \kappa/n$. Hence, with some constant c_4 we have got

$$\left|\frac{1}{t-z}\,\frac{w_n(z)}{q_{n-1}(t)}\,\frac{q_{n-1}(z)}{w_n(t)}\,r_n^*(t)\right| \quad \leq \quad c_4 n^3 \left(\frac{1}{\rho}\right)^{n+1-\nu_n} (\rho-\varepsilon)^{n-\mu_n}.$$

Inserting the last bound in the error function (6.8) we have

$$|r_n^*(z) - r_{n-1}(z)| \leq c_5 n^3 \left(\frac{\rho-\varepsilon}{\rho}\right)^{n-\mu_n}$$

for $n \geq n_2$ and $z \in [-1,1]$ or

$$\limsup_{n\to\infty} \|r_n^* - r_{n-1}\|^{1/n} \leq \left(\frac{\rho-\varepsilon}{\rho}\right)^{1-1/c} = \delta < 1$$

and Lemma 4 is proved. □

7. Proof of Theorem 1

Let us assume first that Green´s function $G(z,\infty)$ is not exact harmonic majorant of $\{H_n\}_{n=1}^{\infty}$. By Lemma 4, there exists $0 < \delta < 1$ and $r_{n-1} \in \mathscr{R}_{n-1,m(n-1)}$ such that

$$\limsup_{n\to\infty} \|r_n^* - r_{n-1}\|^{1/n} = \delta < 1. \tag{7.1}$$

For abbreviation, we set $E_n = E_{n,m(n)}(f)$. Then because of

$$E_{n-1} = \|f - r_{n-1}^*\| \leq \|f - r_{n-1}\| \leq \|f - r_n^*\| + \|r_n^* - r_{n-1}\|,$$

we obtain

$$E_{n-1} - E_n \leq \|r_n^* - r_{n-1}\|.$$

Hence, by (7.1) we get

$$\limsup_{n\to\infty} (E_{n-1}-E_n)^{1/n} \le \delta.$$

That implies that

$$E_{n-1}-E_n \le K(\delta+\varepsilon)^n$$

with $\varepsilon = (1-\delta)/2$ and some constant K. For $\nu \in \mathbb{N}$, $\nu \ge 1$, we get

$$\begin{aligned} E_n &= E_n - E_{n+1} + E_{n+1} - E_{n+2} + \ldots + E_{n+\nu-1} - E_{n+\nu} + E_{n+\nu} \\ &\le K\sum_{\mu=1}^{\nu} (\delta+\varepsilon)^{n+\mu} + E_{n+\nu} \le K(\delta+\varepsilon)^{n+1}\frac{1}{1-\delta-\varepsilon} + E_{n+\nu} \\ &= \frac{K}{\varepsilon}(\delta+\varepsilon)^{n+1} + E_{n+\nu}. \end{aligned}$$

Using Weierstraß´ theorem we have got

$$\limsup_{n\to\infty} E_n^{1/n} \le \delta+\varepsilon = 1-\varepsilon < 1$$

which contradicts the assumptions of the theorem.

Hence, we have proved that $G(z,\infty)$ is an exact harmonic majorant of $\{H_n\}_{n=1}^{\infty}$. To get the sharper statement of the theorem let S be a continuum in $\mathbb{C}\setminus[-1,1]$. According to Lemma 4,

$$\limsup_{n\to\infty} \max_{z\in S} H_n(z) = \max_{z\in S} G(z,\infty) = \|G(\cdot,\infty)\|_S.$$

That yields

$$\limsup_{n\to\infty} \|r_n^*\|_S^{1/(n-\mu_n)} \ge \exp(\|G(\cdot,\infty)\|_S)$$

or

$$\begin{aligned} \limsup_{n\to\infty} \|r_n^*\|_S^{1/(n+m(n))} &\ge \exp\left(\frac{n-\mu_n}{n+m(n)}\|G(\cdot,\infty)\|_S\right) \\ &\ge \exp\left(\frac{n-m(n)}{n+m(n)}\|G(\cdot,\infty)\|_S\right) \\ &= \exp\left(\frac{c-1}{c+1}\|G(\cdot,\infty)\|_S\right) \end{aligned}$$

and Theorem 1 is proved.

8. How restrictive is the pole condition?

Let $f(x) = |x|^\alpha$ and $\alpha \in \mathbb{R}_+\setminus 2\mathbb{N}$, $N_c = \{(n,m)\}$, $c > 1$. Saff and Stahl [6] have proved that in this case all poles of the best approximants $r_{n,m}^*$ to f converge to $z=0$ and all zeros cluster on $[-1,1]$. Hence, the pole condition (P_τ) does not hold for any $\tau > 1$. Moreover, if $r_{n,m}^* = p_n^*/q_m^*$ then the polynomials p_n^* resp. q_m^* have exact degree n resp. m for all sufficiently big (n,m) and n,m even (cf. Lemma 2.1 in [6]).

Motivated by this example we introduce

Condition (PZ):

f satisfies the *condition (PZ)* for $N_c = \{(n,m)\}$, $c > 1$, if the best approximations $r_{n,m}^* = p_n^*/q_m^*$ to f satisfy the following conditions:

(i) p_n^* and q_m^* have no common factor, degree $p_n^* = n$, degree $q_m^* = m$,

(ii) $r_{n,m}^*(z) = a_{n,m} \dfrac{\prod_{i=1}^{n} (z - y_i^{(n)})}{\prod_{i=1}^{m} (z - z_i^{(n)})}$,

(iii) $0 = \lim\limits_{\substack{n+m\to\infty \\ (n,m)\in N_c}} \dfrac{1}{n-m} \sum_{i=1}^{n} G(y_i^{(n)}, \infty) = \lim\limits_{\substack{n+m\to\infty \\ (n,m)\in N_c}} \dfrac{1}{n-m} \sum_{i=1}^{m} G(z_i^{(n)}, \infty)$.

If f satisfies (PZ) for $N_c = \{(n,m)\}$, then the generalized Bernstein Lemma implies

$$\frac{1}{n-m} \log|r_{n,m}^*(z)| \le \frac{1}{n-m} \left[\sum_{i=1}^{m} G(z, z_i^{(n)}) + \log(2\|f\|) \right] + G(z, \infty)$$

and for $z \to \infty$ we obtain

$$\limsup_{\substack{n+m\to\infty \\ (n,m)\in N_c}} \frac{1}{n-m} \log|a_{n,m}| \le \log 2 + \limsup_{\substack{n+m\to\infty \\ (n,m)\in N_c}} \frac{1}{n-m} \sum_{i=1}^{m} G(z_i^{(n)}, \infty) \le \log 2. \tag{8.1}$$

Since $(n,m) \in N_c$, we may assume w.l.o.g. that the function f is strictly positive, i.e. there exists $\kappa > 0$ such that

$$f(x) \ge \kappa \text{ for all } x \in [-1,1]. \tag{8.2}$$

In this case we may write

$$\frac{f - r_{n,m}^*}{f \cdot r_{n,m}^*} = \frac{1}{r_{n,m}^*} - \frac{1}{f}. \tag{8.3}$$

Since $\lim\limits_{\substack{n+m\to\infty \\ (n,m)\in N_c}} \|f - r_{n,m}^*\| = 0$, there exists $n_0 \in \mathbb{N}$ such that

$$r_{n,m}^*(x) \ge \frac{\kappa}{2} \text{ for } x \in [-1,1] \text{ and } n \ge n_0.$$

Consequently, for $n \ge n_0$

$$\left\| \frac{1}{r_{n,m}^*} - \frac{1}{f} \right\| \le \frac{2}{\kappa^2} \|f - r_{n,m}^*\|.$$

Using Lemma 1 again for $r = 1/r_{n,m}^*$ we get

$$\liminf_{\substack{n+m\to\infty \\ (n,m)\in N_c}} \frac{1}{n-m} \log|a_{n,m}| \ge \log 2 - \limsup_{\substack{n+m\to\infty \\ (n,m)\in N_c}} \sum_{i=1}^{n} G(y_i^{(n)}, \infty). \tag{8.4}$$

Condition (PZ) and (8.1), (8.4) imply

$$\lim_{\substack{n+m\to\infty \\ (n,m)\in N_c}} \frac{1}{n-m} \log|a_{n,m}| = \log 2. \tag{8.5}$$

Now let us fix $K > 2\|f\|$ and define

$$S_{n,m,K} := \{z \in \mathbb{C} : |r_{n,m}^*(z)| \le K\}. \tag{8.6}$$

Consider

$$H_{n,m,K}(z) := \frac{1}{n-m} \left[\log \left| \frac{r_{n,m}^*(z)}{K} \right| - \sum_{i=1}^{m} G_{n,m,K}(z, z_i^{(n)}) \right], \tag{8.7}$$

where $G_{n,m,K}(z, z_i^{(n)})$ is Green´s function of the domain $\overline{\mathbb{C}} \setminus S_{n,m,K}$ with pole at $z_i^{(n)}$, $1 \le i \le m$. Then $H_{n,m,K}(z)$ is Green´s function of $\overline{\mathbb{C}} \setminus S_{n,m,K}$ with pole at ∞ and it is well known that

$$\lim_{z\to\infty} (H_{n,m,K}(z) - \log|z|) = -\log\operatorname{cap} S_{n,m,K}, \tag{8.8}$$

where $\operatorname{cap} S_{n,m,K}$ is the logarithmic capacity of $S_{n,m,K}$. Then the left-hand side of (8.8) leads to

$$\begin{aligned}\limsup_{\substack{n+m\to\infty\\(n,m)\in N_c}} \log\operatorname{cap} S_{n,m,K} &= \limsup_{\substack{n+m\to\infty\\(n,m)\in N_c}} \frac{1}{n-m}\left[\log\frac{1}{|a_{n,m}|} + \sum_{i=1}^{m} G_{n,m,K}(z_i^{(n)},\infty)\right]\\ &\le \limsup_{\substack{n+m\to\infty\\(n,m)\in N_c}} \frac{1}{n-m}\left[\log\frac{1}{|a_{n,m}|} + \sum_{i=1}^{m} G(z_i^{(n)},\infty)\right]\\ &= \log\frac{1}{2} = \log\operatorname{cap}[-1,1]. \end{aligned}\tag{8.9}$$

Since $[-1,1] \subset S_{n,m,K}$ for sufficiently big n we have got

$$\lim_{\substack{n+m\to\infty\\(n,m)\in N_c}} \log\operatorname{cap} S_{n,m,K} = \log\frac{1}{2}.$$

Hence, we have obtained

Theorem 2. *If f satisfies condition (PZ) for $N_c = \{(n,m)\}$, then the absolute value of the best rational approximants $r^*_{n,m}$ converge quasi-everywhere to ∞ in $\overline{\mathbb{C}} \setminus [-1,1]$, i. e. the set of exceptional points, where $\{r^*_{n,m}\}$ is bounded, is of capacity* 0.

References

[1] V. V. Andrievskii and H.-P. Blatt, *Discrepancy of signed measures and polynomial approximation.* Springer Verlag New York, Berlin, Heidelberg, 2002.

[2] H.-P. Blatt and E. B. Saff, Behavior of zeros of polynomials of near best approximation. *J. Approx. Theory* **46** (1986), 323 - 344.

[3] H.-P. Blatt, A. Iserles and E. B. Saff, Remarks on the behaviour of zeros and poles of best approximating polynomials and rational functions. In: J.C. Mason and M. G. Cox (Eds.), *Algorithms for Approximation,* pp. 437-445. Inst. Math. Appl. Conf. Ser. New Ser. 10, Clarendon Press, Oxford, 1987.

[4] A. A. Gončar, Properties of functions related to their rate of approximability by rational functions. *Amer. Math. Soc. Transl.* **91** (1970), 99 - 127.

[5] A. A. Gončar, A local condition of single-valuedness of analytic functions. *Mat. Sb. USSR* **18** (1972), 151 - 167.

[6] E. B. Saff nad H. Stahl, Ray sequences of best rational approximants for $|x|^\alpha$. *Can. J. Math.* **49** (1997), 1034 - 1065.

[7] J. L. Walsh, Overconvergence, degree of convergence, and zeros of sequences of analytic functions. *Duke Math. J.* **13** (1946), 195 - 234.

[8] J. L. Walsh, *Interpolation and approximation by rational functions in the Complex Plane,* 5th edition. American Mathematical Society, 1969.

Katholische Universitaet Eichstaett-Ingolstadt, Mathematisch-Geographische Fakultaet, Lehrstuhl für Mathematik - Angewandte Mathematik, D-85071 Eichstaett, Germany

E-mail address: hans.blatt@ku-eichstaett.de

Potential Theory and Stochastics in Albac, 65–84
Aurel Cornea Memorial Volume

NON-SYMMETRIC RESISTANCE FORMS

NICU BOBOC AND GHEORGHE BUCUR

ABSTRACT. We introduce the notion of (non-symetric) resistance form on a set and we develop a potential theory associated with this form. The notion of symmetric resitance form was introduced by J. Kigami in connection with the so called "analysis of fractals".

This paper contains some preliminaries concerning the (non-symmetric) resistance forms, the excessive and co-excessive functions with respect to such forms and particularly the associated Green function: other aspects, as sub Markovian resolvents in duality associated with such a (non-symmetric) resistance form, will be treated later.

1. PRELIMINARIES

In the sequel we consider a system $(X, D(\mathcal{E}), \mathcal{E})$ where X is an arbitrary set which contains at least two points, $D(\mathcal{E})$ is a real vector space of real functions on X which contains any real constant function on X and separates the points of X and $\mathcal{E}$ is a bilinear functional on $D(\mathcal{E}) \times D(\mathcal{E})$.

A system $(X, D(\mathcal{E}), \mathcal{E})$ is called a *resistance form on the set* X if the following properties hold:

RF1: For any $u \in D(\mathcal{E})$ we have:

a) $\mathcal{E}(u,u) \geq 0$ and $\mathcal{E}(u,u) = 0 \Leftrightarrow u$ is constant;

b) There exists $M \in \mathbf{R}_+$ such that

$$(\mathcal{E}(u,v))^2 \leq M\mathcal{E}(u,u)\mathcal{E}(v,v), \quad \forall u,v \in D(\mathcal{E}).$$

RF2: For any $u, v \in D(\mathcal{E})$ the function

$$x \longmapsto (u \wedge v)(x) := inf(u(x), v(x))$$

belongs to $D(\mathcal{E})$ and we have

$$u \wedge v = 0 \Rightarrow \mathcal{E}(u,v) \leq 0.$$

RF3: There exists $x_0 \in X$ such that the real vector space

$$D_{x_0}(\mathcal{E}) := \{u \in D(\mathcal{E}) | u(x_0) = 0\}$$

is a Hilbert space with respect to the scalar product

$$\langle u, v \rangle := \frac{1}{2}(\mathcal{E}(u,v) + \mathcal{E}(v,u))$$

2000 *Mathematics Subject Classification.* 60J45, 60J57, 31C15, 60J80, 35B27, 47D08.
Key words and phrases. Resistance form, excessive function, fine topology.

and the convex cone

$$D^+_{x_0}(\mathcal{E}) := \{u \in D_{x_0}(\mathcal{E}) | u \geq 0\}$$

is a closed subset of this Hilbert space.

If moreover

$$\mathcal{E}(u,v) = \mathcal{E}(v,u), \quad u,v \in D(\mathcal{E})$$

the resistance form $(X,\mathcal{E},D(\mathcal{E}))$ is called *symmetric*: In fact $(X,D(\mathcal{E}),\mathcal{E})$ will be symmetric iff the constant M which appear in the above definition is equal 1. This case was introduced and studied by J. Kigami ([2],[3]) in connection with the so called analyse of fractals. The notion of non-symmetric resistance form is analogous, in this context, of the non-symmetric Dirichlet form (see [4]).

Remark. Let $(X,D(\mathcal{E}),\mathcal{E})$ be a resistance form on a set X. Then the systems $(X,D(\mathcal{E}^*),\mathcal{E}^*)$, $(X,D(\tilde{\mathcal{E}}),\tilde{\mathcal{E}})$ are also resistance forms on the set X where

$$D(\mathcal{E}^*) = D(\tilde{\mathcal{E}}) = D(\mathcal{E})$$

and for any $u,v \in D(\mathcal{E})$,

$$\mathcal{E}^*(u,v) = \mathcal{E}(v,u), \quad \tilde{\mathcal{E}}(u,v) = \frac{1}{2}(\mathcal{E}(u,v) + \mathcal{E}(v,u)).$$

The resistance form $(X,D(\mathcal{E}^*),\mathcal{E}^*)$ is called the adjoint of $(X,D(\mathcal{E}),\mathcal{E})$ and the resistance form $(X,D(\tilde{\mathcal{E}}),\tilde{\mathcal{E}})$ is symmetric.

We have for any $u,v \in D(\mathcal{E})$

$$\mathcal{E}^*(u,u) = \mathcal{E}(u,u) = \tilde{\mathcal{E}}(u,u).$$

Proposition 1.1. *Assume that $(X,D(\mathcal{E}),\mathcal{E})$ is a resistance form on the set X. Then we have:*

a) $\mathcal{E}(|u|,|u|) \leq \mathcal{E}(u,u), \ \forall u \in D(\mathcal{E})$;
b) $\mathcal{E}(u \wedge \alpha, u \wedge \alpha) \leq \mathcal{E}(u,u), \ \forall u \in D(\mathcal{E}), \ \alpha \in \mathbf{R}$;
c) *Any linear functional $\varphi : D_{x_0}(\mathcal{E}) \to \mathbf{R}$ such that $\varphi(u) \geq 0$ for all $u \in D^+_{x_0}(\mathcal{E})$ is continuous on the Hilbert space $(D_{x_0}(\mathcal{E}), \langle,\rangle)$.*

Proof. a) If $u \in D(\mathcal{E})$ we have $|u| = u^+ + u^-$, $u = u^+ - u^-$ and since $u^+ \wedge u^- = 0$ we get $\mathcal{E}(u^-,u^+) \leq 0$ and $\mathcal{E}(u^+,u^-) \leq 0$. Therefore

$$\begin{aligned}\mathcal{E}(|u|,|u|) \leq & \ \mathcal{E}(u^+,u^+) + \mathcal{E}(u^-,u^-) + \mathcal{E}(u^+,u^-) + \mathcal{E}(u^-,u^+) \\ \leq & \ \mathcal{E}(u^+,u^+) + \mathcal{E}(u^-,u^-) - \mathcal{E}(u^+,u^-) - \mathcal{E}(u^-,u^+) = \mathcal{E}(u,u).\end{aligned}$$

b) If $u \in D(\mathcal{E})$ and $\alpha \in \mathbf{R}$ we get

$$\mathcal{E}(u,\alpha) = 0, \ \ \mathcal{E}(\alpha,u) = 0,$$

$$\mathcal{E}(\alpha - u \wedge \alpha, u - u \wedge \alpha) \leq 0, \ \ \mathcal{E}(u - u \wedge \alpha, \alpha - u \wedge \alpha) \leq 0$$

and so

$$\langle \alpha - u \wedge \alpha, u - u \wedge \alpha \rangle \leq 0, \ \ \langle u \wedge \alpha, u \wedge \alpha - u \rangle \leq 0,$$

$$\langle u \wedge \alpha, u \wedge \alpha \rangle \leq \langle u \wedge \alpha, u \rangle \leq \langle u \wedge \alpha, u \wedge \alpha \rangle^{\frac{1}{2}} \langle u,u \rangle^{\frac{1}{2}},$$

$$\mathcal{E}(u \wedge \alpha, u \wedge \alpha) = \langle u \wedge \alpha, u \wedge \alpha \rangle \leq \langle u,u \rangle = \mathcal{E}(u,u).$$

c) Let $\varphi : D_{x_0}(\mathcal{E}) \to \mathbf{R}$ be a linear functional such that $\varphi(u) \geq 0$ for all $u \in D^+_{x_0}(\mathcal{E})$ and assume that φ is not continuous. Then there exists a sequence $(u_n)_n$ in $D_{x_0}(\mathcal{E})$ such that $\langle u_n, u_n\rangle \leq 1$ and $\varphi(u_n) \geq 2^n$ for all $n \in \mathbf{N}$. From the above property we get

$$\langle |u_n|, |u_n|\rangle \leq \langle u_n, u_n\rangle \leq 1, \quad \forall n \in \mathbf{N}$$

and so the series $\sum_n \frac{1}{2^n}|u_n|$ is convergent in $D_{x_0}(\mathcal{E})$ to an element $u \in D^+_{x_0}(\mathcal{E})$.

We have, for any $p \in \mathbf{N}$

$$\sum_{n=1}^{p} \frac{1}{2^n}|u_n| \leq u,$$

$$p \leq \sum_{n=1}^{p} \frac{1}{2^n}\varphi(u_n) \leq \sum_{n=1}^{p} \frac{1}{2^n}\varphi(|u_n|) \leq \varphi(u)$$

which is a contradiction. □

Corollary 1.2. *If $(X, D(\mathcal{E}), \mathcal{E})$ is a resistance form on the set X then for any $y \in X$ there exists $M_y \in \mathbf{R}_+$ such that*

$$(u(y))^2 \leq M_y \mathcal{E}(u, u), \quad \forall u \in D_{x_0}(\mathcal{E}).$$

In particular for any $y, z \in X$ we have

$$\sup\{\frac{(u(y) - u(z))^2}{\mathcal{E}(u, u)} \mid u \in D(\mathcal{E}), \mathcal{E}(u, u) > 0\} < \infty.$$

Corollary 1.3. *If $(X, D(\mathcal{E}), \mathcal{E})$ is a resistance form on the set X then for any $y \in X$ the real vector space*

$$D_y(\mathcal{E}) := \{u \in D(\mathcal{E}) \mid u(y) = 0\}$$

is a Hilbert space with respect to the scalar product

$$\langle u, v\rangle := \frac{1}{2}(\mathcal{E}(u, v) + \mathcal{E}(v, u))$$

and the convex cone

$$D^+_y(\mathcal{E}) := \{u \in D_y(\mathcal{E}) \mid u \geq 0\}$$

is a closed subset of this Hilbert space.

We remember (see [2]) the folowing result:

Proposition 1.4. *If $(X, D(\mathcal{E}), \mathcal{E})$ is a resistance form on the set X then the map $d^{\mathcal{E}}$: $X \times X \to \mathbf{R}$ where*

$$d^{\mathcal{E}}(x, y) := \sup\{\frac{(u(x) - u(y))^2}{\mathcal{E}(u, u)} \mid u \in D(\mathcal{E}),\ \mathcal{E}(u, u) > 0\}$$

is a distance on X called the distance associated with $(X, D(\mathcal{E}), \mathcal{E})$.

In the sequel $(X, D(\mathcal{E}), \mathcal{E})$ will be a given resistance form on a set X, x_0 is a fixed point in X, $D_0(\mathcal{E}) := \{u \in D(\mathcal{E}),\ u(x_0) = 0\}$, $\langle,\rangle$ is the scalar product on $D_0(\mathcal{E})$ given by $\langle u, v\rangle := \frac{1}{2}(\mathcal{E}(u, v) + \mathcal{E}(v, u))$ and $||\cdot||$ the norm on $D_0(\mathcal{E})$ associated with this scalar product.

Remark. There exist two linear continuous and bijective operators on the Hilbert space $(D_0(\mathcal{E}), \langle , \rangle)$ such that

$$\mathcal{E}(u,v) = \langle Tu, v\rangle = \langle u, T^* v\rangle$$

for any $u, v \in D_0(\mathcal{E})$. In particular we have

$$||u||^2 = \langle Tu, u\rangle$$

$$u, v \in D_0(\mathcal{E}),\ u \wedge v = 0 \Rightarrow \langle Tu, v\rangle \leq 0$$

i.e. the system $(D_0(\mathcal{E}), D_0^+(\mathcal{E}), T)$ is a Dirichlet space (see [1], pag. 256).

Proposition 1.5. a) *For any $y \in X$ there exists $g^y, {}^*g^y,\ g_0^y \in D_0(\mathcal{E})$ uniquely determined such that*

$$\langle g_0^y, u\rangle = \mathcal{E}(g^y, u) = \mathcal{E}(u, {}^*g^y) = u(y)$$

for all $u \in D_0(\mathcal{E})$. In particular we have

$$g_0^y = 0 \Leftrightarrow g^y = 0 \Leftrightarrow {}^*g^y = 0 \Leftrightarrow y = x_0.$$

b) *We have for any $y, z \in X$*

$$||g^y - g^z|| = ||{}^*g^y - {}^*g^z|| = ||g_0^y - g_0^z|| =: d_{\mathcal{E}}(y,z)$$

and

$$d_{\mathcal{E}}(y,z) \leq (d^{\mathcal{E}}(y,z))^{\frac{1}{2}} \leq M d_{\mathcal{E}}(y,z).$$

In particular if $(X, D(\mathcal{E}), \mathcal{E})$ is symmetric we get

$$d^{\mathcal{E}}(y,z) = (d_{\mathcal{E}}(y,z))^2$$

$$g^y(z) = \frac{1}{2}(d^{\mathcal{E}}(y,x_0) + d^{\mathcal{E}}(z,x_0) - d^{\mathcal{E}}(y,z)).$$

The function on $X \times X$ given by $(x,y) \to g(x,y) := g^y(x) = {}^*g^x(y)$ is called the Green function associated with $(\mathcal{E}, D(\mathcal{E}), X)$ and x_0.

Proof. a) If $y \in X$ then by Corollary 1.3 the linear functional on $D_0(\mathcal{E})$ given by $u \longrightarrow u(y)$ is continuous and so there exists $g_0^y \in D_0(\mathcal{E})$ uniquely determined such that

$$u(y) = \langle g_0^y, u\rangle,\ \forall u \in D_0(\mathcal{E}).$$

Also if we put $g^y := T^{-1}(g_0^y), {}^*g^y = (T^*)^{-1}(g_0^y)$ we get

$$\mathcal{E}(g^y, u) = \langle u, T(g^y)\rangle = \langle u, g_0^y\rangle = u(y),$$

$$\mathcal{E}({}^*g^y, u) = \langle u, T^*({}^*g^y)\rangle = \langle u, g_0^y\rangle = u(y).$$

Since $D_0(\mathcal{E})$ separates the points of X we get

$$g_0^y = 0 \Leftrightarrow g^y = 0 \Leftrightarrow {}^*g^y = 0 \Leftrightarrow y = x_0.$$

b) If $y, z \in X$ we get

$$\begin{aligned} ||g^y - g^z||^2 = &\ \mathcal{E}(g^y - g^z, g^y - g^z) = (g^y - g^z)(y) - (g^y - g^z)(z) \\ = &\ \mathcal{E}(g^y - g^z, {}^*g^y - {}^*g^z) = ({}^*g^y - {}^*g^z)(y) - ({}^*g^y - {}^*g^z)(z) \\ = &\ \mathcal{E}({}^*g^y - {}^*g^z, {}^*g^y - {}^*g^z) = ||{}^*g^y - {}^*g^z||^2. \end{aligned}$$

From a) we get

$$||g^y - g^z||^2 = \langle g^y - g^z, g_0^y - g_0^z\rangle \leq ||g^y - g^z|| \cdot ||g_0^y - g_0^z||$$

and so

$$||g^y - g^z||^2 \le ||g_0^y - g_0^z||^2 = (g_0^y - g_0^z)(y) - (g_0^y - g_0^z)(z)$$
$$= \frac{((g_0^y - g_0^z)(y) - (g_0^y - g_0^z)(z))^2}{\mathcal{E}(g_0^y - g_0^z, g_0^y - g_0^z)} \le d^{\mathcal{E}}(y,z).$$

On the other hand if $u \in D_0(\mathcal{E})$ is such that $\mathcal{E}(u,u) > 0$ we get

$$\frac{(u(y)-u(z))^2}{\mathcal{E}(u,u)} = \frac{(\mathcal{E}(g^y - g^z, u))^2}{\mathcal{E}(u,u)} \le M\frac{\mathcal{E}(u,u)\mathcal{E}(g^y - g^z, g^y - g^z)}{\mathcal{E}(u,u)} = M||g^y - g^z||^2$$

i.e.

$$d^{\mathcal{E}}(y,z) \le M||g^y - g^z||^2.$$

If $(X, D(\mathcal{E}), \mathcal{E})$ is symmetric then $M = 1$ and so

$$d^{\mathcal{E}}(y,z) = ||g^y - g^z||^2.$$

Hence

$$\frac{1}{2}(d^{\mathcal{E}}(y,x_0) + d^{\mathcal{E}}(x,x_0) - d^{\mathcal{E}}(y,z)) = \frac{1}{2}(||g^y||^2 + ||g^z||^2 - ||g^y - g^z||^2)$$
$$= \langle g^y, g^z \rangle = g^y(z).$$

□

Proposition 1.6. *Let $(u_n)_n$ be a sequence in the Hilbert space $D_0(\mathcal{E})$ and $u \in D_0(\mathcal{E})$.*

a) *$(u_n)_n$ is weakly convergent to u iff $(u_n)_n$ is bounded in $D_0(\mathcal{E})$ and*

$$\lim_{n\to\infty} u_n(x) = u(x), \quad \forall x \in X$$

b) *If u_n is weakly convergent to u then the sequences $(u_n^+)_n$, $(u_n^-)_n$, $(|u_n|)_n$ are weakly convergent to u^+, u^- and $|u|$ respectively.*

Proof. a) If $(u_n)_n$ is weakly convergent in $(D_0(\mathcal{E}), \langle,\rangle)$ then $(u_n)_n$ is bounded in $(D_0(\mathcal{E}), \langle,\rangle)$ and for any $x \in X$ we have

$$\lim_{n\to\infty} u_n(x) = \lim_{n\to\infty} \langle u_n, g_0^x \rangle = \langle u, g_0^x \rangle = u(x).$$

Conversely, assume that $(u_n)_n$ is bounded in $D_0(\mathcal{E})$ and $u_n(x) \to u(x)$, $\forall x \in X$.

It follows that u is the only weak limit point of the sequence $(u_n)_n$ and so $(u_n)_n$ is weakly convergent to u.

b) The assertion follows using Proposition 1.1 and the above point a). □

Corollary 1.7. *Let $(u_n)_n$ and $(v_n)_n$ be two sequences in $(D_0(\mathcal{E}), \langle,\rangle)$ which are weakly convergent to u and v respectively. Then the sequences $(u_n \wedge v_n)_n$ and $(u_n \vee v_n)_n$ are weakly convergent to $u \wedge v$ and $u \vee v$ respectively.*

We have the folowing assertion (see [1], Theorem 7.2.5):

Proposition 1.8. *The latice operations in $D_0(\mathcal{E})$ are continuous.*

Proof. Let $d^{\mathcal{E}}$ be the distance on X associated with the resistance form $(X, D(\mathcal{E}), \mathcal{E})$ and let $\bar{X}$ the completion of X with respect to this distance.

Since for any $u \in D(\mathcal{E})$ we have

$$(u(x) - u(y))^2 \le \mathcal{E}(u,u)d^{\mathcal{E}}(x,y)$$

for all $x, y \in X$ it follows that any $u \in D(\mathcal{E})$ may be extended uniquely to a continuous function $\bar{u}$ on $\bar{X}$ and we have

$$(\bar{u}(x) - \bar{u}(y))^2 \leq \mathcal{E}(u,u) d^{\mathcal{E}}(x,y), \quad \forall x, y \in \bar{X}.$$ □

Proposition 1.9. *The system* $(\bar{X}, D(\bar{\mathcal{E}}), \bar{\mathcal{E}})$ *where*

$$D(\bar{\mathcal{E}}) = \{\bar{u} \mid u \in D(\mathcal{E})\}$$

$$\bar{\mathcal{E}}(\bar{u}, \bar{v}) := \mathcal{E}(u, v)$$

is a resistance form on the set $\bar{X}$ *such that the distance* $d^{\mathcal{E}}$ *on* $\bar{X}$ *associated with this resistance form coincides with the natural extension of* $d^{\mathcal{E}}$ *to* $\bar{X}$.

Remark. From now on whenever $(X, D(\mathcal{E}), \mathcal{E})$ is a resistance form on a set X we may assume that the metric space $(X, d^{\mathcal{E}})$ is complete.

Example 1. Let $X = [0,1]$, λ the Lebesgue measure on $[0,1]$, $D(\mathcal{E})$ the set of all real functions u on $[0,1]$ which are absolutely continuous with $u' \in L^2(\lambda)$ (u' is the derivative of u) and $\mathcal{E}$ is the bilinear functional on $D(\mathcal{E}) \times D(\mathcal{E})$ given by

$$\mathcal{E}(u,v) = \int_0^1 u'v' d\lambda + \int_0^1 (u - u(0))v' d\lambda.$$

Then the system $([0,1], D(\mathcal{E}), \mathcal{E})$ is a resistance form on $[0,1]$.

Proof. Obviously $D(\mathcal{E})$ is a vector sublattice of the vector lattice of all real functions on $[0,1]$ which contains the real constant functions and moreover separates the point of $[0,1]$. Since for any $u \in D(\mathcal{E})$ we have

$$\int_0^1 (u - u(0))u' d\lambda = \frac{(u(1) - u(0))^2}{2}$$

it follows that for any $u \in D(\mathcal{E})$ we have $\mathcal{E}(u,u) \geq 0$ and $\mathcal{E}(u,u) = 0 \Leftrightarrow u = \text{constant}$.

Also for any $u, v \in D(\mathcal{E})$ we get

$$\mathcal{E}(u,v)^2 \leq 4\mathcal{E}(u,u)\mathcal{E}(v,v)$$

and so the system $([0,1], D(\mathcal{E}), \mathcal{E})$ satisfies the property RF1. Also we have

$$u, v \in D(\mathcal{E}), \ u \wedge v = 0 \Rightarrow \mathcal{E}(u,v) \leq 0$$

i.e. the system $([0,1], D(\mathcal{E}), \mathcal{E})$ satisfies the property RF2.

It is easy to see that real vector space

$$D_0(\mathcal{E}) = \{u \in D(\mathcal{E}) \mid u(0) = 0\}$$

is a Hilbert space with respect to the scalar product

$$\langle u, v \rangle := \frac{1}{2}(\mathcal{E}(u,v) + \mathcal{E}(v,u)) = \int_0^1 u'v' d\lambda + (uv)(1)$$

and the set

$$D_0^+(\mathcal{E}) = \{u \in D_0(\mathcal{E}) \mid u \geq 0\}$$

is closed in this Hilbert space i.e. $([0,1], D(\mathcal{E}), \mathcal{E})$ satisfies the property RF3. □

Example 2. Let $X = \{x_1, x_2, \dots, x_n\}$ be a finite set with $n \geq 3$ and

$$i \neq j \Rightarrow x_i \neq x_j.$$

We denote

$$D(\mathcal{E}) := \{u : X \to \mathbf{R}\}$$

and for any $u, v \in D(\mathcal{E})$

$$\mathcal{E}(u,v) = \sum_{i=1}^{n-1} u(x_i)(v(x_i) - v(x_{i+1})) + u(x_n)(v(x_n) - v(x_1)).$$

Since for any $u \in D(\mathcal{E})$ we have

$$\mathcal{E}(u,u) = \frac{1}{2}(\sum_{i=1}^{n-1}(u(x_i) - u(x_{i+1}))^2 + (u(x_n) - u(x_1))^2$$

it follows

$$\mathcal{E}(u,u) \geq 0 \text{ and } \mathcal{E}(u,u) = 0 \Leftrightarrow u = \text{constant}.$$

Since

$$\mathcal{E}(u,v) = \sum_{i=1}^{n-1}\sum_{j=2}^{i}(u(x_j) - u(x_{j-1}))(v(x_i) - v(x_{i+1})) + (u(x_n) - u(x_1))(v(x_n) - v(x_1))$$

it follows that

$$|\mathcal{E}(u,v)| \leq 2n(\mathcal{E}(u,u))^{\frac{1}{2}}(\mathcal{E}(v,v))^{\frac{1}{2}}$$

i.e. $(X, D(\mathcal{E}), \mathcal{E})$ satisfies the property RF1.

Obviously we get

$$u, v \in D(\mathcal{E}),\ u \wedge v = 0 \Rightarrow \mathcal{E}(u,v) \leq 0$$

i.e. the property RF2.

Since $D(\mathcal{E})$ has a finite dimension, then the property RF3 is also satisfied.

2. Excessive functions associated with a resistance form on a set

In the sequel $(X, D(\mathcal{E}), \mathcal{E})$ is a resistance form on a set X and x_0 is a fixed point in X. We denote

$$X_0 := X \setminus \{x_0\}, D_0(\mathcal{E}) := \{u \in D(\mathcal{E}) / u(x_0) = 0\},$$

$$D_0^+(\mathcal{E}) := \{u \in D_0(\mathcal{E}) / u \geq 0\},$$

$\langle , \rangle$ the scalar product on $D_0(\mathcal{E})$ given by

$$\langle u, v \rangle := \frac{1}{2}(\mathcal{E}(u,v) + \mathcal{E}(v,u)),$$

and T, T^* the linear, bijective and continuous operators on the Hilbert space $(D_0(\mathcal{E}), \langle\rangle)$ such that

$$\mathcal{E}(u,v) = \langle Tu, v \rangle = \langle u, T^* v \rangle.$$

We recall now an important technique introduced by G. Stampacchia (see [1], Theorem 7.1.4).

Proposition 2.1. *Let K be a non-empty closed convex subset of $(D_0(\mathcal{E}), \langle,\rangle)$. Then, for any $u \in D_0(\mathcal{E})$ there exists a unique element $\Pi^K u \in K$ with the property*

$$\langle T(u - \Pi^K u), v - \Pi^K u\rangle \leq 0 \;\forall\; v \in K.$$

The map

$$u \to \Pi^K u$$

is continuous. More precisely we have

$$\left\|\Pi^k u' - \Pi^K u''\right\| \leq M \left\|u' - u''\right\|, \;\forall\; u', u'' \in D_0(\mathcal{E}).$$

Remark. The operators $\Pi^K : D_{x_0}(\mathcal{E}) \to K$ is called the Stampacchia projection on K with respect to $(X, D(\mathcal{E}), \mathcal{E})$. The Stampacchia projection with respect to $(X, D(\mathcal{E}^*), \mathcal{E}^*)$ is denoted by ${}^*\Pi^K$. We have the following results ([1], Corollary 7.1.6).

Proposition 2.2. *Let $(K_i)_{i\in I}$ be an increasing (resp. decreasing and $\bigcap_{i\in I} K_i \neq \emptyset$) family of closed convex non-empty subsets of $D_0(\mathcal{E})$. Then for any $u \in D_0(\mathcal{E})$ the family $\left(\Pi^{K_i} u\right)_{i\in I}$ converges to $\Pi^K u$ where*

$$K = \overline{\bigcup_{i\in I} K_i} \text{ (resp. } K = \bigcap_{i\in I} K_i).$$

Definition. A function $s \in D_0(\mathcal{E})$ is called $\mathcal{E}$ - *excessive* (or T - *potential*) if for any $h \in D_0^+(\mathcal{E})$ we have

$$\mathcal{E}(s, h) \geq 0 \text{ (or } \langle Ts, h\rangle \geq 0).$$

We denote by $S := S(\mathcal{E})$ the set of all $\mathcal{E}$ - excessive functions and by $S^* := S^*(\mathcal{E})$ the set of all $\mathcal{E}^*$ - excessive functions.

It is easy to see that S and S^* are convex cones which are closed in the Hilbert space.

Proposition 2.3. *For any $s \in S$ (resp. $t \in S^*$) we have $s \geq 0$ (resp. $t \geq 0$) and $S - S$ (resp. $S^* - S^*$) is dense in $(D_0(\mathcal{E}), \langle,\rangle)$.*

Proof. If $s \in S$ then we have

$$0 \leq \mathcal{E}(s, s_-) = \mathcal{E}(s_+, s_-) - \mathcal{E}(s_-, s_-) \leq -\mathcal{E}(s_-, s_-)$$

i.e.

$$\mathcal{E}(s_-, s_-) = 0, s_- = 0, s \geq 0.$$

From the definition of S we have

$$h \in D_0^+(\mathcal{E}) \Leftrightarrow \langle Ts, h\rangle \geq 0 \;\forall\; s \in S$$

and so

$$h \in D_0(\mathcal{E}), \langle s, h\rangle = 0 \Rightarrow T^* h = 0 \Rightarrow h = 0$$

i.e. $S - S$ is dense in $(D_0(\mathcal{E}), \langle,\rangle)$. □

Definition. For any $u \in D_0(\mathcal{E})$ we denote by K_u the non-empty closed convex subset of $(D_0(\mathcal{E}), \langle,\rangle)$ given by

$$K_u := \{v \in D_0(\mathcal{E}) / v \geq u\}$$

and

$$Ru := \Pi^{K_u}(0).$$

The function Ru is called *the reduite of u*.

Proposition 2.4. *For any $u \in D_0(\mathcal{E})$ Ru is the unique element of $D_0(\mathcal{E})$ with the following properties*

$$a)\ Ru \geq u,$$

$$b)\ Ru \in S,$$

$$c)\ \mathcal{E}(Ru, Ru - u) = 0.$$

Proof. a) We have $Ru \in K_u$ and so

$$Ru \geq u.$$

b) If $h \in D_0^+(\mathcal{E})$ then $Ru + h \in K_u$ and so

$$\mathcal{E}(-Ru, h) = \langle -TRu, h\rangle = \langle -TRu, Ru + h - Ru\rangle \leq 0$$

i.e.

$$\mathcal{E}(Ru, h) \geq 0,\ Ru \in S.$$

c) From the points a) we get

$$\mathcal{E}(Ru, Ru - u) \geq 0.$$

Since $u \in K_u$ it follows

$$\mathcal{E}(-Ru, u - Ru) \leq 0$$

i.e.

$$\mathcal{E}(Ru, Ru - u) \leq 0$$

and so

$$\mathcal{E}(Ru, Ru - u) = 0.$$

Conversely let $v \in D_0(\mathcal{E})$ be such that we have $v \geq u$, $v \in S$ and $\mathcal{E}(v, v - u) = 0$. Hence $v \in K_u$ and for any $w \in K_u$ we get

$$\langle T(-v), w - v\rangle = -\mathcal{E}(v, w - v) = -\mathcal{E}(v, w - u) + \mathcal{E}(v, v - u) =$$

$$= -\mathcal{E}(v, w - v) \leq 0.$$

i.e.

$$v = \Pi^{K_u}(0) = Ru.$$ □

Corollary 2.5. *If $u \in D_0(\mathcal{E})$ we have*

$$u \in S \Leftrightarrow u = Ru.$$

Proposition 2.6. *a) For any $u \in D_0(\mathcal{E})$ we have*

$$Ru = \inf\{s \in S / s \geq u\}.$$

b) For any $s_1, s_2 \in S$ we have

$$s_1 \wedge s_2 \in S.$$

Proof. a) Let $u \in D_0(\mathcal{E})$. For any $s \in S$ with $s \geq u$ we have

$$u \leq s \wedge Ru \leq s.$$

and therefore

$$\begin{aligned}\mathcal{E}(Ru - s \wedge Ru, Ru - s \wedge Ru) &= \mathcal{E}(Ru - s, Ru - s \wedge Ru) + \mathcal{E}(s - s \wedge Ru, Ru - s \wedge Ru)\\ &\leq \mathcal{E}(Ru - s, Ru - s \wedge Ru) \leq \mathcal{E}(Ru, Ru - s \wedge Ru)\\ &\leq \mathcal{E}(Ru, Ru - u) = 0,\end{aligned}$$

$$Ru = s \wedge Ru,\ Ru \leq s.$$

Hence

$$Ru = \inf\{s \in S / s \geq u\}.$$

b) Let $s_1, s_2 \in S$. We have $u := s_1 \wedge s_2 \in D_0(\mathcal{E})$ and by the preceding point a) it follows

$$u = s_1 \wedge s_2 \geq Ru \geq u,$$

$$s_1 \wedge s_2 = Ru,\ s_1 \wedge s_2 \in S. \qquad \square$$

Proposition 2.7. *If $(s_i)_{i\in I}$ is a decreasing (resp. increasing and dominated) family in S then $(s_i)_{i\in I}$ is convergent to an element $s \in S$ in the Hilbert space $(D_0(\mathcal{E}), \langle,\rangle)$ and we have*

$$s(x) = \inf_{i\in I} s_i(x)\ \forall\ x \in X$$

$$(\text{resp. } s(x) = \sup_{i\in I} s_i(x)\ \forall\ x \in X).$$

Proof. If $(s_i)_{i\in I}$ is decreasing then the family $(K_{s_i})_{i\in I}$ of convex closed non-empty subsets of $D_0(\mathcal{E})$ given by

$$K_{s_i} := \{u \in D_0(\mathcal{E}) / u \geq s_i\}$$

is increasing,

$$s_i = R(s_i) = \Pi^{K_i}(0)\ \forall\ i \in I$$

and therefore, by Proposition 2.2, the family $(s_i)_{i\in I}$ converges in the Hilbert space $(D_0(\mathcal{E}), \langle,\rangle)$ to s. Since S is closed in this Hilbert space it follows that $s \in S$ and from Corollary 1.2 we get

$$s(x) = \inf_{i\in I} s_i(x)\ \forall\ x \in X.$$

If $(s_i)_{i\in I}$ is increasing and dominated in S then the family $(K_{s_i})_{i\in I}$ is decreasing and $\bigcap_{i\in I} K_{s_i} \neq 0$. Therefore, from Proposition 2.2, it follows that $(s_i)_{i\in I}$ is convergent in $(D_0(\mathcal{E}), \langle,\rangle)$ to a function $s \in S$. Moreover, from Corollary 1.2 we get

$$s(x) = \sup_{i\in I} s_i(x)\ \forall\ x \in X. \qquad \square$$

Proposition 2.8. *a) For any $s \in S$ and any $\alpha \in \mathbb{R}_+$ we have $s \wedge \alpha \in S$.*

b) For any $s_1, s_2 \in S$ there exists $s \in S$ such that

$$R(s_2 - s_1) + s = s_2.$$

Proof. a) Let $u := \alpha \wedge s$. Since

$$\mathcal{E}(Ru - \alpha \wedge Ru, Ru - \alpha \wedge Ru) \le \mathcal{E}(-\alpha \wedge Ru, Ru - \alpha \wedge Ru) =$$

$$= \mathcal{E}(\alpha - \alpha \wedge Ru, Ru - \alpha \wedge Ru) \le 0,$$

it follows that

$$Ru = \alpha \wedge Ru,$$

and so

$$Ru \le \alpha \wedge s, Ru = u, u \in S.$$

We show that

$$s_2 - R(s_2 - s_1) = R(s_2 - R(s_2 - s_1)).$$

Indeed, we have

$$\mathcal{E}(R(s_2 - R(s_2 - s_1)) - (s_2 - R(s_2 - s_1)), R(s_2 - R(s_2 - s_1)) - (s_2 - R(s_2 - s_1))) =$$

$$= \mathcal{E}(R(s_2 - s_1) - s_2, R(s_2 - R(s_2 - s_1)) - (s_2 - R(s_2 - s_1))) \le$$

$$\le \mathcal{E}(R(s_2 - s_1), s_1 - (s_2 - R(s_2 - s_1))) =$$

$$= \mathcal{E}(R(s_2 - s_1), R(s_2 - s_1) - (s_2 - s_1)) = 0$$

and therefore

$$R(s_2 - R(s_2 - s_1)) = s_2 - R(s_2 - s_1).$$

□

Corollary 2.9. *For any $s, s_1, s_2 \in S$ such that $s \le s_1 + s_2$ there exists $s_1', s_2' \in S$ with*

$$s = s_1' + s_2',\ s_1' \le s_1,\ s_2' \le s_2.$$

Proof. If we put

$$s_1' := R(s - s_2)$$

we get from Proposition 2.8 b)

$$s_2' := s - R(s - s_2) \in S$$

and

$$s = s_1' + s_2',\ s_1' \le s_1,\ s_2' \le s_2.$$

□

Remark. From Proposition 2.7 and Corollary 2.9 it follows that the ordered convex cone $(S, \le)$, where $\le$ is the pointwise order relation in S, is an H-cone and moreover

a) for any family $(s_i)_{i\in I}$ in S the function $\inf_{i\in I} s_i$ belongs to S

b) for any increasing and dominated family $(s_i)_{i\in I}$ in S the function $\sup_{i\in I} s_i$ belongs to S.

Proposition 2.10. *The linear space $S - S$ is dense in the Hilbert space $(D_0(\mathcal{E}), \langle,\rangle)$ and the set*

$$(S - S) \cap D_0^+(\mathcal{E})$$

is dense in $D_0^+(\mathcal{E})$. Moreover, for any $u \in D_0^+(\mathcal{E})$ there exists a sequence $(u_n)_n$ in $(S - S) \cap D_0^+(\mathcal{E})$ which converges to u in $(D_0(\mathcal{E}), \langle,\rangle)$ and $u_n \le u$, $\forall\, n \in \mathbb{N}$.

Proof. By the definition of S it follows that an element $u \in D_0(\mathcal{E})$ belongs to $-S$ iff for any $h \in D_0^+(\mathcal{E})$ we have

$$\langle u, T^* h\rangle \leq 0$$

i.e. $-S$ is the polar of the set $T^*(D_0^+(\mathcal{E}))$ and so we deduce that an element $h \in D_0(\mathcal{E})$ will be in $D_0^+(\mathcal{E})$ iff

$$\langle -s, T^*(h)\rangle \leq 0 \ \forall\ s \in S,$$

or equivalently

$$\mathcal{E}(s,h) \geq 0 \ \forall\ s \in S.$$

Let now $h \in D_0(\mathcal{E})$ be such that

$$\langle u, h\rangle = 0 \ \forall\ u \in S - S$$

or equivalently

$$\langle s, h\rangle = 0 \ \forall\ s \in S.$$

Since T^* is a bijection on $D_0(\mathcal{E})$ it follows

$$\langle T^* h'\rangle = 0 \ \forall\ s \in S$$

where $h' = (T^*)^{-1}(h)$. We deduce using the preceding considerations that $h' = 0$ and so $h = 0$ i.e. $S - S$ is dense in $D_0(\mathcal{E})$.

Let now $u \in D_0^+(\mathcal{E})$ and let $(u_n)_n$ be a sequence in $S - S$ which converges to u in $(D_0(\mathcal{E}), \langle,\rangle)$. From Proposition 1.8 it follows that the sequence $(u_n^+)_n$ converges in $(D_0(\mathcal{E}), \langle,\rangle)$ to $u^+ = u$. Since the operator $u \to Ru$ is continuous it follows that the sequence $(R(u_n - u))_n$ converges to zero in $(D_0(\mathcal{E}), \langle,\rangle)$ and so the sequence $(v_n)_n$ where

$$v_n = u_n - R(u_n - u)$$

converges to u in $(D_0(\mathcal{E}), \langle,\rangle)$ and we have

$$v_n \in S - S,\ v_n \leq u \ \forall\ n \in \mathbb{N}.$$

It follows that the sequence $(v_n^+)_n$ converges to u and we have

$$v_n^+ \in (S - S) \cap D_0^+(\mathcal{E}),\ v_n^+ \leq u \ \forall\ n \in \mathbb{N}. \qquad \square$$

Definition. The coarsest topology on X_0 for which any $s \in S$ is continuous on X_0 is called *the fine topology* on X_0. The coarsest topology on X_0 for which any $t \in S^*$ is continuous is called *the co-fine topology* on X_0.

Proposition 2.11. *The fine topology on X_0 coincide with the co-fine topology on X_0 and each of them coincide with the coarsest topology on X_0 for which any function $u \in D_0(\mathcal{E})$ is continuous on X_0.*

Proof. It is sufficient to show that any $u \in D_0^+(\mathcal{E})$ is fine and co-fine continuous.

Let $u \in D_0^+(\mathcal{E})$ and $(u_n)_n$ a sequence in $(S - S) \cap D_0^+(\mathcal{E})$ which converges to u in $(D_0(\mathcal{E}), \langle,\rangle)$ and moreover $u_n \leq u \ \forall\ n \in \mathbb{N}$.

If we put, for any $n \in \mathbb{N}$, $x \in X$

$$v_n(x) = \sup(u_1(x), u_2(x), \ldots, u_n(x)),$$

it follows that $v_n \in S - S$, $(v_n)_n$ is increasing

$$u_n \leq v_n \leq u \ \forall\ n \in \mathbb{N}$$

$$\lim_{n\to\infty} u_n(x) = u(x) \;\; \forall\, x \in X$$

and so the function u is lower semicontinuous with respect to the fine topology.

Since $Ru - u \in D_0^+(\mathcal{E})$ and $Ru \in S$ we deduce that $Ru - u$ is also lower semicontinuous with respect to the fine topology and so u is fine continuous.

Analogously it follows that any $u \in D_0^+(\mathcal{E})$ is co-fine continuous and so any $u \in D_0(\mathcal{E})$ is co-fine continuous. □

Notation. We denote by $\preccurlyeq$ the *specific order relation* in S i.e.

$$s', s'' \in S,\; s' \preccurlyeq s'' \overset{\text{def}}{\Leftrightarrow} \exists s \in S \text{ with } s' + s = s''.$$

Proposition 2.12. *a) For any $y \in X_0$ the function g^y given in Proposition 1.5 belongs to S. Moreover for any $y \in X_0$ we have*

$$s \in S, s \preccurlyeq g^y \Rightarrow \exists\, \alpha \ge 0 \textit{ with } s = \alpha g^y.$$

$$s \in S, s(y) \ge g^y(y) \Rightarrow s \ge g^y \textit{ on } X_0.$$

b) If A is a dense subset of X_0 with respect to the fine topology, then for any $s \in S$ and any finite set $F \subset A$, $F \neq \emptyset$ there exists $(\alpha_y)_{y\in F}$, $\alpha_y \in \mathbb{R}_+$ with

$$\sum_{y\in F} \alpha_y g^y \le s, \;\; \sum_{y\in F} \alpha_y g^y = s \textit{ on } F.$$

Proof. a) If $h \in D_0^+(\mathcal{E})$ we have

$$\mathcal{E}(g^y, h) = h(y) \ge 0$$

and so $g^y \in S$. Let now $y \in X_0$ and $s \in S$ be such that $s \preccurlyeq g^y$. It follows that there exists $s' \in S$ with $s' + s = g^y$. For any $h \in D_0(\mathcal{E})$ be such that $h(y) = 0$. We get $\mathcal{E}(s,h) = \mathcal{E}(s',h) = 0$.

Indeed, we have

$$\mathcal{E}(s,h^+) + \mathcal{E}(s',h^+) = \mathcal{E}(g^y,h^+) = h^+(y) = 0,$$

$$\mathcal{E}(s,h^-) + \mathcal{E}(s',h^-) = \mathcal{E}(g^y,h^-) = h^-(y) = 0,$$

and so

$$\mathcal{E}(s,h) = \mathcal{E}(s',h) = 0.$$

Let now $h_0 \in D_0^+(\mathcal{E})$ with $h_0(x) = 1$. We have for any $h \in D_0(\mathcal{E})$

$$\mathcal{E}(s,h) = \mathcal{E}(s, h(y), h_0) = h(y)\mathcal{E}(s,h_0),$$

$$\mathcal{E}(s',h) = \mathcal{E}(s', h(y), h_0) = h(y)\mathcal{E}(s',h_0),$$

$$\mathcal{E}(s',h_0) + \mathcal{E}(s,h_0) = \mathcal{E}(g^y,h_0) = 1.$$

If we put $\alpha = \mathcal{E}(s,h_0)$ we have

$$0 \le \alpha \le 1$$

and

$$\mathcal{E}(s,h) = h(y)\mathcal{E}(s,h_0) = \alpha h(y) = \alpha\mathcal{E}(g^y,h)$$

for any $h \in D_0(\mathcal{E})$ and so

$$s = \alpha g^y.$$

Let now $s \in S$ be such that $s(y) \geq g^y(y)$. We get $s \geq g^y$. Indeed, from the preceding considerations and $R(g^y - s) \preccurlyeq g^y$ it follows that there exists $\alpha \in \mathbb{R}_+$, $\alpha \leq 1$ with

$$R(g^y - s) = \alpha g^y.$$

We get

$$\mathcal{E}(\alpha g^y, \alpha g^y - (g^y - s)) = \mathcal{E}(R(g^y - s), R(g^y - s) - (g^y - s)) = 0$$

i.e.

$$\alpha s(y) = \alpha(1-\alpha) g^y(y),$$

$$\alpha g^y(y) \leq \alpha(1-\alpha) g^y(y),\ \alpha = 0$$

and so $R(g^y - s) = 0$, $g^y \leq s$.

b) Let $s \in S$ and $F \subset A$ be a non-empty finite set. The assertion is trivial if F has a single point. Assume that card$F = n$ and the assertion holds for F. Let us denote $F_0 = F \cup \{z\}$ where $z \in A$, $z \notin F$. By hypothetis there exists $(\alpha_y)_{y\in F}$, $\alpha_y \geq 0$ such that

$$\sum_{g\in F} \alpha_g g^y \leq s \text{ and } \sum_{y\in F} \alpha_y g^y = s \text{ on } F.$$

Let further $\beta \in R_+$ be such that $\beta g^z(z) = s(z)$ and

$$\beta g^z \leq s \text{ on } X_0.$$

If we put

$$u := \sum_{y\in F} \alpha_y g^y + \beta g^z - \left(\sum_{y\in F} \alpha_y g^y\right) \wedge \beta g^z,$$

it follows

$$u = \sup\left(\sum_{y\in F} \alpha_y g^y, \beta g^z\right) \leq s,$$

$$u = s \text{ on } F_0,$$

and so

$$u \leq Ru \leq s,$$

$$Ru = s \text{ on } F_0.$$

On the other hand, since

$$Ru \leq \sum_{y\in F} \alpha_y g^y + \beta g^z,$$

there exists $(\alpha'_y)_{y\in F}$, $0 \leq \alpha'_y \leq \alpha_y$ and $\beta' \in R_+$, $\beta' \leq \beta$ such that

$$Ru = \sum_{y\in F} \alpha'_y g^y + \beta' g^z.$$

Hence the assertion from b) holds for F_0. □

Corollary 2.13. *For any dense subset A of X_0, with respect to the fine topology, the set*

$$\mathscr{A} = \left\{\sum_{i\in I} \alpha_i g^{y_i} / \alpha_i \in R_+, y_i \in A, I \text{ finite}\right\}$$

is increasingly dense in S.

Proposition 2.14. *Let $y, z \in X_0$ be such that $g^y(z) > 0$ and $g^z(y) > 0$. Then, there exists $\alpha, \beta > 0$ such that*

$$\alpha g^y \leq g^z \leq \beta g^y.$$

Particulary, for any $x \in X_0$ we get

$$g^y(x) > 0 \Leftrightarrow g^z(x) > 0.$$

Proof. From proposition 2.12 b) we have

$$g^z(y) g^y \leq g^y(y) g^z, \; g^y(z) g^z \leq g^z(z) g^y$$

and so

$$\alpha g^y \leq g^z \leq \beta g^y$$

where

$$\alpha = g^z(y)/g^y(y), \; \beta = g^z(z)/g^y(z).$$ □

Definition. A function

$$s : X_0 \to \overline{\mathbb{R}}_+ \text{ (resp. } t : X_0 \to \overline{\mathbb{R}}_+)$$

is called *excessive* (resp. *co-excessive*) with respect to the resistance form $(X, D(\mathcal{E}), \mathcal{E})$ and the point $x_0 \in X$ if it is finite on a dense subset of X_0 with respect to the fine topology and moreover

$$s' \in S \Rightarrow \inf(s', s) \in S$$

$$\text{(resp. } t' \in S^* \Rightarrow \inf(t', t) \in S^*).$$

We denote by $\bar{S} = \bar{S}(\mathcal{E})$ (resp. $\bar{S}^* = \bar{S}^*(\mathcal{E})$) the set of all excessive (resp. co-excessive) functions with respect to $(X, D(\mathcal{E}), \mathcal{E})$ and $x_0 \in X$.

It is easy to see that $\bar{S}$ (resp. $\bar{S}^*$) is a convex cone, $1 \in \bar{S}$ (resp. $1 \in \bar{S}^*$), $S \subset \bar{S}$ (resp. $S^* \subset \bar{S}^*$) and S (resp. S^*) is solid and increasingly dense in $\bar{S}$ (resp. $\bar{S}^*$) with respect to pointwise order relation.

Also if $(s_i)_{i \in I}$ (resp. $(t_i)_{i \in I}$) is a family in $\bar{S}$ (resp. $\bar{S}^*$) then we have

$$\inf_{i \in I} s_i \in \bar{S} \text{ (resp. } \inf_{i \in I} t_i \in \bar{S}^*).$$

Moreover, if $(s_i)_{i \in I}$ (resp. $(t_i)_{i \in I}$) is an increasing and dominated family in $\bar{S}$ (resp. $\bar{S}^*$) then we have

$$\sup_{i \in I} s_i \in \bar{S} \text{ (resp. } \sup_{i \in I} t_i \in \bar{S}^*).$$

Proposition 2.15. *Any $s \in \bar{S}$ (resp. $t \in \bar{S}^*$) is finite and fine continuous.*

Proof. Let $s \in \bar{S}$ and assume that there exists $x \in X_0$ with $s(x) = +\infty$. From Proposition 2.12 a) we have

$$s' \in \bar{S}, \; s'(x) \geq g^x(x) \Rightarrow s' \geq g^x$$

and so for any $\alpha > 0$ we get

$$\alpha s \geq g^x.$$

Since s is finite, on a fine dense subset of X_0 we deduce $g = 0$, contradiction.

Let now $x \in X_0$, $s \in \bar{S}$ and $s' \in S$ be such that

$$s(x) < s'(x).$$

Since the set

$$U := \{y \in X_0 / s(y) < s'(y)\} = \{y \in X_0 / \inf(s', s)(y) < s'(y)\}$$

is a fine neighbourhood of x and

$$\inf(s', s) = s \text{ on } U$$

it follows that S is fine continuous at x. □

Proposition 2.16. *The ordered convex cone $(\bar{S}, \leq)$, where $\leq$ is the pointwise order relation, is an H-cone (see [1]) such that*

a) for any family $(s_i)_{i\in I}$ in $\bar{S}$ the function $\inf\limits_{i\in I} s_i$ belongs to $\bar{S}$

b) for any increasing and dominated family $(s_i)_{i\in I}$ in $\bar{S}$ the function $\sup\limits_{i\in I} s_i$ belongs to $\bar{S}$.

Proof. It remains to show that for any $s, s_1, s_2 \in \bar{S}$ with $s \leq s_1 + s_2$ there exist $s_1', s_2' \in \bar{S}$ with $s = s_1' + s_2'$, $s_1' \leq s_1$, $s_2' \leq s_2$.

Let $s' \in S$. We have

$$s \wedge s' \leq s_1 \wedge s' + s_2 \wedge s'.$$

If we put

$$f_1^{s'} := R(s \wedge s' - s_2 \wedge s')$$

we have

$$s \wedge s' - f_1^{s'} := h_{s'} \in S,$$

$$f_1^{s'} \leq s_1 \wedge s',\ h_{s'} \leq s_2 \wedge s'.$$

Let further

$$f_2^{s'} := \inf\{h_{s''} / s'' \in S,\ s'' \geq s'\}.$$

We have

$$s' \leq s'' \Rightarrow f_1^{s'} \leq f_1^{s''},\ f_2^{s'} \leq f_2^{s''}$$

and

$$f_1^{s'} \leq s_1 \wedge s' \leq s_1,\ f_2^{s'} \leq s_2.$$

By the definition of $f_1^{s'}, f_2^{s'}$ it follows that

$$s_1' := \sup_{s'\in S} f_1^{s'},\ s_2' := \sup_{s'\in S} f_2^{s'}$$

are elements in $\bar{S}$ and from

$$s \wedge s' = f_1^{s'} + h_{s'} \geq f_1^{s'} + f_2^{s'}$$

it follows

$$s \geq s_1' + s_2'.$$

On the other hand from $s', s'' \in S$, $s' \leq s''$ we get

$$s \wedge s' \leq s \wedge s'' = f_1^{s''} + h_{s''} \leq s_1' + h_{s''}$$

and so

$$s \wedge s' \leq s_1' + f_2^{s'} \leq s_1' + s_2'.$$

Since

$$\sup_{s'}(s \wedge s') = s,$$

it follows that

$$s = s_1' + s_2'. \qquad \square$$

Corollary 2.17. *The ordered convex cone $(S, \leq)$, where $\leq$ is the pointwise order relation, is an H-cone such that S is solid in $\bar{S}$ and increasingly dense with respect to the order relation $\leq$.*

Definition. A functional

$$\varphi : \bar{S} \to \bar{\mathbb{R}}_+$$

is called an *H-integral* on $\bar{S}$ if we have:

a) φ is additive and increasing

b) φ is continuous in order from below i.e.

$$s_i \uparrow s \Rightarrow \varphi(s_i) \uparrow \varphi(s)$$

c) φ is sufficiently finite i.e. the set

$$\{s \in \bar{S} / \varphi(s) < \infty\}$$

is increasingly dense in $\bar{S}$.

Notation. We denote by $\bar{\mathcal{E}}$ the functional

$$\bar{\mathcal{E}} : \bar{S} \times \bar{S}^* \to \bar{\mathbb{R}}_+$$

given by

$$\bar{\mathcal{E}}(s,t) = \sup\{\mathcal{E}(s',t') / s' \in S,\ t' \in S^*,\ s' \leq s,\ t' \leq t\}.$$

and for any $t \in \bar{S}^*$ we denote by φ_t the functional

$$\varphi_t : \bar{S} \to \bar{\mathbb{R}}_+$$

given by

$$\varphi_t(s) = \bar{\mathcal{E}}(s,t).$$

Remarks. It is easy to see that:

a) For any $t \in S$, φ_t is an H-integral on $\bar{S}$.

b) $t_1, t_2 \in \bar{S}^* \Rightarrow t_1 \leq t_2 \Leftrightarrow \varphi_{t_1}(s) \leq \varphi_{t_2}(s)\ \forall\ s \in \bar{S}$.

c) For any $t \in \bar{S}^*$ we have

$$t \in S^* \Leftrightarrow \varphi_t(s) < \infty\ \forall\ s \in S.$$

Proposition 2.18. *For any functional*

$$\varphi : S \to \mathbb{R}_+$$

such that φ is additive and increasing there exists a unique $t \in S^*$ with

$$\varphi(s) = \varphi_t(s).$$

Proof. Obviously, φ can be extended by linearity as a linear functional $\widetilde{\varphi}$ on the linear space $S-S$ which is positive. This functional $\widetilde{\varphi}$ is continuous with respect to the norm of the Hilbert space $(D_0(\mathcal{E}), \langle , \rangle)$. Indeed, in the contrary case, there exists a sequence $(u_n)_n$ in $S-S$ with $\|u_n\| \leq 1/2^n$ and $\widetilde{\varphi}(u_n) \geq 1$ $\forall n \in \mathbb{N}$.

If we put

$$s_n = Ru_n$$

we have

$$\|s_n\| \leq M\|u_n\| \leq M1/2^n \quad \forall n \in \mathbb{N}$$

and so the series $\sum\limits_{n=1} s_n$ is convergent in $(D_0(\mathcal{E}), \langle , \rangle)$ to an element $s \in S$. We have

$$n \in \mathbb{N} \Rightarrow \varphi(s) \geq \sum_{k=1}^{n} \varphi(s_k) \geq n$$

which is a contradiction.

Since $\widetilde{\varphi}$ is continuous, this functional can be extended by continuity to a linear continuous functional on $(D_0(\mathcal{E}), \langle , \rangle)$ and so there exists $t \in D_0(\mathcal{E})$ with

$$\widetilde{\varphi}(u) = \mathcal{E}(u,t) \quad \forall u \in D_0(\mathcal{E}).$$

Since

$$u \in D_0(\mathcal{E}),\ u \geq 0 \Rightarrow \widetilde{\varphi}(u) \geq 0$$

we deduce that $t \in S^*$. □

Proposition 2.19. *For any H-integral φ on $\bar{S}$ there exists $t \in \bar{S}^*$ uniquely determined such that*

$$\varphi(s) = \varphi_t(s) \quad \forall s \in \bar{S}.$$

In fact we have

$$t(x) := \varphi(g^x) \quad \forall x \in X_0.$$

Proof. Let t be the function

$$t : X_0 \to \bar{\mathbb{R}}_+$$

given by

$$t(x) := \varphi(g^x).$$

We show that $t(x) < \infty$ $\forall x \in X_0$. Indeed, let $(s_i)_{i \in I}$ be an increasing family in S such that $s_i \uparrow g^x$ and $\varphi(s_i) < \infty$ $\forall i \in I$.

Obviously, there exists $i_0 \in I$ with $s_i(x) \geq 1/2\ g^x(x)$ and so $s_{i_0} \geq 1/2\ g^x$, $\varphi(g^x) \leq 2\varphi(s_{i_0}) < \infty$.

We show now that $t \in \bar{S}^*$. Let $t' \in S^*$ and let ψ the functional

$$\psi : S \to \mathbb{R}_+$$

given by

$$\psi(s) = \inf\{\varphi_{t'}(s') + \varphi(s'')/ s', s'' \in S,\ s' + s'' = s\}.$$

It is easy to see that ψ is additive and increasing and so there exists $t'' \in S^*$ with

$$\psi(s) = \varphi_{t''}(s) \quad \forall s \in S.$$

For any $x \in X$ we have

$$\psi(g^x) = \varphi_{t''}(g^x) = t''(x),$$

$$\psi(g^x) = \inf\{\varphi_{t'}(s') + \varphi(s'') \,/\, s', s'' \in S,\ s' + s'' = g^x\}$$
$$= \inf\{\varphi_{t'}(\alpha g^x) + \varphi((1-\alpha)g^x) \,/\, \alpha \in \mathbb{R}_+,\ 0 \le \alpha \le 1\}$$
$$= \inf\{\alpha t'(x) + (1-\alpha)t(x) \,/\, \alpha \in \mathbb{R}_+,\ 0 \le \alpha \le 1\} = \inf(t'(x), t(x)),$$

and so

$$t''(x) = \varphi_{t''}(g^x) = \psi(g^x) = \inf(t'(x), t(x))$$

i.e.

$$t' \wedge t = t'' \in S^*.$$

From the above considerations it follows that $t \in \bar{S}^*$ and

$$\varphi(g^x) = \varphi_t(g^x).$$

Since the set

$$\mathscr{A} = \left\{ \sum_{i\in I} \alpha_i g^{x_i} \,/\, I \text{ finite},\ \alpha_i \in \mathbb{R}_+,\ x_i \in X_0 \right\}$$

is increasingly dense in S it follows

$$\varphi(s) = \varphi_t(s)\ \forall s \in S$$

and so

$$\varphi = \varphi_t. \qquad \square$$

Definition. A *balayage* on S (resp. on S^*) is a map

$$B : S \to S \text{ (resp. } B : S^* \to S^*)$$

which is additive, increasing, continuous in order from below, contractive (i.e. $Bs \le s$ (resp. $Bt \le t$) for all $s \in S$ (resp. $t \in S^*$)) and idempotent.

Remarks. a) Using Proposition 2.18 it follows that a map

$$B : S \to S \text{ (resp. } B : S^* \to S^*)$$

which is additive, increasing, contractive and idempotent is continuous in order from below and so it is a balayage on S (resp. S^*).

b) If $F \subset X_0$, $F \neq \emptyset$ the map

$$B^F : S \to S \text{ (resp. } {}^*B^F : S^* \to S^*)$$

given by

$$B^F s = \inf\{s' \in S / s' \ge s \text{ on } F\} \text{ (resp } {}^*B^F t = \inf\{t' \in S^* / t' \ge t \text{ on } F\})$$

is a balayage on S (resp. S^*).

Proposition 2.20. *For any balayage B on S there exists a unique balayage B^* on S^* such that*

$$\mathcal{E}(Bs, t) = \mathcal{E}(s, B^* t)\ \ \forall s \in S,\ t \in S^*.$$

The balayage B^* on S^* is called the adjoint of B.

Proof. For any $t \in S^*$ the functional

$$S \ni s \to \mathcal{E}(Bs, t) \in \mathbb{R}_+$$

is additive and increasing and therefore for Proposition 2.19 there exists $B^* t \in S^*$, uniquely determined, such that

$$\mathcal{E}(Bs, t) = \mathcal{E}(s, B^* t).$$

Obviously, we get for any $t, t_1, t_2 \in S^*$

$$B^*(t_1 + t_2) = B^* t_1 + B^* t_2$$

$$t_1 \leq t_2 \Rightarrow B^* t_1 \leq B^* t_2$$

$$B^* t \leq t,\ B^*(B^* t) = B^* t$$

and so $t \to B^* t$ is a balayage on S^*. □

Proposition 2.21. *For any $F \subset X_0$, $F \neq \emptyset$ the balayage ${}^*B^F$ in S^* is the adjoint of the balayage B^F on S.*

Proof. Let $s \in S$, $t \in S^*$. We have to show that

$$\mathcal{E}(B^F s, t) = \mathcal{E}(s, {}^*B^F t).$$

Using Proposition 2.12 it is sufficient to assume that

$$s = \sum_{i \in I} \alpha_i g^{x_i},\ I \text{ finite},\ \alpha_i \in \mathbb{R}_+,\ x_i \in F.$$

In this case we get

$$B^F s = s$$

and

$$\mathcal{E}(B^F s, t) = \mathcal{E}(s, t) = \sum_{i \in I} \alpha_i t(x_i) = \sum_{i \in I} \alpha_i\, {}^*B^F t(x_i) = \mathcal{E}(s, {}^*B^F t).$$ □

ACKNOWLEDGMENTS

Work supported by the Romanian Ministry of Education, Research and Inovation (PN II Program, CNCSIS code: 1186/2008).

REFERENCES

[1] N. Boboc, Gh. Bucur and A. Cornea, *Order and convexity in potential theory: H-cones* (Lecture Notes in Math. **853**). Springer-Verlag, 1981.
[2] J. Kigami, *Analysis on Fractals.* Cambridge University Press, 2001.
[3] J. Kigami, Harmonic analysis for resistance forms. *J. Funct. Anal* **204** (2003), 399-444.
[4] Z.M. Ma and M. Röckner, *Introduction to the Theory of (Non-Symmetric) Dirichlet Forms.* Springer-Verlag, 1992.

FACULTY OF MATHEMATICS AND COMPUTER SCIENCES, UNIVERSITY OF BUCHAREST, STR. ACADEMIEI 14, RO-010014 BUCHAREST, ROMANIA.
E-mail address: nicu_boboc@yahoo.com

"SIMION STOILOW" INSTITUTE OF MATHEMATICS OF THE ROMANIAN ACADEMY, P.O.BOX 1-764, RO-014700 BUCHAREST AND FACULTY OF MATHEMATICS AND COMPUTER SCIENCES, UNIVERSITY OF BUCHAREST, STR. ACADEMIEI 14, RO-010014 BUCHAREST, ROMANIA.

Potential Theory and Stochastics in Albac, 85–100
Aurel Cornea Memorial Volume

HYPERBOLIC HARMONIC FUNCTIONS AND THEIR FUNCTION THEORY

SIRKKA-LIISA ERIKSSON AND HEINZ LEUTWILER

Dedicated to the memory of Professor Aurel Cornea

ABSTRACT. There is a rich interplay between potential theory in the plane and complex function theory. In higher dimensions potential theory is very well developed but extensions of one variable complex function theory to higher dimensions has plenty of open problems. A higher dimensional generalization of the algebra of complex numbers is a Clifford algebra which is the smallest extension of the Euclidean space $\mathbb{R}^n$ to an associative algebra that inherits the algebraic, geometric and metric properties of the Euclidean space $\mathbb{R}^n$. It is a generalization of the algebra of quaternions introduced by Hamilton around 1843.

In 1935 Fueter defined regular functions in the algebra of quaternions and was able to prove the main theorems as Cauchy theorem, Cauchy formula and Laurent power series expansions. The key idea is that a regular function is a conjugate gradient of a harmonic function. Delanghe in 1970 was able to prove similar results in $\mathbb{R}^n$ using Clifford algebras. A generalization of the Laplacian is the Laplace-Beltrami operator defined on manifolds. The second author noticed around 1990 that if the usual Euclidean metric is changed to a hyperbolic one on the upper half space model then the power function, calculated using Clifford algebra, is a solution of the hyperbolic Dirac equation. Another important, but not so well known, fact is that in $\mathbb{R}^3$ and in higher dimensions, hyperbolic distance is Möbius invariant without any conformal factor.

We present an overview to the theory of hyperbolic harmonic functions and Clifford algebra valued hypermonogenic functions.

1. HYPERBOLIC HARMONIC FUNCTIONS

We consider harmonic functions in the upper half space, denoted by

$$\mathbb{R}^{n+1}_{+} = \left\{ (x_0, ..., x_n) \in \mathbb{R}^{n+1} \mid x_n > 0 \right\},$$

with respect to the Poincaré hyperbolic metric

$$ds^2 = \frac{dx_0^2 + ... + dx_n^2}{x_n^2}.$$

They satisfy the Laplace-Beltrami equation, which in the usual coordinates $(x_0, ..., x_n)$ of the the upper half space is of the form

$$\Delta_h h = x_n^2 \Delta h - (n-1)\, x_n \frac{\partial h}{\partial x_n} = 0 \tag{1}$$

2000 *Mathematics Subject Classification.* Primary 30G35. Secondary 30A05, 30F45.
Key words and phrases. Hyperbolic harmonic, hypermonogenic, Dirac operator, hyperbolic metric.

([1, p. 56]). This equation has been studied for example by A. Weinstein [26], A. Huber [21], Loo-keng Hua [20], L. Ahlfors [1] and H. Leutwiler [16], [3] and Ö. Akin [3]. A differentiable solution $f:\Omega\to\mathbb{R}$ of (1) on an open subset Ω of $\mathbb{R}_+^{n+1}$ is called hyperbolic harmonic.

Hyperbolic metric and hyperbolic harmonic functions have an important property with respect to the conformal transformations defined next.

Definition 1. *The group $M(\mathbb{R}^{n+1})$ is the group of sense preserving Möbius transformations on $\overline{\mathbb{R}}^{n+1}=\mathbb{R}^{n+1}\cup\{\infty\}$, generated by translations, dilatations, orthogonal mappings and the reflection map. The subgroup $M(\mathbb{R}_+^{n+1})$ of $M(\mathbb{R}^{n+1})$ is the Möbius transformations mappping the upper half space onto itself.*

The hyperbolic metric is invariant under $M(\mathbb{R}_+^{n+1})$, as noticed for example by L. Ahlfors [1] and Loo-keng Hua [20].

Theorem 2 ([1]). *The group $M(\mathbb{R}_+^{n+1})$ is the group of isometries of the upper half space model, that is mappings f satisfying*

$$d(f(v),f(w))=d(v,w)$$

where $d=d_h$ denotes the hyperbolic distance.

Hyperbolic harmonic functions are also invariant under $M(\mathbb{R}_+^{n+1})$.

Theorem 3 ([16]). *Let Ω be an open subset of the upper half space $\mathbb{R}_+^{n+1}$. If $f:\Omega\to\mathbb{R}$ is hyperbolic harmonic then $f\circ T$ is hyperbolic harmonic on $T^{-1}(\Omega)$ for any $T\in M(\mathbb{R}_+^{n+1})$.*

Hyperbolic harmonic functions are forming also a harmonic space as stated next.

Theorem 4. *Let U be an open subset of $\mathbb{R}_+^{n+1}$. If $\mathcal{H}(U)$ is a set of hyperbolic harmonic functions on U then $\mathcal{H}:U\to\mathcal{H}(U)$ is a harmonic sheaf and the pair $\left(\mathbb{R}_+^{n+1},\mathcal{H}\right)$ is a Brelot space in which regular sets are the same as with respect to the usual Laplace equation.*

This result is a special case of the theorem concerning elliptic differential equations, proved for example by M. Brelot [4] or R.-M. Herve [19] (see also [6, p. 79]). We are interested in building a function theory based on hyperbolic harmonic functions.

2. WHAT HAPPENS IN HIGHER DIMENSIONS?

The theory of complex holomorphic functions and potential theory in $\mathbb{R}^2$ have a rich interplay. For example, if $f:\Omega\to\mathbb{C}$ is holomorphic then $\operatorname{Re} f$ and $\operatorname{Im} f$ are harmonic functions. If $f:\Omega\to\mathbb{C}$ is holomorphic, then f can be written locally in the form $f=\frac{\partial H}{\partial x}-i\frac{\partial H}{\partial y}$ for some harmonic function H. Moreover a function $f:\Omega\to\mathbb{C}$ is holomorphic if and only if f and zf are harmonic. Holomorphic functions have more properties that harmonic functions. They yield conformal mappings between domains and their products are also holomorphic.

In higher dimensions in $\mathbb{R}^3$ or generally in $\mathbb{R}^n$ the function theory is more complicated. W. Hamilton tried to multiply vectors in $\mathbb{R}^3$. He finally realized that in order to multiply vectors geometrically in $\mathbb{R}^3$ the space has to be extended by one dimension. According to his own words

> "we must admit, in some sense, a fourth dimension of space for the purpose of calculating with triples ... An electric circuit seemed to close, and a spark flashed forth".

He introduced quaternions in 1843.

The algebra $\mathbb{H}$ of quaternions is the real associative algebra generated by e_1, e_2 satisfying

$$e_1^2 = e_2^2 = -1,$$
$$e_1 e_2 = -e_2 e_1.$$

Setting

$$e_{12} = e_1 e_2,$$

any element in $\mathbb{H}$ may be written as

$$x = x_0 + x_1 e_1 + x_2 e_2 + x_{12} e_{12}$$

for $x_0, x_1, x_2, x_{12} \in \mathbb{R}$.

Any quaternion x may be presented as

$$x = z_1 + z_2 e_2$$

where z_1 and z_2 are complex numbers $\mathbb{C}$ identified with

$$\{x_0 + x_1 e_1 \mid x_0, x_1 \in \mathbb{R}\}.$$

Note that if $z \in \mathbb{C}$ then

$$z e_2 = e_2 \bar{z}.$$

More generally, if $z_1, ..., z_4 \in \mathbb{C}$ then

$$(z_1 + z_2 e_2)(z_3 + z_4 e_2) = z_1 z_3 - z_2 \bar{z}_4 + (z_1 z_4 + z_2 \bar{z}_3) e_2.$$

3. Clifford algebra

Clifford algebras were introduced by William Clifford in 1878. We are using the (universal) Clifford algebra $C\ell_{0,n}$ that is the algebra generated by $e_1, ..., e_n$ satisfying the relation

$$e_i e_j + e_j e_i = -2\delta_{ij},$$

where δ_{ij} is the usual Kronecker delta. The Clifford algebra $C_{0,n}$ is formed by the elements presented as

$$x = \sum_A x_A e_A,$$

where x_A are real numbers,

$$A = \{i_1, i_2, \cdots, i_k \mid 1 \le i_1 < i_2 < \cdots < i_k \le n\},$$

and

$$e_A = e_{i_1} e_{i_2} \cdots e_{i_k}, \qquad e_\emptyset = e_0 = 1.$$

The special cases are:

- $C\ell_{0,0} = \mathbb{R}$
- $C\ell_{0,1} = \mathbb{C}$,
- $C\ell_{0,2} = \mathbb{H}$.

In the complex plane there is one involution, conjugation,

$$\overline{x+yi} = x - yi$$

In Clifford algebras there are several involutions and anti-involutions. We define in the following table involutions and anti-involutions used subsequently

involutions	$i = 1, \dots, n$	products
main or grade $'$	$e_i' = -e_i$	$(ab)' = a'b'$
reversion *	$e_i^* = e_i$	$(ab)^* = b^* a^*$
conjugation $-$	$\overline{e}_i = -e_i$	$\overline{ab} = \overline{b}\,\overline{a}$
$\widehat{\ }$	$\hat{e}_i = (-1)^{\delta_{in}} e_i$	$\widehat{ab} = \hat{a}\hat{b}$

Note that in the complex plane the main and the conjugation are the usual complex conjugation.

Any element $a \in C\ell_{0,n}$ may be uniquely decomposed as

$$a = b + ce_n$$

for $b, c \in C\ell_{0,n-1}$(the Clifford algebra generated by $e_1, \dots, e_{n-1}$). The mappings $P : C\ell_{0,n} \to C\ell_{0,n-1}$ and $Q : C\ell_{0,n} \to C\ell_{0,n-1}$ are defined by

$$Pa = b, \qquad Qa = c.$$

The mappings P and Q are generalizations of real and imaginary parts, since in case $n = 1$, the mapping P is the real part and Q the imaginary part of the complex number.

The following calculation rules [11] are useful

$$P(ab) = (Pa)\,Pb + (Qa)\,Q\left(b'\right),$$
$$Q(ab) = aQb + (Qa)\,b'.$$

We abbreviate $\left(Qf\right)' = Q'f$ and $\left(Pf\right)' = P'f$.

Not all elements of the Clifford algebra are invertible. However, elements $x = x_0 + x_1 e_1 + \dots + x_n e_n$ for $x_1, \dots, x_n \in \mathbb{R}$, called *paravectors,* are invertible and

$$x^{-1} = \frac{\overline{x}}{|x|^2}$$

where $|x| = \sqrt{x_0^2 + \dots + x_n^2}$. The set of paravectors is identified with $\mathbb{R}^{n+1}$. The product of invertible elements is also invertible and therefore the products of non zero paravectors are also invertible. The *Clifford group* Γ_{n-1} consists of all elements which can be written as products of non-zero paravectors $x = x_0 + x_1 e_1 + \dots + x_{n-1} e_{n-1}$ in $\mathbb{R}^n$.

Lemma 5 ([2])**.** *For any $a \in \Gamma_{n-1}$ there exists a matrix $\rho(a) \in SO(n)$ such that $axa'^{-1} = \rho(a)\,x$. The mapping $a \to \rho(a)$ is a homeomorphism from Γ_{n-1} onto $SO(n)$ with the center* $\mathbb{R} \setminus \{0\}$.

Clifford algebras may be used to characterize Möbius transformations. As Poincaré noticed any Möbius transformation in $\mathbb{R}^n \cup \{\infty\}$ may be extended to $\mathbb{R}^{n+1} \cup \{\infty\}$. We are interested in Möbius transformation in $M\left(\mathbb{R}_+^{n+1}\right)$. We say that the mapping T :

$\mathbb{R}^{n+1} \cup \{\infty\} \to \mathbb{R}^{n+1} \cup \{\infty\}$ is induced by the matrix $\begin{pmatrix} a & b \\ c & d \end{pmatrix}$ with the coefficients a, b, c and d in $C\ell_{n-1}$ if

$$Tx = (ax+b)(cx+d)^{-1}.$$

They may be characterized as follows.

Theorem 6 ([2]). *A mapping $T \in M(\mathbb{R}^{n+1}_+)$ if and only if T is induced by the matrix $\begin{pmatrix} a & b \\ c & d \end{pmatrix}$ where*

(1) $a, b, c, d \in \Gamma_{n-1} \cup \{0\}$,
(2) $\triangle = ad^* - bc^* \in \mathbb{R} \setminus \{0\}$
(3) ab^*, cd^*, c^*a *and* d^*b *in* $\mathbb{R}^n$.

Hyperbolic harmonic functions are easy to transform to Poincaré's unit ball model as follows.

Proposition 7 ([3]). *Let T be the Möbius transformation mapping the open unit ball B_{n+1} onto $\mathbb{R}^{n+1}_+$ defined by*

$$T(y) = (y+e_n)(e_n y+1)^{-1}$$

A function u is hyperbolic harmonic on $\mathbb{R}^{n+1}_+$ if and only if the function

$$v(x) = u(T(y))$$

satisfies the equation

$$\Delta_h^B v(y) = \left(1-\|y\|^2\right)^2 \Delta v(y) + 2(n-1)\left(1-\|y\|^2\right)\sum_{i=0}^{n} y_i \frac{\partial v}{\partial y_i} = 0 \tag{2}$$

on the unit ball.

Using the invariance of the hyperbolic distance $d_h(x,a)$ between the points $x = x_0 + e_1 x_2 + \ldots + x_n e_n$ and $a = a_0 + e_1 a_1 + \ldots + a_n e_n$ it may be calculated from

$$R_h = d_h(x,a) = \operatorname{arcosh} \delta(x,a),$$

where

$$\delta(x,a) = \frac{|x-a|^2 + 2a_n x_n}{2x_n a_n}.$$

Moreover

$$\frac{|x-a|^2}{|x-\hat{a}|^2} = \tanh^2\left(\frac{R_h}{2}\right).$$

Hence making simple calculations we note that the hyperbolic sphere with the hyperbolic center a and the radius R_h is an euclidean sphere with the euclidean center

$$a_e = Pa + e_n a_n \cosh R_h$$

and the euclidean radius

$$R_e = a_n \sinh R_h.$$

Note that the hyperbolic surface measure and the volume measure are

$$d\sigma_h = \frac{d\sigma}{x_n^n}, dx_h = \frac{dx}{x_n^{n+1}}.$$

and the hyperbolic normal derivative is

$$\frac{\partial}{\partial n_h} = x_n \frac{\partial}{\partial n}.$$

in the usual coordinates $x = (x_0, ..., x_n)$ of the upper half space. In the hyperbolic unit ball model, they are

$$d\sigma_h = \frac{2^n d\sigma}{\left(1 - |y|^2\right)^n}, dy_h = \frac{2^{n+1} dy}{\left(1 - |y|^2\right)^{n+1}},$$

where the usual coordinates are denoted by $y = \left(y_0, ..., y_n\right)$.

Lemma 8. *If $f \in \mathscr{C}^2(U)$ and $\overline{U} \subset \mathbb{R}^{n+1}_+$, then*

$$\frac{d}{dR_h}\left(\frac{1}{\sigma_n \sinh^n R_h}\int_{\partial B_h(a,R_h)} f\, d\sigma_h\right) = \frac{1}{\sigma_n \sinh^n R_h}\int_{B_h(e_n,R_h)} \Delta_h(f)\, dx_h.$$

Proof. Applying Möbius transformations of $M\left(\mathbb{R}^{n+1}_+\right)$ and Theorem 3 we infer that it is enough to prove the result for $B_h(e_n, R_h)$. Since the surface measure is the coordinate invariant applying the mapping T given in the preceedig proposition we obtain

$$\begin{aligned}\frac{1}{\sigma_n}\int_{\partial B_h(e_n,R_h)} f\, d\sigma_h(x) &= \frac{1}{\sigma_n}\int_{\partial B_e(0,r)} f\circ T\, d\sigma_h(y)\\ &= \frac{2^n}{\sigma_n\left(1-\tanh^2\frac{R_h}{2}\right)^n}\int_{\partial B_e(0,r)} f\circ T(y)\, d\sigma(y)\end{aligned}$$

where $r = \left|T^{-1}(x)\right| = \frac{|x-e_n|}{|x+e_n|} = \tanh\left(\frac{R_h}{2}\right)$. Since

$$\begin{aligned}\int_{\partial B_e(0,r)} f\circ T(y)\, d\sigma(y) &= r^n \int_{\partial B_e(0,1)} f\circ T(r\theta)\, d\sigma(\theta)\\ &= \tanh^n\left(\frac{R_h}{2}\right)\int_{\partial B_e(0,1)} f\circ T(r\theta)\, d\sigma(\theta)\end{aligned}$$

and $\sinh R_h = 2\sinh\frac{R_h}{2}\cosh\frac{R_h}{2}$ we obtain

$$\frac{1}{\sigma_n \sinh^n R_h}\int_{\partial B_h(e_n,R_h)} f\, d\sigma_h = \int_{\partial B_e(0,1)} f\circ T(r\theta)\, d\sigma.$$

Hence

$$\frac{d}{dR_h}\left(\frac{1}{\sigma_n \sinh^n R_h}\int_{\partial B_h(e_n,R_h)} f\, d\sigma_h\right) = \frac{dr}{dR_h}\frac{d}{dr}\int_{\partial B_e(0,1)} f\circ T(r\theta)\, d\sigma.$$

Since $\frac{dr}{dR_h} = = \frac{1}{2}\left(1-|y|^2\right)$ we infer using ([1, p. 63]) that

$$\begin{aligned}
\frac{d}{dR_h}\left(\frac{1}{\sinh^n R_h}\int_{\partial B_h(e_n,R_h)} f\,d\sigma_h\right) &= \frac{1}{2}\left(1-|r|^2\right)\int_{\partial B_e(0,1)} \frac{d}{dr} f\circ T(r\theta)\,d\sigma(\theta)\\
&= \frac{1}{2}\left(1-|r|^2\right)\int_{\partial B_e(0,1)} \operatorname{grad} f\circ T(r\theta)\cdot\theta\,d\sigma(\theta)\\
&= \frac{\left(1-|r|^2\right)}{2r^n}\int_{\partial B_e(0,r)} \operatorname{grad} f\circ T(y)\cdot\frac{y}{r}\,d\sigma(y)\\
&= \frac{1}{r^n}\int_{\partial B_e(0,r)} \frac{1}{2}\left(1-|y|^2\right)\frac{\partial f\circ T}{\partial n}\,d\sigma(y)\\
&= \frac{\left(1-r^2\right)^n}{2^n r^n}\int_{\partial B_e(0,r)} \frac{\partial f\circ T}{\partial n_h}\,d\sigma_h(y)\\
&= \frac{\left(1-r^2\right)^n}{2^n r^n}\int_{B_e(0,r)} \Delta_h^B(f\circ T)\,dy_h\\
&= \frac{1}{\sinh^n R_h}\int_{B_h(e_n,R_h)} \Delta_h(f)\,dx_h
\end{aligned}$$

where Δ_h^B is the hyperbolic Laplace operator in the unit ball given by (2). □

Using the above lemma we obtain the usual characterizations of hyperbolic harmonic functions.

Theorem 9. *Let $U\subset\mathbb{R}_+^{n+1}$ be open. The following properties are equivalent*

(1) *h is hyperbolic harmonic on U.*
(2) *$h\in\mathscr{C}^2(U)$ and*

$$h(a) = \frac{1}{\sigma_n \sinh^n R_h}\int_{\partial B_h(a,R_h)} h\,d\sigma_h$$

for all hyperbolic balls satisfying $\overline{B_h(a,R_h)}\subset U$.
(3) *$h\in\mathscr{C}^2(U)$ and*

$$h(a) = \frac{1}{V_h(B_h(a,R_h))}\int_{B_h(a,R_h)} f\,dx_h,$$

for all hyperbolic balls satisfying $\overline{B_h(a,R_h)}\subset U$, where the hyperbolic volume of the ball $B_h(a,R_h)$ is given by $V_h(B_h(a,R_h)) = \sigma_n\int_0^{R_h}\sinh^n t\,dt$.

Proof. We modify the methods of [22] for the hyperbolic case. Assume that h is hyperbolic harmonic on U. By virtue of the the preceeding lemma the function $R_h\to \frac{1}{\sigma_n \sinh^n R_h}.\int_{\partial B_h(a,R_h)} h\,d\sigma_h$ is a constant. Since

$$\lim_{R_h\to 0}\frac{1}{\sigma_n\sinh^n R_h}\int_{\partial B_h(a,R_h)} h\,d\sigma_h = h(a)$$

we obtain the assertion (2).

Assume that (2) holds. Applying Möbius transformations of $M\left(\mathbb{R}_+^{n+1}\right)$ and Theorem 3 we infer that it is enough to prove the result for $B_h(e_n, R_h)$. Then

$$\begin{aligned}\frac{1}{V_h(B_h(e_n,R_h))}\int_{B_h(e_n,R_h)} h\,dx_h &= \frac{1}{V_h(B_h(e_n,R_h))}\int_{B(0,r)} h\circ T\,dy_h \\ &= \frac{1}{V_h(B_h(e_n,R_h))}\int_0^{\tanh\frac{R_h}{2}}\int_{\partial B_h(0,r)} h\circ T\,d\sigma dr \\ &= \frac{f(e_n)}{V_h(B_h(e_n,R_h))}\int_0^{R_h}\sigma_n \sinh^n t\,dt = h(e_n).\end{aligned}$$

Hence (3) holds.

Assume that (3) holds. Then

$$\begin{aligned}h(e_n) &= \frac{1}{V_h(B_h(e_n,R_h))}\int_{B_h(e_n,R_h)} h\,dx_h = \frac{1}{V_h(B_h(e_n,R_h))}\int_{B_h(0,r)} h\circ T\,dy_h \\ &= \frac{1}{V_h(B_h(e_n,R_h))}\int_0^{\tanh\frac{R_h}{2}}\frac{2}{1-r^2}\int_{\partial B_h(0,r)} h\circ T\,d\sigma_h dr \\ &= \frac{1}{V_h(B_h(e_n,R_h))}\int_0^{R_h}\int_{\partial B_h(e_n,s)} h\,d\sigma_h ds.\end{aligned}$$

Hence we have

$$\sigma_n\int_0^{R_h}\sinh^n t\,dt\,h(e_n) = V_h(B_h(e_n,R_h))\,h(e_n) = \int_0^{R_h}\int_{\partial B_h(e_n,s)} h\,d\sigma_h ds.$$

Differentiating with respect to R_h we obtain

$$h(e_n) = \frac{1}{\sigma_n\sinh^n R_h}\int_{\partial B_h(e_n,s)} h\,d\sigma_h.$$

Applying Lemma 8 and usual measure theoretic arguments we note that the property (2) implies (1). □

4. Hyperbolic function theory

Hyperbolic function theory was initiated by the second author around 1990 ([17], [18]). He noticed that for any $m\in\mathbb{Z}$, the power function

$$f(x) = x^m,$$

which is paravector-valued for any paravector x, has the property that

$$x^m = \frac{\partial h}{\partial x_0} - \sum_{i=1}^n e_i\frac{\partial h}{\partial x_i} \tag{3}$$

where $h(x) = \frac{1}{m+1}\operatorname{Re} x^{m+1}$ is a hyperbolic harmonic function.

The second author started the research of the paravector valued functions f, called *H-solutions,* admitting locally the a representation of the form (3) in terms of some hyperbolic harmonic function h.

Another characterization of H-solutions is that they are paravector-valued solutions $f = u_0 + u_1 e_1 + ... + u_n e_n$ of the following generalized Cauchy-Riemann equation

$$\begin{aligned} & x_n \left(\frac{\partial u_0}{\partial x_0} - \sum_{i=1}^{n} \frac{\partial u_i}{\partial x_i} \right) + (n-1)\, u_n = 0, \\ & \frac{\partial u_i}{\partial x_k} = \frac{\partial u_k}{\partial x_i}, \quad i,k = 1,...,n, \\ & \frac{\partial u_0}{\partial x_k} = -\frac{\partial u_k}{\partial x_0}, \quad k = 1,...,n. \end{aligned} \tag{H}$$

Let Ω be an open subset of $\mathbb{R}^{n+1}_+$. We consider function

$$f : \Omega \to C\ell_{0,n},$$

whose components are continuously differentiable.

The left and right Dirac operators in $C\ell_{0,n}$ are defined by

$$D_l f = \sum_{i=0}^{n} e_i \frac{\partial f}{\partial x_i} \qquad \text{and} \qquad D_r f = \sum_{i=0}^{n} \frac{\partial f}{\partial x_i} e_i,$$

respectively. Their conjugate operators $\overline{D_l}$ and $\overline{D_r}$ are defined by

$$\overline{D_l} f = \sum_{i=0}^{n} \overline{e_i} \frac{\partial f}{\partial x_i} \qquad \text{and} \qquad \overline{D_r} f = \sum_{i=0}^{n} \frac{\partial f}{\partial x_i} \overline{e_i}.$$

Let Ω be an open subset of $\mathbb{R}^{n+1}_+$. The modified Dirac operators M_k^l and M_k^r are introduced by

$$\begin{aligned} M_k^l f(x) &= D_l f(x) + k \frac{Q' f}{x_n}, \\ M_k^r f(x) &= D_r f(x) + k \frac{Q f}{x_n}. \end{aligned}$$

where $f \in \mathscr{C}^1 (\Omega, C\ell_{0,n})$. Conjugate modified Dirac operators $\overline{M}_k^l$ and $\overline{M}_k^r$ are introduced by

$$\begin{aligned} \overline{M}_k^l f(x) &= \overline{D}_l f(x) - k \frac{Q' f}{x_n}, \\ \overline{M_k}^r f(x) &= \overline{D}_r f(x) - k \frac{Q f}{x_n}. \end{aligned}$$

The operator M_{n-1}^l is also denoted by M.

Definition 10. *Let $\Omega \subset \mathbb{R}^{n+1}_+$ be open. A function $f : \Omega \to C\ell_{0,n}$ is a left k-hypermonogenic function, if $f \in \mathscr{C}^1 (\Omega)$ and*

$$M_k^l f(x) = 0$$

for any $x \in \Omega \setminus \{x_n = 0\}$. The right k-hypermonogenic functions are defined similarly. If $k = n-1$ left hypermonogenic functions are called hypermonogenic functions.

Hypermonogenic functions were introduced by the authors in [11]. Note that hypermonogenic functions have values in the total Clifford algebra $C\ell_{0,n}$. Subsequently, their domain of definition is for the sake of simplicity the upper half space although they may also be defined in the whole space $\mathbb{R}^{n+1}$. Paravector-valued hypermonogenic functions are H-solutions.

Lemma 11. *Let $\Omega \subset \mathbb{R}^{n+1}_+$ be open. A function $f : \Omega \to C\ell_{0,n}$ is left k−hypermonogenic function if and only if f^* is right hypermonogenic.*

The theory is notably studied in addition to the authors by J. Cnops ([5]), P. Cerejeiras, Th. Hempfling, P. Zeilinger, L. Pernas, Junxia Li, J. Kettunen, K. Hirvonen, H. Orelma, J. Ryan ([23]), Qiao ([23]), Yuying ([23])), S. Krausshar, I. Ramadanoff, Sw. Bernstein, G. Laville, Eric Lehman and Xiaoli Bian.

Operating with P and Q to $M_k f = 0$ we obtain the generalized Cauchy-Riemann equations as follows

Theorem 12 ([11]). *Let Ω be an open subset of $\mathbb{R}^{n+1}_+$ and $f : \Omega \to C\ell_{0,n}$ be a mapping with continuous partial derivatives. The equation $M_k f = 0$ is equivalent with the following system of equations*

$$D_{n-1}(Pf) - \frac{\partial(Q'f)}{\partial x_n} + k\frac{Q'f}{x_n} = 0,$$
$$D_{n-1}(Qf) + \frac{\partial P'(f)}{\partial x_n} = 0.$$

Hyperbolic harmonic functions and k-hypermonogenic functions are related.

Theorem 13 ([11]). *Let $f : \Omega \to C\ell_{0,n}$ be twice continuously differentiable. Then*

$$P\left(M_k \overline{M}_k f\right) = \Delta Pf - \frac{k}{x_n}\frac{\partial Pf}{\partial x_n},$$
$$Q\left(M_k \overline{M}_k f\right) = \Delta Qf - \frac{k}{x_n}\frac{\partial Qf}{\partial x_n} + k\frac{Qf}{x_n^2}.$$

If f is k-hypermonogenic, then Pf satisfies the equation

$$\Delta Pf - \frac{k}{x_n}\frac{\partial Pf}{\partial x_n} = 0$$

and Qf satisfies the equation

$$\Delta Qf - \frac{k}{x_n}\frac{\partial Qf}{\partial x_n} + k\frac{Qf}{x_n^2} = 0.$$

These are Laplace-Beltrami equations with respect to the Riemannian metric

$$ds^2 = \frac{\sum_{i=0}^n dx_i^2}{x_n^{\frac{2k}{n-1}}}.$$

Real hyperbolic harmonic functions may be extended to total Clifford algebra valued.

Definition 14. *A twice continuously differentiable function $f : \Omega \to Cl_{0,n}$ is called k-hyperbolic harmonic if $M_k \overline{M}_k f = 0$.*

Theorem 15 ([11]). *Let $f : \Omega \to Cl_{0,n}$ be twice continuously differentiable. Then f is k-hypermonogenic if and only if f and xf are k-hyperbolic harmonic functions.*

Theorem 16 ([11]). *Let Ω be an open subset of $\mathbb{R}^{n+1}_+$ and $f : \Omega \to C\ell_{0,n}$ be twice continuously differentiable. Then f is k-hypermonogenic if and only if there exists locally a k-hyperbolic harmonic mapping H with values in $C\ell_{0,n-1}$ satisfying $\overline{D}H = f$.*

Example 17 ([17]). *The following functions are hypermonogenic:*

(1) x^m, $m \in \mathbb{Z}$, *where* $x = x_0 + x_1 e_1 + \ldots + x_n e_n$;
(2) $e^x = \sum \frac{1}{k!} x^k$;
(3) $\sin x = \sum \frac{1}{(2k+1)!} (-1)^k x^{2k+1}$;
(4) $\cos x = \sum \frac{1}{(2k)!} (-1)^k x^{2k}$.
(5) *If* $f(z) = \sum a_k z^k$ *is holomorphic and* $a_k \in \mathbb{R}$, *then* $f(x) = \sum a_k x^k$ *is hypermonogenic.*
(6) *(Fueter Construction) If* $f = u + iv$ *is holomorphic in an open set* $\Omega \subset \mathbb{C}$, *then*

$$\widetilde{f}(x) = u\left(x_0, \sqrt{x_1^2 + \ldots + x_n^2}\right) + \frac{x_1 e_1 + \ldots + x_n e_n}{\sqrt{x_1^2 + \ldots + x_n^2}} v\left(x_0, \sqrt{x_1^2 + \ldots + x_n^2}\right)$$

is hypermonogenic.

Proposition 18 ([11]). *Let* f *be a* k*-hypermonogenic function in* Ω. *Then* $\frac{\partial f}{\partial x_i}$, $i = 0, 1, \ldots, n-1$, *is* k*-hypermonogenic. Moreover the function* $\frac{\partial f}{\partial x_n}$ *is* k*-hypermonogenic if and only if* $\frac{\partial f}{\partial x_n} = 0$.

Proposition 19 ([11]). *The* k*-hypermonogenic functions in an open subset* Ω *of* $\mathbb{R}^{n+1}$ *form a right* $C\ell_{0,n-1}$*-module.*

The M-operators satisfy the following algebraic identities.

Theorem 20 ([15]). *Let* Ω *be an open subset of* $\mathbb{R}^{n+1}_+$ *and* f, g *be* $\mathscr{C}^1(\Omega, C\ell_{0,n})$ *functions. Then*

$$\left((M_k^r g) f + g M_k^l f\right) \frac{1}{x_n^k} = D_r\left(\frac{g}{x_n^k}\right) f + \frac{g}{x_n^k} D_l f + \frac{k}{x_n^{k+1}} P(g f') e_n,$$
$$(M_{-k}^r g) f + g M_k^l f = \left(D_r g\right) f + g D_l f - \frac{k}{x_n} Q(g f').$$

The hyperbolic version of the Stokes theorem is stated next.

Theorem 21 ([9]). *For* Ω *an open subset of* $\mathbb{R}^{n+1}_+$ *(or* $\mathbb{R}^{n+1}_-$*),* $K \subset \Omega$ *a smoothly bounded compact set with outer unit normal field* ν, *and* $f, g \in \mathscr{C}^1(\Omega, C\ell_{0,n})$,

$$\int_{\partial K} P(g\nu f) \frac{d\sigma}{x_n^k} = \int_K P\left((M_k^r g) f + g M_k^l f\right) \frac{dx}{x_n^k},$$
$$\int_{\partial K} Q(g\nu f)\, d\sigma = \int_K Q\left((M_{-k}^r g) f + g M_k^l f\right) dx.$$

There exist Borel-Pompeiu theorem for P-part and Q-parts.

Theorem 22 ([15]). *Let* Ω *be an open subset of* $\mathbb{R}^{n+1}_+$ *(or* $\mathbb{R}^{n+1}_-$*) and* $K \subset \Omega$ *a smoothly bounded compact set with outer unit normal field* ν. *If* $f \in \mathscr{C}^1(\Omega, C\ell_n)$, *then for any*

$y \in intK$ *and* $C_1 = \frac{2^n}{\omega_{n+1}}$

$$Pf(y) = C\int_{\partial K} P\left(p(x,y)\,\nu(x)\,f(x)\right) x_n^{1-n}\,d\sigma$$
$$-C\int_K P\left(p(x,y)\,M_{n-1}^l f(x)\right) x_n^{1-n}\,dx,$$

$$\begin{aligned} p(x,y) &= \frac{x_n^{n-1}y_n^{n-1}}{2}\left(\frac{(x-y)^{-1}-(x-\widehat{y})^{-1}}{|x-y|^{n-1}|x-\widehat{y}|^{n-1}}\right) \\ &= -\frac{1}{2^{2n-1}}\overline{D}_l^x\int_{\frac{|x-y|}{|x-\widehat{y}|}}^1 \frac{(1-s^2)^{n-1}}{s^n}\,dt. \end{aligned}$$

Borel-Pompeiu theorem for Q-part formulates as follows.

Theorem 23 ([9]). *Let* Ω *be an open subset of* $\mathbb{R}_+^{n+1}$ *(or* $\mathbb{R}_-^{n+1}$*) and* $K \subset \Omega$ *a smoothly bounded compact set with outer unit normal field* ν*. If* $f \in \mathscr{C}^1(\Omega, C\ell_n)$*, then for any* $y \in intK$ *and* $C = \frac{2^n}{\omega_{n+1}}$

$$Qf(y) = C\int_{\partial K} Q\left(q(x,y)\,\nu(x)\,f(x)\right) d\sigma$$
$$-C\int_K Q\left(q(x,y)\,M_{n-1}^l f(x)\right) dx,$$

$$\begin{aligned} q(x,y) &= y_n^{n-1}\frac{(x-y)^{-1}+(x-\widehat{y})^{-1}}{2|x-y|^{n-1}|x-\widehat{y}|^{n-1}} \\ &= -\frac{y_n^{n-1}}{2(n-1)}\overline{D}_l^x\frac{1}{|x-y|^{n-1}|x-\widehat{y}|^{n-1}}. \end{aligned}$$

The basic integral formulas exists for P- and Q-parts:

Theorem 24 ([9]). *Let* Ω *be an open subset of* $\mathbb{R}_+^{n+1}$ *(or* $\mathbb{R}_-^{n+1}$*) and* $K \subset \Omega$ *a smoothly bounded compact set with outer unit normal field* ν*. If* f *is hypermonogenic in* Ω *and* $y \in intK$*, then*

$$Pf(y) = C\int_{\partial K} P\left(p(x,y)\,\nu(x)\,f(x)\right) x_n^{1-n}\,d\sigma,$$
$$Qf(y) = C\int_{\partial K} Q\left(q(x,y)\,\nu(x)\,f(x)\right) d\sigma,$$

where $C = \frac{2^n}{\omega_{n+1}}$.

If we multiply the second equation with e_n and add them together we obtain the following result.

Theorem 25 ([9]). *Let* Ω *be an open subset of* $\mathbb{R}_+^{n+1}$ *and* $\overline{K} \subset \Omega$ *a smoothly bounded compact set with outer unit normal field* ν*. If* f *is hypermonogenic in* Ω *and* $y \in K$*, then*

$$f(y) = \frac{2^{n-1}}{\omega_{n+1}}\int_{\partial K}\left(K_1(x,y)\,\nu(x)\,f(x) - K_2(x,y)\,\widehat{\nu(x)}\,\widehat{f(x)}\right) d\sigma,$$

where

$$K_1(x,y) = \frac{y_n^{n-1}(x-y)^{-1}}{|x-y|^{n-1}|x-\widehat{y}|^{n-1}},$$
$$K_2(x,y) = \frac{y_n^{n-1}(\widehat{x}-y)^{-1}}{|x-y|^{n-1}|x-\widehat{y}|^{n-1}},$$

and ω_{n+1} *is the surface measure of the unit ball in* $\mathbb{R}^{n+1}$.

The kernel $K_1(x,y)$ is not hypermonogenic. However the following decomposition holds

$$x_n^{n-1}K_1(x,y) = \frac{x_n^{n-1}y_n^{n-1}(x-y)^{-1}}{|x-y|^{n-1}|x-\widehat{y}|^{n-1}} = g_1 + g_2 e_n,$$

where the functions g_1 and g_2 are hypermonogenic functions on $\mathbb{R}^{n+1}\setminus\{y,\widehat{y}\}$ for each y with $y_n \neq 0$ given by

$$g_1(x,y) = y_n p(x,y),$$
$$g_2(x,y) = -x_n^{n-1} q(x,y)\, e_n.$$

Theorem 26 ([15]). *Let* Ω *be an open subset of* $\mathbb{R}_+^{n+1}$ *and* $K \subset \Omega$ *a smoothly bounded compact set with outer unit normal field* ν. *If* f *is hypermonogenic in* Ω *and* $x \in int K$, *then*

$$f(y) = \frac{2^{n-1}}{\omega_{n+1}} \int_{\partial K} \left(x_n k(y,x) P(\nu(x) f(x)) + k_1(y,x) Q'(\nu(x) f(x))\right) d\sigma,$$

where the kernels $k(y,x)$ *and* $k_1(y,x)$ *are both hypermonogenic with respect to* y *in* $\mathbb{R}^{n+1}\setminus\{x,\widehat{x}\}$. *They are defined by*

$$\begin{aligned} k(y,x) &= \frac{y_n^{n-1}}{x_n}\frac{(x-y)^{-1}-(\widehat{x}-y)^{-1}}{|y-x|^{n-1}|\widehat{y}-x|^{n-1}} \\ &= 2y_n^{n-1}\frac{(x-y)^{-1}}{|y-x|^{n-1}} e_n \frac{(y-\widehat{x})^{-1}}{|x-\widehat{y}|^{n-1}} \\ &= \frac{1}{2^{2n-2}x_n^n}\overline{D}^y\left(\int_{\frac{|y-x|}{|x-\widehat{y}|}}^1 \frac{(1-s^2)^{n-1}}{s^n} ds\right), \end{aligned}$$

and

$$\begin{aligned} k_1(y,x) &= y_n^{n-1}\frac{(x-y)^{-1}+(\widehat{x}-y)^{-1}}{|y-x|^{n-1}|\widehat{y}-x|^{n-1}} e_n \\ &= \frac{1}{(n-1)} y_n^{n-1}\overline{D}^y\left(\frac{1}{|x-y|^{n-1}|x-\widehat{y}|^{n-1}}\right) e_n. \end{aligned}$$

Theorem 27 ([15]). *Let* Ω *be an open subset of* $\mathbb{R}_+^{n+1}$ *and* $\overline{K} \subset \Omega$ *a smoothly bounded compact set with outer unit normal field* ν. *If* f *is continuous function in* $\partial\Omega$, *then* *function g defined by*

$$g(y)=\int_{\partial K}\left(x_n k(y,x)P(\nu(x)f(x))+k_1(y,x)Q'(\nu(x)f(x))\right)d\sigma$$

is hypermonogenic in $y\in\mathbb{R}^{n+1}_{+}\setminus\partial K$.

Applying the preceeding theorem we obtain also interesting identities for the $\overline{M}^l_k M^l_k f$ operator.

Theorem 28 ([15]). *For* Ω *an open subset of* $\mathbb{R}^{n+1}_{+}$ *(or* $\mathbb{R}^{n+1}_{-}$*),* $K\subset\Omega$ *a smoothly bounded compact set with outer unit normal field* ν*, and* $f,g\in\mathscr{C}^1(\Omega,C\ell_{0,n-1})$,

$$\begin{aligned}&\int_K P\left(g\overline{M}^l_k M^l_k f-\left(\overline{M}^r_k M^r_k g\right)f\right)x_n^{-k}dx\\&=\int_{\partial K}P\left(g\nu\overline{M}^l_k f-(M^r_k g)\overline{\nu}f\right)x_n^{-k}d\sigma.\end{aligned}$$

and

$$\begin{aligned}&\int_K Q\left(\left(\overline{M}^r_{-k}M^r_{-k}g\right)f-g\overline{M}^l_k M^l_k f\right)dx\\&=\int_{\partial K}Q\left(\overline{M}^r_{-k}g\nu f-g\overline{\nu}M^l_k f\right)d\sigma.\end{aligned}$$

We also find an integral formula for hyperbolic harmonic functions.

Theorem 29 ([15]). *Let* Ω *be an open subset of* $\mathbb{R}^{n+1}_{+}$ *(or* $\mathbb{R}^{n+1}_{-}$*) and* $K\subset\Omega$ *a smoothly bounded compact set with outer unit normal field* ν*. If* $h:\Omega\to C\ell_{0,n-1}$ *is hyperbolic harmonic, with* $k=n-1$*, then*

$$h(y)=c\int_{\partial K}P\left(G(x,y)\nu(x)\overline{D}_l h(x)-\overline{\tilde{p}(x,y)\nu(x)}h(x)\right)x_n^{1-n}d\sigma,$$

for all $y\in intK$*, where* $c=\frac{1}{2^{n-1}\omega_{n+1}}$, $\tilde{p}(x,y)=-2^{2n-1}p(x,y)=\overline{D}^x G(x,y)$ *and*

$$G(x,y)=g\left(\frac{|x-y|}{|x-\widehat{y}|}\right),\quad \textit{where}\quad g(t)=\int_t^1\frac{\left(1-s^2\right)^{n-1}}{s^n}dt.$$

Theorem 30 ([15]). *Let* Ω *be an open subset of* $\mathbb{R}^{n+1}_{+}$ *(or* $\mathbb{R}^{n+1}_{-}$*) and* $K\subset\Omega$ *a smoothly bounded compact set with outer unit normal field* ν*. If* $w:\Omega\to C\ell_{0,n-1}$ *is a* $(1-n)$*-hyperbolic harmonic function, then*

$$\begin{aligned}w(y)=&\,c\int_{\partial K}P\left(H(x,y)\nu(x)\overline{D}_l w(x)\right)x_n^{n-1}d\sigma\\&+c\int_{\partial K}P\left(2y_n^{n-1}\overline{q(x,y)\nu(x)}w(x)\right)x_n^{n-1}d\sigma,\end{aligned}$$

for all $y\in intK$*, where* $c=\frac{2^{n-1}}{\omega_{n+1}}$ *and* $H(x,y)=\frac{1}{(n-1)|x-y|^{n-1}|x-\widehat{y}|^{n-1}}$.

Choose v an arbitrary, $C\ell_{0,n-1}$-valued, solution of the equation

$$\Delta v-\frac{n-1}{x_n}\frac{\partial v}{\partial x_n}+\frac{n-1}{x_n^2}v=0, \tag{4}$$

Theorem 31 ([15]). *Let Ω be an open subset of $\mathbb{R}^{n+1}_+$ (or $\mathbb{R}^{n+1}_-$) and $K \subset \Omega$ a smoothly bounded compact set with outer unit normal field ν. If $v : \Omega \to C\ell_{n-1}$ is a solution of (4), then, for all $y \in intK$,*

$$v(y) = c\int_{\partial K} Q\Big(H(x,y)\overline{\nu(x)}M^l_{n-1}(ve_n)(x)\Big)\,d\sigma + c\int_{\partial K} Q\big(2q(x,y)\nu(x)\,v(x)e_n\big)\,d\sigma,$$

where $c = \frac{2^{n-1}y_n^{n-1}}{\omega_{n+1}}$.

References

[1] L.V. Ahlfors, *Möbius transformations in several dimensions* (Ordway Lectures in Math.). University of Minnesota, 1981.

[2] L.V. Ahlfors, Möbius transformations and Clifford numbers. In: I. Chavel and H. M. Farkas (Eds.), *Differential Geometry and Complex Analysis*, pp. 65-73. Springer-Verlag, Berlin, 1985.

[3] Ö Akin and H. Leutwiler, On the invariance of the solutions of the Weinstein equation under Möbius transformations. In: *Classical and Modern Potential Theory and Applications* (ASI Ser., Ser. C, Math. Phys. Sci. **430**), pp. 19-29. Kluwer, Dordrecht, 1994.

[4] M. Brelot, Une axiomatique générale du probléme de Dirichlet dans les espaces localement compacts. *Sém. Théorie du Potentiel* **1** (1958), 6.01-6.16.

[5] J. Cnops, *Hurwitz pairs and applications of Möbius transformation*. Habilitation thesis, Univ. Gent, 1994.

[6] C. Constantinescu and A. Cornea, *Potential Theory on Harmonic Spaces*. Springer-Verlag, Berlin Heidelberg New York, 1972.

[7] S.-L Eriksson-Bique, On modified Clifford analysis. *Complex Variables* **45** (2001), 11-32.

[8] S.-L. Eriksson-Bique, $k-$hypermonogenic functions. In: *Progress in Analysis* **I** (2003), World Scientific , 337-348.

[9] S.-L. Eriksson, Integral formulas for hypermonogenic functions. *Bull. Bel. Math. Soc.* **11** (2004), 705-717.

[10] S.-L. Eriksson-Bique and H. Leutwiler, On modified quaternionic analysis in $\mathbb{R}^3$. *Arch. Math.* **70** (1998), 228-234.

[11] S.-L. Eriksson-Bique and H. Leutwiler, Hypermonogenic functions. In: *Clifford Algebras and their Applications in Mathematical Physics* **2**, pp. 287-302. Birkhäuser, Boston, 2000.

[12] S.-L. Eriksson-Bique and H. Leutwiler, Hypermonogenic functions and Möbius transformations. *Advances in Applied Clifford algebras*, **11** (S2) (2001), 67-76.

[13] S.-L. Eriksson and H. Leutwiler, Hypermonogenic functions and their Cauchy-type theorems. In: *Trend in Mathematics: Advances in Analysis and Geometry*, pp. 97–112. Birkhäuser, Basel/Switzerland, 2004.

[14] S.-L. Eriksson and H. Leutwiler, Contributions to the theory of hypermonogenic functions. *Complex Variables and elliptic equations* **51** (2006), 547-561.

[15] S.-L. Eriksson and H. Leutwiler, On hyperbolic function theory, to appear.

[16] H. Leutwiler, Best constants in the Harnack inequality for the Weinstein equation. *Aequation Math.* **34** (1987), 304-315.

[17] H. Leutwiler, Modified Clifford analysis. *Complex Variables* **17** (1992), 153-171.

[18] H. Leutwiler, Modified quaternionic analysis in $\mathbb{R}^3$. *Complex Variables* **20**(1992), 19-51.

[19] R.-M. Hervé, Reserches axiomatiques sur la théorie des fonctions surharmoniques et du potentiel. *Ann. Inst. Foiurier* **12** (1962), 415-571.

[20] L.K. Hua, *Starting with the Unit Circle, Background to Higher Analysis*. Springer, New York, 1981.

[21] A. Huber, On the uniqueness of generalized axially symmetric potentials. *Ann. of Math.* **60** (1954), 351-358.

[22] J. Hyvönen, Classical potential theory. Summer school in potenitial theory. *Department of Mathematics and Physics Report Ser.* **5-M** (1983), 3-63.

[23] Y. Qiao, S. Bernstein, S.-L. Eriksson and J. Ryan, Function theory for Laplace and Dirac-Hodge operators in hyperbolic space. *Journal d'Analyse Mathématique* **98** (2006), 43-64.
[24] K. Vahlen, Über Bewegungen und komplexe Zahlen. *Math. Anal.* **55** (1902), 585-593.
[25] P.L. Waterman, Möbius transformations in several dimensions. *Adv. in Math* **101** (1993), 87-113.
[26] A. Weinstein, Generalized axially symmetric potential theory. *Bull. Amer. Math. Soc.* **59** (1953), 20-38.

DEPARTMENT OF MATHEMATICS, TAMPERE UNIVERSITY OF TECHNOLOGY, P.O.BOX 553, FI-33101 TAMPERE, FINLAND
E-mail address: sirkka-liisa.eriksson@tut.fi

UNIVERSITY OF ERLANGEN-NÜRNBERG, BISMARCKSTRASSE $1\frac{1}{2}$, D-91054 ERLANGEN, GERMANY
E-mail address: leutwil@mi.uni-erlangen.de

Potential Theory and Stochastics in Albac, 101–110
Aurel Cornea Memorial Volume

ON EXTENDED DIRICHLET SPACES AND THE SPACE OF BL FUNCTIONS

MASATOSHI FUKUSHIMA

ABSTRACT. We first formulate the notions of Dirichlet form and extended Dirichlet space without assuming any topology on the underlying space. We next study the space of BL functions on an Euclidean domain in relation to the extended Dirichlet space of the Sobolev space of order 1 as well as an extension of the time changed transient Brownian motion by reflection at infinity.

The (symmetric) Dirichlet space and the Dirichlet form were introduced by A. Beurling and J. Deny [1] in 1959 and then the extended Dirichlet space was defined by M.L. Silverstein [15] in 1974. In both cases, the underlying space was assumed to be a locally compact Hausdorff topological space. We first formulate them without assuming any topology on the underlying space using an idea of B. Schmuland [13].

The notion of the extended Dirichlet space serves as a useful tool to give criteria for the transience and recurrence of the associated Markovian semigroup. Besides, it plays the role of an invariant in describing the time changes of the process or equivalently, the exchanges of the underlying measure, and it enables us to formulate its maximal extension called the reflected Dirichlet space in transient case.

The space of BL (Beppo Levi) functions on an Euclidean domain was introduced and studied in J. Deny and J.L. Lions [8] in 1953-4 that preceded [1]. The second aim of the present paper is to examine this space in relation to the above mentioned concepts for the Brownian motion, producing a new extension of it at infinity.

1. EXTENDED DIRICHLET SPACES

Let $(E,\mathscr{B}(E))$ be a measurable space and m be a σ-finite measure on it. Numerical functions f, g on E are said to be m-equivalent ($f = g$ $[m]$ in notation) if $f = g$ m-a.e.

Definition 1.1. For $1 \le p \le \infty$,, a linear operator L on $L^p(E;m)$ is called *Markovian* if

$$0 \le f \le 1\ [m],\ f \in \mathscr{D}(L) \quad \Longrightarrow \quad 0 \le Lf \le 1\ [m].$$

A real function φ, namely, a mapping from $\mathbb{R}$ to $\mathbb{R}$, is said to be a *normal contraction* if

$$\varphi(0) = 0, \quad |\varphi(s) - \varphi(t)| \le |s - t|, \quad \forall s, t \in \mathbb{R}.$$

A function defined by $\varphi(t) = (0 \vee t) \wedge 1,\ t \in \mathbb{R}$, is a normal contraction which is called the *unit contraction*. For any $\epsilon > 0$, a real function φ_ϵ satisfying the next condition is a

2000 *Mathematics Subject Classification*. 31C25, 60J50.
Key words and phrases. Extended Dirichlet space, BL function, time changed Brownian motion.

normal contraction:

$$\varphi_\epsilon(t) = t,\ \forall t \in [0,1];\ -\epsilon \le \varphi_\epsilon(t) \le 1+\epsilon,\ \forall t \in \mathbb{R},$$
$$s < t \to 0 \le \varphi_\epsilon(t) - \varphi_\epsilon(s) \le t - s,\ \forall s,t \in \mathbb{R}. \tag{1.1}$$

Definition 1.2. A symmetric form $(\mathcal{E},\mathcal{D}(\mathcal{E}))$ on $L^2(E;m)$ is called *Markovian* if, for any $\epsilon > 0$, there exists a real function φ_ϵ satisfying (1.1) and

$$f \in \mathcal{D}(\mathcal{E}) \implies g = \varphi_\epsilon \circ f \in \mathcal{D}(\mathcal{E}),\quad \mathcal{E}(g,g) \le \mathcal{E}(f,f). \tag{1.2}$$

A closed symmetric form $(\mathcal{E},\mathcal{F})$ on $L^2(E;m)$ is called a *Dirichlet form* if it is Markovian. In this case, the domain $\mathcal{F}(=\mathcal{D}(\mathcal{E}))$ is said to be a *Dirichlet space.*

Theorem 1.3. *Let $(\mathcal{E},\mathcal{F})$ be a closed symmetric form on $L^2(E;m)$ and $\{T_t\}_{t>0}$, $\{G_\alpha\}_{\alpha>0}$ be the strongly continuous contraction semigroup and resolvent on $L^2(E;m)$ generated by $(\mathcal{E},\mathcal{F})$, respectively. Then the next conditions are mutually equivalent:*

(a) *T_t is Markovian for each $t > 0$.*
(b) *αG_α is Markovian for each $\alpha > 0$.*
(c) *$(\mathcal{E},\mathcal{F})$ is a Dirichlet form on $L^2(E;m)$.*
(d) *The unit contraction operates on $(\mathcal{E},\mathcal{F})$:*

$$f \in \mathcal{F} \implies g = (0 \vee f) \wedge 1 \in \mathcal{F},\quad \mathcal{E}(g,g) \le \mathcal{E}(f,f).$$

(e) *Every normal contraction operates on $(\mathcal{E},\mathcal{F})$: for any normal contraction φ*

$$f \in \mathcal{F} \implies g = \varphi \circ f \in \mathcal{F},\quad \mathcal{E}(g,g) \le \mathcal{E}(f,f).$$

This theorem can be shown in exactly the same way as the proof of [9, Th.1.4.1] except for the implication (a) ⇒ (e), which follows however from the more general theorem formulated below. In what follows, we occasionally use for a Dirichlet form $(\mathcal{E},\mathcal{F})$ on $L^2(E;m)$ the notations $\|f\|_{\mathcal{E}} = \sqrt{\mathcal{E}(f,f)}$, $f \in \mathcal{F}$.

Definition 1.4. Let $(\mathcal{E},\mathcal{F})$ be a closed symmetric form on $L^2(E;m)$. Denote by $\mathcal{F}_e$ the totality of m-equivalence classes of all m-measurable functions f on E satisfying the next condition:

$$|f| < \infty\ [m],\ \exists\{f_n\} \subset \mathcal{F},\ \lim_{n,n'\to\infty} \|f_n - f_{n'}\|_{\mathcal{E}} = 0,\ \lim_{n\to\infty} f_n = f\ [m]. \tag{1.3}$$

$\{f_n\} \subset \mathcal{F}$ in the above is called an *approximating sequence of $f \in \mathcal{F}_e$*. We call the space $\mathcal{F}_e$ the *extended space* attached to $(\mathcal{E},\mathcal{F})$. When the latter is a Dirichlet form on $L^2(E;m)$, the space $\mathcal{F}_e$ will be called its *extended Dirichlet space.*

Theorem 1.5. *Let $(\mathcal{E},\mathcal{F})$ be a closed symmetric form on $L^2(E;m)$ and $\mathcal{F}_e$ be the extended space attached to it. If the semigroup $\{T_t; t > 0\}$ generated by $(\mathcal{E},\mathcal{D}(\mathcal{F}))$ is Markovian, then the followings are true:*

(i) *For any $f \in \mathcal{F}_e$ and for any approximating sequence $\{f_n\} \subset \mathcal{F}$ of f, the limit $\mathcal{E}(f,f) = \lim_{n\to\infty} \mathcal{E}(f_n,f_n)$ exists independently of the choice of an approximating sequence $\{f_n\}$ of f.*
(ii) *Every normal contraction operates on $(\mathcal{F}_e,\mathcal{E})$: for any normal contration φ*

$$f \in \mathcal{F}_e \implies g = \varphi \circ f \in \mathcal{F}_e,\quad \mathcal{E}(g,g) \le \mathcal{E}(f,f).$$

(iii) *$\mathcal{F} = \mathcal{F}_e \cap L^2(E;m)$. In particular, $(\mathcal{E},\mathcal{F})$ is a Dirichlet form on $L^2(E;m)$.*

Assertions (ii) and (iii) of this theorem imply the implication (a) $\Rightarrow$ (e) in Theorem 1.3. Before giving the proof of Theorem 1.5, we shall fix a Markovian contractive symmetric linear operator T on $L^2(E;m)$ and make some preparatory observations.

By the linearity and the Markovian property of T on $L^2(E;m)\cap L^\infty(E;m)$,

$$f_1,\, f_2 \in L^2\cap L^\infty,\ 0<f_1<f_2\ [m] \Longrightarrow 0\le Tf_1\le Tf_2\le \|f_2\|_\infty\ [m].$$

Due to the σ-finitness of m, we can construct a Borel function $\eta\in L^1(E;m)$ which is strictly positive on E. If we put $\eta_n(x)=(n\eta(x))\wedge 1$, then $0<\eta_n\le 1$, $\eta_n\uparrow 1$, $n\to\infty$. Hence we can define an extension of T from $L^2(E;m)\cap L^\infty(E;m)$ to $L^\infty(E;m)$ by

$$\begin{cases} Tf(x) &= \lim_{n\to\infty} T(f\cdot\eta_n)(x),\ m\text{-a.e. } x\in E,\ f\in L^\infty_+(E;m),\\ Tf &= Tf^+ - Tf^-,\quad f\in L^\infty(E;m),\quad f=f^+-f^-. \end{cases} \tag{1.4}$$

By the symmetry of T, $(g, T(f\cdot\eta_n)) = (Tg, f\cdot\eta_n)$, $g\in bL^1(E;m)$, and we let $n\to\infty$ to see that the function Tf, $f\in L^\infty(E;m)$, defined by (1.4) satisfies the identity

$$\langle g, Tf\rangle = \langle Tg, f\rangle,\quad \forall g\in bL^1(E;m), \tag{1.5}$$

where $\langle g,f\rangle$ denotes the integral $\int_E gf\,dm$ for $g\in L^1(E;m)$, $f\in L^\infty(E;m)$. Consequently Tf is uniquely determined up to the m-equivalence for $f\in L^\infty(E;m)$. T becomes a Markovian linear operator on $L^\infty(E;m)$ and satisfies

$$f_n,\, f\in L^\infty_+(E;m),\ f_n\uparrow f\ [m],\ n\to\infty \Longrightarrow \lim_{n\to\infty} Tf_n = Tf\ [m]. \tag{1.6}$$

Further, if a sequence $\{f_n\}\subset L^\infty(E;m)$ is uniformly bounded and converges to f m-a.e. as $n\to\infty$, then

$$\lim_{n\to\infty}\langle g, Tf_n\rangle = \langle g, Tf\rangle,\quad \forall g\in bL^1(E;m). \tag{1.7}$$

Lemma 1.6. (i) *For any $g\in L^\infty(E;m)$,*

$$T(g^2) - 2gTg + g^2T1 \ge 0\quad [m].$$

(ii) *For any $g\in L^\infty(E;m)$, define*

$$\mathscr{A}_T(g) = \frac{1}{2}\int_E [T(g^2) - 2gTg + g^2T1]\,dm + \int_E g^2(1-T1)\,dm. \tag{1.8}$$

It holds for $g\in L^2(E;m)\cap L^\infty(E;m)$ that

$$\mathscr{A}_T(g) = (g - Tg, g). \tag{1.9}$$

(iii) *For any $g\in L^\infty(E;m)$ and for any normal contraction φ,*

$$\mathscr{A}_T(\varphi\circ g) \le \mathscr{A}_T(g). \tag{1.10}$$

Proof. Since $T(g^2)-2sTg+s^2T1\ge 0\quad [m]$ for $g\in L^\infty$ and a simple function s, we get (i) by letting $s\to g$. (ii) can be shown first for $\mathscr{A}_T^k(g) = \langle T(g^2)-2gTg+g^2T1, \eta_k\rangle + \langle g^2(1-T1), \eta_k\rangle$ by approximating g with simple functions and using (1.7). We then let $k\to\infty$ to obtain (ii). □

Let φ^ℓ be a specific sequence of normal contractions defined by

$$\varphi^\ell(t) = [(-\ell)\vee t]\wedge \ell,\quad t\in\mathbb{R}. \tag{1.11}$$

For any m-measurable function g on E with $|g| < \infty$ $[m]$, $\mathscr{A}_T(\varphi^\ell \circ g)$ is increasing as ℓ increases, as is clear from $\varphi^\ell \circ (\varphi^{\ell+1} \circ g) = \varphi^\ell \circ g$ and Lemma 1.6 (iii). We can then extend the definition of $\mathscr{A}_T(g)$ to g by letting

$$\mathscr{A}_T(g) = \lim_{\ell\to\infty} \mathscr{A}_T(\varphi^\ell \circ g) \ (\leq \infty). \tag{1.12}$$

Lemma 1.7. (i) *For* $g \in L^2(E;m)$, $\mathscr{A}_T(g) = (g - Tg, g)$.
(ii) **(Fatou's property)** *For any m-measurable functions g_n, g on E with* $|g_n| < \infty$, $|g| < \infty[m]$, $\lim_{n\to\infty} g_n = g$ $[m]$,

$$\mathscr{A}_T(g) \leq \liminf_{n\to\infty} \mathscr{A}_T(g_n). \tag{1.13}$$

(iii) *For any m-measurable function g on E with $|g| < \infty$ $[m]$ and for any normal contraction φ, $\mathscr{A}_T(\varphi \circ g) \leq \mathscr{A}_T(g)$.*

Proof. (i) follows from Lemma 1.6 (ii) and the contraction property of T on $L^2(E;m)$. Assertion (ii) for uniformly bounded $\{g_n\}$ can be shown in the same manner as the proof of [13, Prop.1] with a help of property (1.6). It then suffices to use the contraction φ^ℓ. For the proof of (iii), we put $f = \varphi \circ g$, $f_\ell = \varphi \circ \varphi_\ell \circ g$. Then $f_\ell \to f$, $\ell \to \infty$, and it suffices to combine (ii) with (1.10). □

More details of the proof of Lemma 1.6 and Lemma 1.7 are being given in [10] and [6].

Proof of Theorem 1.5 (i) For any $f \in \mathscr{F}_e$, take its approximating sequence $\{f_n\} \subset \mathscr{F}$. f_n being $\mathscr{E}$-Cauchy, the triangular inequality guarantees the existence of the limit $\mathscr{E}(f,f) = \lim_{n\to\infty} \mathscr{E}(f_n, f_n)$. Let us prove that

$$\frac{1}{t}\mathscr{A}_{T_t}(f) \uparrow \mathscr{E}(f,f) \quad t \downarrow 0, \tag{1.14}$$

which in particular implies that $\mathscr{E}(f,f)$ does not depend the choice of the approximating sequence.

Since $f - f_\ell \in \mathscr{F}_e$ for each ℓ and $\{f_n - f_\ell\} \subset \mathscr{F}$ is its approximating sequence, we have from Lemma 1.7

$$\frac{1}{t}\mathscr{A}_{T_t}(f - f_\ell) \leq \liminf_{n\to\infty} \frac{1}{t}\mathscr{A}_{T_t}(f_n - f_\ell) \leq \lim_{n\to\infty} \|f_n - f_\ell\|_{\mathscr{E}}^2.$$

Therefore $\lim\limits_{\ell\to\infty} \mathscr{A}_{T_t}(f - f_\ell) = 0$; by the triangular inequality $\mathscr{A}_{T_t}(f) = \lim_{\ell\to\infty} \mathscr{A}_{T_t}(f_\ell)$, which in particular implies the monotonicity of the left hand side of (1.14) in t. Since $\lim_{t\downarrow 0} \frac{1}{t}\mathscr{A}_{T_t}(f_\ell) = \|f_\ell\|_{\mathscr{E}}^2$, we can get from the triangular inequality and the inequality obtained above that

$$\left| \lim_{t\downarrow 0} \sqrt{\frac{1}{t}\mathscr{A}_{T_t}(f)} - \|f_\ell\|_{\mathscr{E}} \right| \leq \lim_{t\downarrow 0} \sqrt{\frac{1}{t}\mathscr{A}_{T_t}(f - f_\ell)} \leq \lim_{n\to\infty} \|f_n - f_\ell\|_{\mathscr{E}}.$$

The last term tends to 0 as $\ell \to \infty$, yielding (1.14).
(ii) By Lemma 1.7 and (1.14), we have

$$\frac{1}{t}\mathscr{A}_{T_t}(\varphi \circ f) \leq \frac{1}{t}\mathscr{A}_{T_t}(f) \leq \mathscr{E}(f,f), \quad \forall t > 0,$$

and consequently, it is enough to prove that $\varphi \circ f \in \mathscr{F}_e$. But this is an easy consequence of a Banach-Saks type theorem. □

Remark 1.8. Theorem 1.3 has been formulated in Bouleau-Hirsch [2] and in Ma-Röckner [12] without any topology on E, while Theorem 1.5 was shown in Silverstein [15] and Fukushima-Oshima-Takeda [9] under a certain topological assumption on E.

Let $(\mathscr{E},\mathscr{F})$ be a Dirichlet form on $L^2(E;m)$, $(\mathscr{F}_e,\mathscr{E})$ be its extended Dirichlet space and $\{T_t; t>0\}$ be the associated strongly continuous contraction semigroup of Markovian symmetric operators on $L^2(E;m)$. We put

$$S_t f = \int_0^t T_s f ds, \quad t>0, \quad f \in L^2(E;m), \tag{1.15}$$

the integral being taken in Bochner's sense. S_t is a linear operator on $L^2(E;m)$ and satisfies the boundedness $\|S_t f\|_2 \le t\|f\|_2$, $f \in L^2(E;m)$.

Take $f \in L^2(E;m)\cap L^1(E;m)$. Choose $B_n \in \mathscr{B}(E)$ with $m(B_n)<\infty$, $B_n \uparrow E$. By the symmetry and the Markov property of T_t,

$$\int_{B_n} |T_t f(x)| m(dx) \le (T_t|f|, 1_{B_n}) = (|f|, T_t 1_{B_n}) \le \int_E |f(x)| m(dx).$$

Letting $n\to\infty$, we get $\|T_t f\|_1 \le \|f\|_1$. Similarly, we get $\|S_t f\|_1 \le t\|f\|_1$ and hence both $\{T_t\}$, $\{S_t\}$ can be extended to linear operators on $L^1(E;m)$ satisfying

$$T_s T_t f = T_{s+t} f,\ \|T_t f\|_1 \le \|f\|_1,\ \|S_t f\|_1 \le t\|f\|_1,\ f \in L^1(E;m).$$

Further T_t and $\frac{1}{t} S_t$ are Markovian.

Since S_t so extended satisfies for $f \in L^1_+(E;m)$ the positivity and the monotonicity $0 \le S_s f \le S_t f$, $[m]$, $0<s<t$, we can define a function $Gf(x) (\le \infty)$ by

$$Gf(x) = \lim_{N\to\infty} S_N f(x)\ [m],\ f \in L^1_+(E;m), \tag{1.16}$$

uniquely up to the m-equivalence.

Definition 1.9. (i) $\{T_t; t>0\}$ is called *transient* if $Gg(x)<\infty\ [m]$, for some $g \in L^1_+(E;m)$ with $g>0\ [m]$.

(ii) $\{T_t; t>0\}$ is called *recurrent* if, for any $f \in L^1_+(E;m)$, $Gf(x)$ is either ∞ or $0\ [m]$, namely, $m\{x \in E : 0 < Gf(x) < \infty\} = 0$.

It is known that the transience is equivalent to the condition that $Gf(x)<\infty\ [m]$ for all $f \in L^1_+(E;m)$, while the recurrence is equivalent to $Gf(x)=\infty\ [m]$ for all $f \in L^1(E;m)$ with $f>0\ [m]$. We take the following theorem from [9] (cf. [15]).

Theorem 1.10. (i) *The transience of* $\{T_t; t>0\}$ *is equivalent to one of the following conditions:*

$$u \in \mathscr{F}_e, \quad \mathscr{E}(u,u)=0 \Longrightarrow u=0. \tag{1.17}$$

$$(\mathscr{F}_e,\mathscr{E}) \text{ is a real Hilbert space.} \tag{1.18}$$

(ii) $\{T_t; t>0\}$ *is recurrent if and only if*

$$1 \in \mathscr{F}_e, \qquad \mathscr{E}(1,1)=0. \tag{1.19}$$

2. THE SPACE OF BL FUNCTIONS AND BROWNIAN MOTION

For a non-empty domain D of the Euclidean n-space $\mathbb{R}^n$, we let

$$\mathbf{D}(u,v) = \sum_{i=1}^{n} \int_D \frac{\partial u}{\partial x_i} \frac{\partial v}{\partial x_i} dx, \tag{2.1}$$

$$H^1(D) = \{u \in L^2(D) : \frac{\partial u}{\partial x_i} \in L^2(D),\ 1 \le i \le n.\} \tag{2.2}$$

Here the derivatives $\frac{\partial u}{\partial x_i}$, $1 \le i \le n$, are taken in Schwartz distribution sense and $L^2(D)$ denotes the L^2-space on D based on the Lebesgue measure dx.

We forcus our attention on a related space

$$\mathrm{BL}(D) = \left\{ T : \frac{\partial T}{\partial x_i} \in L^2(D),\ 1 \le i \le n \right\} \tag{2.3}$$

of Schwartz distributions T. It is known that any distribution $T \in \mathrm{BL}(D)$ can be identified with a function in $L^2_{loc}(D)$ (cf. L. Schwartz [14], J. Deny and J.L. Lions [8]) so that

$$\mathrm{BL}(D) = \{u \in L^2_{loc}(D) : \frac{\partial u}{\partial x_i} \in L^2(D),\ 1 \le i \le n,\} \tag{2.4}$$

where again the derivatives are taken in Schwartz distribution sense. Members in BL(D) are called BL (Beppo-Levi) *functions* on D. The space BL(D) is known to enjoy the following properties (cf. [8]):

(BL.1) The quotient space $\dot{\mathrm{BL}}(D)$ of BL(D) by the subspace of constant functions is a Hilbert space with inner product **D**. Any **D**-Cauchy sequence $u_n \in \mathrm{BL}(D)$ admits $u \in \mathrm{BL}(D)$ and constants c_n such that u_n is **D**-convergent to u and $u_n + c_n$ is L^2_{loc}-convergent to u.

(BL.2) A function u on D is in BL(D) if and only if, for each i ($1 \le i \le n$), there is a version $u^{(i)}$ of u such that it is absolutely continuous on almost all straight lines parallel to x_i-axis and the derivative $\partial u^{(i)}/\partial x_i$ in the ordinary sense (which exists a.e. on D) is in $L^2(D)$. In this case, the ordinary derivatives coincide with the distribution derivatives of u.

Notice that the version $u^{(i)}$ in the above for $u \in \mathrm{BL}(D)$ depends on the choice of the coordinate x_i. However each $u \in \mathrm{BL}(D)$ admits a quasi continuous version $\tilde{u}$ called a BLD (Beppo Levi-Deny) *function*, which enjoys the absolute continuity property (BL.2) no matter how the coordinates are choosen (cf. [8]).

The space $H^1(D) = \mathrm{BL}(D) \cap L^2(D)$ is called the *Sobolev space* of order 1 on D.

$$(\mathcal{E}, \mathcal{F}) = \left(\frac{1}{2}\mathbf{D}, H^1(D)\right) \tag{2.5}$$

is a closed symmetric form on $L^2(D)$. To see this, suppose $\{u_n\} \subset \mathcal{F}$ is $\mathcal{E}_1$-Cauchy. Then $\frac{\partial u_n}{\partial x_i}$ is $L^2(D)$-convergent to some $v_i \in L^2(D)$ for each $1 \le i \le n$ and u_n is $L^2(D)$-convergent to some $u \in L^2(D)$. Then, for any $f \in C_0^\infty$,

$$(v_i, f) = \lim_{n\to\infty} (\frac{\partial u_n}{\partial x_i}, f) = -\lim_{n\to\infty} (u_n, \frac{\partial f}{\partial x_i}) = -(u, \frac{\partial f}{\partial x_i})$$

and hence $v_i = \dfrac{\partial u}{\partial x_i}$, $1 \le i \le n$. It is a Dirichlet form on $L^2(D)$ because its Markov property (1.2) can be verified by a direct use of the property (BL.2).

Let us denote by $H^1_e(D)$ the extended Dirichlet space of the Dirichlet form (2.5). When $D = \mathbb{R}^n$, the Dirichlet form (2.5) on $L^2(\mathbb{R}^n)$ is associated with the transition density $g_t(x) = (2\pi t)^{-n/2}\exp(-|x|^2/(2t))$ of the n-dimensional standard Brownian motion. Since $\int_0^\infty g_t(x)dt$, $x \ne 0$, is divergent when $n = 1, 2$, but convergent and equal to the Newtonian kernel $\mathcal{N}(x)$ when $n \ge 3$, the corresponding L^2-semigroup is recurrent in the former case and transient in the latter case.

When $n \ge 3$, the extended Dirichlet space $(H^1_e(\mathbb{R}^n), \mathcal{E})$ of (2.5) for $D = \mathbb{R}^n$ is a real Hilbert space and

$$u \in H^1_e(\mathbb{R}^n),\ \mathcal{E}(u,u) = 0 \implies u = 0, \tag{2.6}$$

in view of Theorem 1.10.

Theorem 2.1. *Assume that $n \ge 3$. $\mathrm{BL}(\mathbb{R}^n)$ is the linear space spanned by $H^1_e(\mathbb{R}^n)$ and constant functions. The space $(H^1_e(\mathbb{R}^n), \mathcal{E})$ is isometric with the space $(\dot{\mathrm{BL}}(\mathbb{R}^n), \frac{1}{2}\mathbf{D})$ by the canonical map $\mathrm{BL}(\mathbb{R}^n) \mapsto \dot{\mathrm{BL}}(\mathbb{R}^n)$.*

Proof. For $u \in H^1_e(\mathbb{R}^n)$, there is a sequence $\{u_n\} \subset H^1(\mathbb{R}^n)$ which is $\mathbf{D}$-Cauchy and convergent to u a.e. $\mathbf{D}(u_n, u_n)$ then converges to $\mathcal{E}(u,u)$. By (BL.1), there exist $v \in \mathrm{BL}(\mathbb{R}^n)$ and constants c_n such that $\{u_n\}$ is $\mathbf{D}$-convergent to v and the sequence $\{u_n + c_n\}$ is convergent to v in $L^2_{loc}(\mathbb{R}^n)$. By choosing a subsequence if necessary, we may assume that the latter sequence converges to v a.e. Then $\lim_{n\to\infty} c_n = c$ exists, $u = v - c$ and consequently, $u \in \mathrm{BL}(\mathbb{R}^n)$ and $\mathcal{E}(u,u) = \mathbf{D}(u,u)$. Further we see from (2.6) that the Hilbert space $(H^1_e(\mathbb{R}^n), \mathcal{E})$ is isometrically imbedded into a closed subspace of $(\dot{\mathrm{BL}}(\mathbb{R}^n), \frac{1}{2}\mathbf{D})$ by the canonical map $\mathrm{BL}(\mathbb{R}^n) \mapsto \dot{\mathrm{BL}}(\mathbb{R}^n)$. We denote by $\dot{u}$ the equivalence class represented by $u \in \mathrm{BL}(\mathbb{R}^n)$.

If $\dot{u} \in \dot{\mathrm{BL}}(\mathbb{R}^n)$ is $\mathbf{D}$-orthogonal to this closed subspace, then, since $C_0^\infty(\mathbb{R}^n) \subset H^1(\mathbb{R}^n)$, we have

$$(\Delta u, f) = -\mathbf{D}(u, f) = 0, \quad \forall f \in C_0^\infty(\mathbb{R}^n),$$

which implies that $\Delta u = 0$, namely, (a version of) u is harmonic on $\mathbb{R}^n$. Since the ordinary derivatives $\dfrac{\partial u}{\partial x_i}$, $1 \le i \le n$, are also harmonic on $\mathbb{R}^n$, we get from the mean-value theorem the estimate

$$\left|\frac{\partial u}{\partial x_i}(x)\right| \le \frac{1}{|B_r(x)|}\int_{B_r(x)}\left|\frac{\partial u}{\partial x_i}\right| dx \le \left(\frac{1}{|B_r(x)|}\mathbf{D}(u,u)\right)^{1/2}, \ x \in \mathbb{R}^n,$$

where $B_r(x)$ is the ball of radius r centered at x and $|B_r(x)|$ denotes it volume. By letting $r \to \infty$, we see that all derivatives of u vanish and hence u is constant, and consequently $\dot{u}$ is the 0 element of $\dot{\mathrm{BL}}(\mathbb{R}^n)$. □

Remark 2.2. Example 1.5.3 and Example 6.2.1 of Fukushima-Oshima-Takeda [9] contained incorrect statements that, when $n \ge 3$, the space $H^1_e(\mathbb{R}^n)$ is obtained from $\mathrm{BL}(\mathbb{R}^n)$ by removing non-zero constant functions. Taking this opportunity, we would like mention that they should be corrected as the statement of Theorem 2.1 of the present paper.

We next consider, on a general domain $D \subset \mathbb{R}^n$, a measure $m(dx) = m(x)dx$ with a density function $m(x)$ satisfying

$$m(x) > 0 \quad \forall x \in D, \quad m \in bC(D) \cap L^1(D) \tag{2.7}$$

and an associated form

$$(\mathscr{E}, \mathscr{F}) = \left(\frac{1}{2}\mathbf{D}, \mathrm{BL}(D) \cap L^2(D; m)\right), \tag{2.8}$$

which is obtained just by replacing $L^2(D)$ with $L^1(D; m)$ in (2.5). Since the convergence in $L^2(D; m)$ implies the convergence in $L^2_{loc}(D)$, (2.8) can be readily seen to be a Dirichlet form on $L^2(D; m)$.

Theorem 2.3. *The Dirichlet form* (2.8) *on* $L^2(D; m)$ *is recurrent. Its extended Dirichlet space* $(\mathscr{F}_e, \mathscr{E})$ *coincides with the space* $(\mathrm{BL}(D), \frac{1}{2}\mathbf{D})$.

Proof. Since m is assumed to be a finite measure on D, the present Dirichlet form enjoys the recurrence condition (1.19). We further have

$$u \in \mathscr{F}_e, \ \mathscr{E}(u, u) = 0 \Longrightarrow u \text{ is constant } a.e. \tag{2.9}$$

To see this, suppose $u \in \mathscr{F}_e$, $\mathscr{E}(u, u) = 0$ and put $u_\ell = \varphi^\ell \circ u$ by the contraction φ^ℓ of (1.11). Then $u_\ell \in L^2(E; m) \cap \mathscr{F}_e = \mathscr{F}$ and hence $\frac{1}{2}\mathbf{D}(u_\ell, u_\ell) = \mathscr{E}(u_\ell, u_\ell) = 0$. Therefore u_ℓ is a constant and we get (2.9) by letting $\ell \to \infty$.

Denote by $\dot{\mathscr{F}}_e$ the quotient space of $\mathscr{F}_e$ by the subspace of constant functions. Just as in the proof of the preceding theorem but using (2.9) in place of (2.6), we conclude that the space $(\dot{\mathscr{F}}_e, \mathscr{E})$ is isometrically embedded into the space $(\dot{\mathrm{BL}}(D), \frac{1}{2}\mathbf{D})$.

Take any $u \in \mathrm{BL}(D)$ and put $u_\ell = \varphi^\ell \circ u$ as above. By (BL.2), $u_\ell \in \mathrm{BL}(D)$ and

$$\mathbf{D}(u - u_\ell, u - u_\ell) = \int_{\{x : |u(x)| > \ell\}} |\nabla u|^2 dx \ \to \ 0, \quad \ell \to \infty.$$

Since $u_\ell \in \mathscr{F}$ and u_ℓ converges to u pointwise, u must be an element of $\mathscr{F}_e$. Hence the above isometric embedding is an onto map and $\mathscr{F}_e = \mathrm{BL}(D)$. □

When the Lebesgue measure of the domain D is finite, then we can take $m(dx)$ to be the Lebesgue measure in (2.8) in reducing $\mathscr{F}$ to $H^1(D)$. Hence

Corollary 2.4. *If the domain D is of finite Lebesgue measure, then*

$$H^1_e(D) = \mathrm{BL}(D). \tag{2.10}$$

Finally, when $D = \mathbb{R}^n$ and m is a finite measure on $\mathbb{R}^n$ with density satisfying (2.7) on $\mathbb{R}^n$, we designate the Dirichlet form (2.8) on $L^2(\mathbb{R}^n; m)$ as

$$(\mathscr{E}^*, \mathscr{F}^*) = \left(\frac{1}{2}\mathbf{D}, \mathrm{BL}(\mathbb{R}^n) \cap L^2(\mathbb{R}^n; m)\right). \tag{2.11}$$

In parallel to this, let us consider the symmetric form on $L^2(\mathbb{R}^n, m)$ defined by

$$(\mathscr{E}^{(0)}, \mathscr{F}^{(0)}) = \left(\frac{1}{2}\mathbf{D}, H^1_e(\mathbb{R}^n) \cap L^2(\mathbb{R}^n; m)\right), \tag{2.12}$$

which is known to be the Dirichlet form associated with the n-dimensional Brownian motion being time changed by its additive functional $A_t = \int_0^t m(X_s)ds$ (cf. [9, Example 6.2.1]).

Assume that $n \geq 3$. Then they are different; (2.11) is recurrent while (2.12) is transient. Further we see from Theorem 2.1 that

$$\mathscr{F}^* = \{u = u_0 + c : u_0 \in \mathscr{F}^{(0)},\ c \text{ is constant}\}, \quad \mathscr{E}^*(u,u) = \mathscr{E}^{(0)}(u_0,u_0). \tag{2.13}$$

Let $\mathbb{R}^n_* = \mathbb{R}^n \cup \{\infty\}$ be the one-point compactification of $\mathbb{R}^n$ and m^* be the extension of m from $\mathbb{R}^n$ to $\mathbb{R}^n_*$ with $m^*(\{\infty\}) = 0$. By identifying $L^2(\mathbb{R}^n; m)$ with $L^2(\mathbb{R}^n_*; m^*)$, we can regard $(\mathscr{E}^*, \mathscr{F}^*)$ as a Dirichlet form on $L^2(\mathbb{R}^n_*; m^*)$.

Theorem 2.5. *When $n \geq 3$, $(\mathscr{E}^*, \mathscr{F}^*)$ is a strongly local regular Dirichlet form on $L^2(\mathbb{R}^n_*, m^*)$.*

Indeed, if we let $\mathscr{C} = \{u + c : u \in C_0^\infty(\mathbb{R}^n),\ c \text{ is constant}\}$, then $\mathscr{C} \subset C(\mathbb{R}^n_*)$ and $\mathscr{C}$ is readily seen to be a core of the Dirichlet form $(\mathscr{E}^*, \mathscr{F}^*)$. The strong locality of $(\mathscr{E}^*, \mathscr{F}^*)$ can be proved in the same way as in the proof of Theorem 3.2 of [11], where a Dirichlet form quite similar to (2.13) was studied in a rather general context.

The m^*-symmetric diffusion process X^* on $\mathbb{R}^n_*$ associated with the Dirichlet form of Theorem 2.5 is an extension of the time changed Brownian motion X^0 on $\mathbb{R}^n$ associated with $(\mathscr{E}^{(0)}, \mathscr{F}^{(0)})$. X^0 approaches to ∞ at a finite life time ζ^0 with mean $\mathbf{E}_x[\zeta^0]$ being equal to the Newtonian potential $\mathscr{N} * m(x)$ of the function m. X is obtained by prolonging the path of X^0 after ζ^0 with a specific reflection at ∞ by piecing together the excursions of X^0 around ∞. Such a probabilistic construction has been studied in [11] and [4]. The relationship between $(\mathscr{E}^*, \mathscr{F}^*)$ and $(\mathscr{E}^{(0)}, \mathscr{F}^{(0)})$ will be studied in terms of the reflected Dirichlet space in [5].

Acknowledgments

The author would like to express his hearty thanks to Professors Lucian Beznea, Nicu Boboc and Gheorghe Bucur for kindly inviting him to the International Conference on Potential Theory and Stochastics dedicated to the memory of Aurel Cornea, held at Albac, Romania in September, 2007.

The research has been supported by Grant-in-Aid for Scientific Research of MEXT No.19540125.

References

[1] A. Beurling and J. Deny, Dirichlet spaces. *Proc. Nat. Acad. Sci. U.S.A.* **45** (1959), 208-215.

[2] N. Bouleau and F. Hirsch, *Dirichlet forms and analysis on Wiener space.* Walter de Gruyter, 1991.

[3] Z.-Q. Chen, On reflected Dirichlet spaces. *Probab. Theory Relat. Fields* **94** (1992), 135-162.

[4] Z.-Q. Chen and M. Fukushima, One-point extensions of symmetric Markov processes by darning. *Probab. Theory Relat. Fields* **141** (2008), 61-112.

[5] Z.-Q. Chen and M. Fukushima, On a unique extension of time changed reflecting Brownian motions, *Ann. Inst. Henri Poincaré Probab. Statist.*, to appear.

[6] Z.-Q. Chen and M. Fukushima, *Symmetric Markov Processes, Time Change and Boundary Theory.* Princeton University Press, in preparation.

[7] J. Deny, Méthods Hilberitiennes en théorie du potentiel. In: *Potential Theory, Centro Internazionale Matematico Estivo*, pp.121-201. Cremonese, Roma, 1970.

[8] J. Deny and J.L. Lions, Les espaces du type de Beppo Levi. *Annales Inst. Fourier* **5** (1953-54), 305-370.
[9] M. Fukushima, Y. Oshima and M. Takeda, *Dirichlet Forms and Symmetric Markov Processes*. Walter de Gruyter, 1994.
[10] M. Fukushima and M. Takeda, *Markov Processes* (in Japanese). Baifukan, 2008.
[11] M. Fukushima and H. Tanaka, Poisson point processes attached to symmetric diffusions. *Ann. Inst. Henri Poincaré Probab. Statist.* **41** (2005), 419-459.
[12] Z.M. Ma and M. Röckner, *Introduction to the Theory of (Non-Symmetric) Dirichlet Forms*. Springer-Verlag, 1992.
[13] B. Schmuland, Positivity preserving forms have the Fatou property. *Potential Analysis* **10** (1999), 373-378.
[14] L. Schwartz, *Théorie des Distributions* **I, II**. Hermann, Paris, 1950, 1951.
[15] M. L. Silverstein, *Symmetric Markov Processes* (Lecture Notes in Math. **426**). Springer-Verlag, 1974.

OSAKA UNIVERSITY, OSAKA, JAPAN
E-mail address: fuku2@mx5.canvas.ne.jp

Potential Theory and Stochastics in Albac, 111–123
Aurel Cornea Memorial Volume

PARTIAL BALAYAGE AND THE EXTERIOR INVERSE PROBLEM OF POTENTIAL THEORY

STEPHEN J. GARDINER AND TOMAS SJÖDIN

ABSTRACT. This paper presents a concise account of partial balayage and how it has recently been used to make progress on the exterior inverse problem of potential theory. It also includes a new proof of the singularity of harmonic measure with respect to Lebesgue measure and a general version of the Hele-Shaw law for fluid flow.

1. INTRODUCTION

Let B be an open ball in Euclidean space $\mathbb{R}^N$ $(N \geq 2)$ and let σ^* denote normalised surface area measure on ∂B. For any function h that is harmonic on a neighbourhood of $\overline{B}$ we have the spherical mean value property

$$h(x_0) = \int_{\partial B} h d\sigma^*, \tag{1}$$

where x_0 denotes the centre of B. Let

$$U_x(y) = \begin{cases} -\log|x-y| & (N=2) \\ |x-y|^{2-N} & (N \geq 3) \end{cases}$$

and (for a suitable measure μ) $U\mu(x) = \int U_x d\mu$. Using (1) with $h = U_x$ $(x \in \mathbb{R}^N \setminus \overline{B})$, we obtain the familiar fact that $U\delta_{x_0} = U\sigma^*$ outside $\overline{B}$, where δ_{x_0} is the Dirac measure at x_0. In this context we regard σ^* as the result of sweeping δ_{x_0} out of B.

Now let λ denote Lebesgue measure on $\mathbb{R}^N$ and let $\lambda^* = (\lambda|_B)/\lambda(B)$. For any integrable harmonic function h on B we have the volume mean value property $h(x_0) = \int_B h d\lambda^*$ and, arguing as before, this yields $U\delta_{x_0} = U\lambda^*$ outside B. By analogy with the previous paragraph we can regard λ^* as the result of a "partial sweeping" (or partial balayage) of δ_{x_0}, in which a certain density of measure is allowed to remain. Such a notion has been developed by Gustafsson and Sakai (see, for example, [8]) in connection with the construction of certain types of quadrature domains for harmonic functions. To be more specific, let μ be a measure with compact support, and suppose we have a procedure for arriving at a bounded open set Ω containing the support of μ such that $U\mu = U(\lambda|_\Omega)$ outside Ω. (Not every measure μ will have such an associated set Ω.) Arranging, by the addition to Ω of a null set if necessary, that the Lebesgue points of

---

2000 *Mathematics Subject Classification.* 31B20.

Key words and phrases. Potentials, partial balayage, harmonic measure, Hele-Shaw flow, inverse problem.

$\mathbb{R}^N\backslash\Omega$ are dense in $\mathbb{R}^N\backslash\Omega$, it follows that $\nabla U\mu = \nabla U(\lambda|_\Omega)$ outside Ω. Thus

$$\int_\Omega h\,d\lambda = \int h\,d\mu \text{ whenever } h \in \{U_x, \partial_i U_x : x \in \mathbb{R}^N\backslash\Omega,\ i = 1,...,N\}.$$

An approximation result of Sakai [11] now shows that $\int_\Omega h\,d\lambda = \int h\,d\mu$ for every λ-integrable harmonic function h on Ω; that is, Ω is a quadrature domain with respect to the measure μ.

Partial balayage is closely related to the obstacle problem, and to the abstract theory of potential cones (see, for example, [2]). However, in this article, we will not rely on these connections. Instead we will present a concise, self-contained exposition of partial balayage from the viewpoint of classical potential theory. Along the way we will obtain a new proof of the singularity of harmonic measure with respect to Lebesgue measure and a general version of the Hele-Shaw law for fluid flow. The paper concludes with an outline of how partial balayage has recently been employed to make progress on the exterior inverse problem of potential theory [6].

2. Partial reductions of superharmonic functions

Now let $\Omega \subseteq \mathbb{R}^N$ be an open set that possesses a Green function (that is, Ω must have nonpolar complement in the case where $N = 2$). Further, let u and v be superharmonic functions on Ω, where $u > 0$ and $v < +\infty$, and let $\mathscr{S}(\Omega)$ denote the collection of all subharmonic functions on Ω. We write

$$^{\Omega}\Phi_u^v = \{s \in \mathscr{S}(\Omega) : s \le u - v\}$$

(the prefix Ω may be omitted where confusion is unlikely to occur), and will use this class to define the "partial reduction" of u with respect to v.

If B is an open ball such that $\overline{B} \subset \Omega$, we denote by $\tau_B u$ the Poisson integral of u in B, and extend it to all of Ω by putting $\tau_B u = u$ elsewhere.

Lemma 1. $-v \le \sup \Phi_u^v \in \Phi_u^v$.

Proof. Clearly $-v \in \Phi_u^v$. The family Φ_u^v is locally uniformly bounded above since, if B is any open ball such that $\overline{B} \subset \Omega$, then $s \le \tau_B s \le \tau_B u - \inf_B v$ on B for every $s \in \Phi_u^v$. Let w denote the upper semicontinuous regularization of $\sup \Phi_u^v$. Then $w \in \mathscr{S}(\Omega)$ and $w \ge \sup \Phi_u^v$, with equality λ-almost everywhere (see Theorem 3.7.5 or 5.7.1 of [1]). It follows that $w \le u - v$ almost everywhere, and hence, by consideration of means over arbitrarily small balls, everywhere on Ω. Thus $w \in \Phi_u^v$, and the result follows. □

Next we define the *partial reduction of u relative to v* by

$$^{\Omega}P_u^v = \sup \Phi_u^v + v,$$

where, again, the prefix Ω may be omitted. It is clear that $0 \le P_u^v \le u$.

Remark 2. If (v_n) is an increasing sequence of superharmonic functions on Ω with limit v, then $\sup \Phi_u^{v_n} \downarrow \sup \Phi_u^v$ and $P_u^{v_n} \to P_u^v$. To see this, we observe that the sequence $\left(\sup \Phi_u^{v_n}\right)$ of subharmonic functions decreases to a subharmonic limit w satisfying $w \ge \sup \Phi_u^v$ and, since $w \le u - v_n$ for each n, we also have $w \le \sup \Phi_u^v$.

Let $B(x,r)$ denote the open ball of centre x and radius r in $\mathbb{R}^N$.

Lemma 3. *If the superharmonic function v is continuous, then*
(a) P_u^v is continuous, and
(b) $\sup \Phi_u^v$ is harmonic on the open set $\omega = \{x \in \Omega : u(x) > P_u^v(x)\}$.

Proof. (a) Let $y \in \Omega$ and $\varepsilon > 0$. Since $\tau_{B(y,\delta)} v \uparrow v$ as $\delta \downarrow 0$, it follows from Dini's theorem that we can choose $\eta > 0$ such that $\overline{B(y,\eta)} \subset \Omega$ and

$$v - \tau_{B(y,\eta)} v < \varepsilon \text{ on } \Omega. \tag{2}$$

Let $w = \sup \Phi_u^v$. Then

$$\tau_{B(y,\eta)} w \leq \tau_{B(y,\eta)}(u - v) \leq u - \tau_{B(y,\eta)} v < u - v + \varepsilon \text{ on } \Omega,$$

so $\tau_{B(y,\eta)} w - \varepsilon \in \Phi_u^v$ and hence

$$\liminf_{x \to y} w(x) \geq \liminf_{x \to y} \tau_{B(y,\eta)} w(x) - \varepsilon = \tau_{B(y,\eta)} w(y) - \varepsilon \geq w(y) - \varepsilon.$$

Since ε and y are arbitrary, w is lower (as well as upper) semicontinuous and so P_u^v is continuous. (The same argument shows more generally that, if v is continuous on an open subset U of Ω, then P_u^v is also continuous on U.)

(b) Let B be any open ball such that $\overline{B} \subset \omega$ and let $\varepsilon = \inf_{\overline{B}}(u - P_u^v)$. It follows from (a) that $\varepsilon > 0$. Let $w = \sup \Phi_u^v$ and $y \in B$, and choose $\eta > 0$ such that $B(y,\eta) \subset B$ and (2) holds. Since $u - v - w = u - P_u^v \geq \varepsilon$ on $\overline{B}$, we have

$$w \leq \tau_{B(y,\eta)} w \leq \tau_{B(y,\eta)}(u - v) - \varepsilon \leq u - \tau_{B(y,\eta)} v - \varepsilon < u - v \text{ on } \Omega.$$

Since $\tau_{B(y,\eta)} w \in \mathcal{S}(\Omega)$, it follows that $\tau_{B(y,\eta)} w = w$. Hence w is harmonic on $B(y,\eta)$, and (b) follows in view of the arbitrary choice of B and y. □

Theorem 4. *The function P_u^v is superharmonic on Ω.*

Proof. In the case where v is continuous, we note from Lemma 3(b) that P_u^v is superharmonic on ω. Since $P_u^v \leq u$ on Ω with equality on $\Omega \backslash \omega$, we now see that P_u^v is supermeanvalued at points of $\Omega \backslash \omega$. This, together with Lemma 3(a), shows that P_u^v is superharmonic on all of Ω.

In the general case we can find an increasing sequence (v_n) of continuous superharmonic functions on Ω with limit v. The nonnegative functions $P_u^{v_n}$ are superharmonic, by the preceding paragraph, and $P_u^{v_n} \to P_u^v$ by Remark 2. Hence the lower semicontinuous regularization $\widehat{P}_u^v$ is superharmonic on Ω and coincides with P_u^v almost everywhere (λ) (see Corollary 5.7.2 in [1]). Since P_u^v is the sum of a subharmonic and a superharmonic function, consideration of means over arbitrarily small balls now shows that $\widehat{P}_u^v = P_u^v$ everywhere, whence P_u^v is superharmonic on Ω. □

Lemma 5. *(a) If (u_n) is an increasing sequence of superharmonic functions on Ω with limit u and, for each n, there is a superharmonic function w_n such that $u = u_n + w_n$, then $P_{u_n}^v \uparrow P_u^v$.*
(b) If (u_n) is a decreasing sequence of superharmonic functions on Ω such that $u_n \downarrow u$ quasi-everywhere, then $P_{u_n}^v \downarrow P_u^v$.
(c) If (Ω_n) is an exhaustion of Ω by open sets, then ${}^{\Omega_n}P_u^v \downarrow {}^{\Omega}P_u^v$.

Proof. (a) Since $P_u^v - v - w_n \in \Phi_{u_n}^v$, we see that

$$P_u^v \geq P_{u_n}^v \geq P_u^v - w_n \uparrow P_u^v \text{ quasi-everywhere on } \Omega.$$

The superharmonic functions $\lim_{n\to\infty} P_{u_n}^v$ and P_u^v thus agree quasi-everywhere, and hence everywhere, on Ω.

(b) The sequence $(\sup \Phi_{u_n}^v)$, of subharmonic functions on Ω, decreases to a subharmonic function w on Ω. Since $\sup \Phi_{u_n}^v \leq u_n - v$, we see that $w \leq u - v$ quasi-everywhere, and hence everywhere, on Ω. Thus

$$w \leq \sup \Phi_u^v \leq \sup \Phi_{u_n}^v \to w \text{ as } n \to \infty,$$

so

$$P_{u_n}^v \downarrow P_u^v \text{ on } \Omega.$$

(c) If $n \geq m$ and $s \in {}^{\Omega_n}\Phi_u^v$, then it is clear that $s|_{\Omega_m} \in {}^{\Omega_m}\Phi_u^v$. Hence ${}^{\Omega_n}P_u^v$ is decreasing in n. Since $\left({}^{\Omega_n}P_u^v - v\right)_{n\geq m}$ is a decreasing sequence of subharmonic functions on Ω_m, it is clear that the limit s is subharmonic and $s \leq u - v$ on Ω, which yields the desired result. □

Remark 6. (a) If $u' = u + h$, where h is positive and harmonic on Ω, then $\Phi_{u'}^v = \{w + h : w \in \Phi_u^v\}$ and so $P_{u'}^v = P_u^v + h$.
(b) If $v' = v + h$, where h is a harmonic function on Ω, then $\Phi_u^{v'} = \{w - h : w \in \Phi_u^v\}$ and so $P_u^{v'} = P_u^v$. Thus P_u^v depends only on the Riesz measure associated with v.

We will build on this last observation in the next section.

3. Partial balayage of measures

Let $G_\Omega\mu$ be the potential on Ω of a measure μ, and let ν denote the Riesz measure associated with v. (Thus, for any ball B satisfying $\overline{B} \subset \Omega$, the functions v and $U(\nu|_B)$ differ on B by a harmonic function.) We denote the Riesz measure associated with the superharmonic function $P_{G_\Omega\mu}^v$ by B_μ^ν and call it the *partial balayage* of μ onto ν (in Ω). It is well-defined, in view of Remark 6(b). Since, by definition, P_u^v is the sum of v and a subharmonic function, we see that $B_\mu^\nu \leq \nu$ on Ω.

We will denote the (classical) sweeping of a measure η onto a set A by η^A; that is, η^A is defined by the equation $G_\Omega\eta^A = \widehat{R}_{G_\Omega\eta}^A$. In particular, if U is an open set compactly contained in Ω, then $\delta_x^{\Omega\setminus U}$ is harmonic measure for U and x.

The next result concerns the structure of the measure B_μ^ν.

Theorem 7. *Suppose v is continuous and let ω be the open set defined by*

$$\omega = \{x \in \Omega : G_\Omega\mu(x) > P_{G_\Omega\mu}^v(x)\}. \tag{3}$$

Then $(\mu|_\omega)^{\Omega\setminus\omega} \geq (\nu|_\omega)^{\Omega\setminus\omega}$ *and*

$$B_\mu^\nu = \nu|_\omega + \mu|_{\Omega\setminus\omega} + (\mu|_\omega)^{\Omega\setminus\omega} - (\nu|_\omega)^{\Omega\setminus\omega}. \tag{4}$$

Proof. Since $P_{G_\Omega\mu}^v < \infty$ on Ω, by definition, it follows that μ does not charge the polar set of irregular boundary points of ω. Hence (see Theorem 7.5.2 in [1]) the greatest harmonic minorant of $P^{\nu}_{G_\Omega\mu}$ on ω is given by $R^{\Omega\setminus\omega}_{G_\Omega\mu}$. We know from Lemma 3(b) that the Riesz measure associated with $P^{\nu}_{G_\Omega\mu}$ on ω is $\nu|_\omega$. Hence, by the Riesz decomposition,

$$P^{\nu}_{G_\Omega\mu} = G_\omega(\nu|_\omega) + R^{\Omega\setminus\omega}_{G_\Omega\mu} \text{ on } \omega. \tag{5}$$

Thus $P^{\nu}_{G_\Omega\mu} = G_\Omega(\nu|_\omega) - \widehat{R}^{\Omega\setminus\omega}_{G_\Omega(\nu|_\omega)} + \widehat{R}^{\Omega\setminus\omega}_{G_\Omega\mu}$ quasi-everywhere on Ω, and (4) follows.

Finally we note from (5) that

$$P^{\nu}_{G_\Omega\mu} = G_\omega(\nu|_\omega) + R^{\Omega\setminus\omega}_{G_\Omega(\mu|_\omega)} + G_\Omega(\mu|_{\Omega\setminus\omega}) \text{ on } \omega.$$

Since the function $w = P^{\nu}_{G_\Omega\mu} - G_\Omega(\mu|_{\Omega\setminus\omega})$ satisfies $w \le G_\Omega(\mu|_\omega)$ on Ω with equality on $\Omega\setminus\omega$, we have

$$R^{\Omega\setminus\omega}_{G_\Omega(\mu|_\omega)} \le w \le G_\Omega(\mu|_\omega) \text{ on } \Omega.$$

Noting that w is superharmonic on ω, we see that w is locally supermeanvalued on Ω, and also lower semicontinuous except perhaps at a polar subset of $\Omega\cap\partial\omega$. Thus $\widehat{w}$ is superharmonic, and it follows from (4) that $(\mu|_\omega)^{\Omega\setminus\omega} \ge (\nu|_\omega)^{\Omega\setminus\omega}$. □

Our next aim is to show that (4) can be simplified when $\nu \ll \lambda$. To this end, we need to borrow some results from the theory of Sobolev spaces. We will do this by proving what we need directly using the same methods. The interested reader can consult [7] for further information about Sobolev spaces. We write U^f in place of $U(f\lambda)$ when f is a bounded Borel measurable function with compact support.

Lemma 8. *Let B_1, B_2, B_3 be open balls such that $\overline{B}_1 \subset B_2$ and $\overline{B}_2 \subset B_3$. There exist constants C_1, C_2, C_3 such that, for any $g \in C_c^\infty(B_1)$,*

$$\left\|D^2U^g\right\|_{L^2(B_2)} \le C_1\left\|g\right\|_{L^2(B_1)} + C_2\left\|\nabla U^g\right\|_{L^2(B_3)} + C_3\left\|U^g\right\|_{L^2(B_3)}. \tag{6}$$

Proof. For any $\psi \in C_c^3(\mathbb{R}^N)$ we have

$$\begin{aligned}\left\|D^2\psi\right\|^2_{L^2(\mathbb{R}^N)} &= \int\sum_{i=1}^N\sum_{j=1}^N\left(\frac{\partial^2\psi}{\partial x_i\partial x_j}\right)^2 d\lambda = -\int\sum_{i=1}^N\sum_{j=1}^N\frac{\partial^3\psi}{\partial x_i^2\partial x_j}\frac{\partial\psi}{\partial x_j}d\lambda \\ &= \int\sum_{i=1}^N\frac{\partial^2\psi}{\partial x_i^2}\sum_{j=1}^N\frac{\partial^2\psi}{\partial x_j^2}d\lambda = \left\|\Delta\psi\right\|^2_{L^2(\mathbb{R}^N)}.\end{aligned}$$

Thus, if $\phi \in C_c^\infty(B_3)$ and $\phi = 1$ on B_2, we obtain

$$\begin{aligned}\left\|D^2U^g\right\|_{L^2(B_2)} &\le \left\|D^2\left(U^g\phi\right)\right\|_{L^2(B_3)} = \left\|\Delta\left(U^g\phi\right)\right\|_{L^2(B_3)} \\ &= \left\|-a_Ng\phi + 2\left\langle\nabla U^g,\nabla\phi\right\rangle + U^g\Delta\phi\right\|_{L^2(B_3)} \quad (g \in C_c^\infty(B_1)),\end{aligned}$$

where a_N is a dimensional constant. It follows that (6) holds with $C_1 = a_N\sup|\phi|$, $C_2 = 2\sup|\nabla\phi|$ and $C_3 = \sup|\Delta\phi|$. □

Theorem 9. *Let h be a harmonic function on an open ball B and let $A = \{x \in B : U^f(x) = h(x)\}$, where f is a bounded Borel function with compact support. Then $f = 0$ almost everywhere (λ) on A.*

Proof. We choose B_1, B_2, B_3 as in the preceding lemma, with the additional condition that $\overline{B} \subset B_1$. There is no loss of generality in assuming that h is harmonic on B_1 and that $f = 0$ outside B_1. Further, since there exists f_1 in $L^\infty(\mathbb{R}^N)$ such that $f_1 = 0$ outside $B_1 \backslash B$ and $h = U^{f_1}$ in B, we may assume that $h = 0$.

We will first show that, for each i in $\{1, ..., N\}$, there exists v_i in $L^2(B_2)$ such that $\partial^2 U^f / \partial x_i^2 = v_i$ in the sense of distributions. To see this, we use smoothing to find a sequence (g_n) of functions in $C_c^\infty(B_1)$ such that $g_n \to f$ almost everywhere (λ) and $\sup |g_n| \le \|f\|_\infty$ for each n. Let x_0 denote the centre of B_1. Since

$$\begin{aligned}\int_{B_1} |\nabla_x U_x(y)| |g_n(y)| \, d\lambda(y) &\le \|f\|_\infty \int_{B_1} |\nabla_x U_x(y)| \, d\lambda(y) \\ &\le \|f\|_\infty \max\{N-2, 1\} \int_{B_1} |x_0 - y|^{1-N} \, d\lambda(y) \\ &< +\infty,\end{aligned}$$

the sequence (∇U^{g_n}) is uniformly bounded and, by dominated convergence, pointwise convergent to ∇U^f. In particular, $\nabla U^{g_n} \to \nabla U^f$ in $L^2(B_1)$, and similar reasoning shows that $U^{g_n} \to U^f$ in $L^2(B_2)$. By the preceding lemma, we know that $\|\partial^2 U^{g_n - g_m} / \partial x_i^2\|_{L^2(B_2)} \to 0$ as $m, n \to \infty$ and hence, by completeness, $(\partial^2 U^{g_n} / \partial x_i^2)$ converges in $L^2(B_2)$ to some v_i. Thus, as claimed,

$$\int_{B_2} U^f \frac{\partial^2 \phi}{\partial x_i^2} \, d\lambda = \lim_{n\to\infty} \int_{B_2} U^{g_n} \frac{\partial^2 \phi}{\partial x_i^2} \, d\lambda = \int_{B_2} \phi v_i \, d\lambda \quad (\phi \in C_c^\infty(B_2)).$$

Let $\varepsilon > 0$ and $\Phi_\varepsilon(t) = \operatorname{sgn}(t)\left(\sqrt{t^2 + \varepsilon^2} - \varepsilon\right)$ for $t \in \mathbb{R}$. Then $\Phi_\varepsilon \in C^1(\mathbb{R})$ and

$$\begin{aligned}\int \Phi_\varepsilon\left(\frac{\partial U^f}{\partial x_i}\right) \frac{\partial \phi}{\partial x_i} \, d\lambda &= \lim_{n\to\infty} \int \Phi_\varepsilon\left(\frac{\partial U^{g_n}}{\partial x_i}\right) \frac{\partial \phi}{\partial x_i} \, d\lambda \\ &= -\lim_{n\to\infty} \int \Phi'_\varepsilon\left(\frac{\partial U^{g_n}}{\partial x_i}\right) \frac{\partial^2 U^{g_n}}{\partial x_i^2} \phi \, d\lambda \\ &= -\int \Phi'_\varepsilon\left(\frac{\partial U^f}{\partial x_i}\right) v_i \phi \, d\lambda \quad (\phi \in C_c^\infty(B_2)).\end{aligned}$$

Letting $\varepsilon \to 0+$ we see that

$$\int U^f \frac{\partial^2 \phi}{\partial x_i^2} \, d\lambda = -\int \frac{\partial U^f}{\partial x_i} \frac{\partial \phi}{\partial x_i} \, d\lambda = \int_{\left\{\frac{\partial U^f}{\partial x_i} \neq 0\right\}} v_i \phi \, d\lambda,$$

and summing over i yields

$$\int U^f \Delta \phi \, d\lambda = \int \phi \sum v_i \chi_{\left\{\frac{\partial U^f}{\partial x_i} \neq 0\right\}} \, d\lambda \quad (\phi \in C_c^\infty(B_2)).$$

Since $\nabla U^f = \nabla h = 0$ almost everywhere on A (a C^1 function has zero gradient at all Lebesgue points of its zero set), we have $f = 0$ almost everywhere on A, as required. □

We will now use partial balayage to give a new proof of the singularity of harmonic measure with respect to n-dimensional measure. For planar domains this was originally proved by Øksendal [10]. In the case of higher dimensions it is contained in work of Bourgain [3] (see also Hansen and Hueber [9]).

Theorem 10. *If U is an open set and $x_0 \in U$, then $\delta_{x_0}^{U^c} \perp \lambda$.*

Proof. We may assume that U is bounded, since the case of an unbounded set U will then follow by consideration of the intersection of U with arbitrarily large balls centred at x_0 We choose r_0 such that $U \subset B(x_0, r_0)$, and define $\Omega = \mathbb{R}^N$ if $N \geq 3$, or $\Omega = B(x_0, r_0 + 1)$ if $N = 2$. We may assume that U is connected.

Let (U_n) be a sequence of open sets with union U such that $x_0 \in U_1$ and $\overline{U}_n \subset U_{n+1}$ for all n. For $t > 0$ we define

$$\mu_t = \lambda|_U + t\delta_{x_0}, \quad \nu_{t,n} = \lambda|_U + t\delta_{x_0}^{U_n^c} \quad \text{and} \quad \nu_t = \lambda|_U + t\delta_{x_0}^{U^c}.$$

We now consider the partial balayage of these measures onto λ (so that the function v may be taken to be a suitable multiple of $-|x|^2$). Since $B_{\mu_t}^{\lambda} \leq \lambda$, we can write $B_{\mu_t}^{\lambda} = f_t \lambda$, where $0 \leq f_t \leq 1$. Let ω_t denote the open set $\{G_\Omega \mu_t > G_\Omega f_t\}$. Then $f_t = 1$ almost everywhere (λ) on ω_t, by Lemma 3(b). Since $G_\Omega \mu_t - G_\Omega f_t$ is nonnegative and superharmonic, but not harmonic, on U, we see that $U \subset \omega_t$. Since

$$G_\Omega(f_t - \chi_U) = G_\Omega(\mu_t - \lambda|_U) = tG_\Omega(x_0, \cdot) \text{ outside } \omega_t,$$

we see from Theorem 9 that $f_t = \chi_U = 0$ almost everywhere (λ) on $\mathbb{R}^N \backslash \omega_t$, and so $B_{\mu_t}^{\lambda} = \lambda|_{\omega_t}$.

Similarly, writing $B_{\nu_{t,n}}^{\lambda} = f_{t,n}\lambda$ and $\omega_{t,n} = \{G_\Omega \nu_{t,n} > G_\Omega f_{t,n}\}$, we see that $U \subset \omega_{t,n}$ and $B_{\nu_{t,n}}^{\lambda} = \lambda|_{\omega_{t,n}}$. Since $G_\Omega \nu_{t,n} = G_\Omega \mu_t$ on $\mathbb{R}^N \backslash \overline{U}_n$ and $G_\Omega f_{t,n} \leq G_\Omega f_t$ we now have

$$\omega_{t,n} = U \cup \{x \in \mathbb{R}^N \backslash U : G_\Omega \mu_t(x) > G_\Omega f_{t,n}(x)\} \supseteq \omega_t$$

and so, in fact, $\lambda|_{\omega_{t,n}} = \lambda|_{\omega_t}$. Hence $B_{\nu_{t,n}}^{\lambda} = \lambda|_{\omega_t}$ for each n, and thus, by Lemma 5(b), $B_{\nu_t}^{\lambda} = \lambda|_{\omega_t}$. From Theorem 7 we now see that $\nu_t(\Omega \backslash \omega_t) = 0$, so $\delta_{x_0}^{U^c} = 0$ outside $(\cap_{t>0}\omega_t) \backslash U$.

Finally, let $A_t = B(x_0, r_t) \backslash B(x_0, r_0)$, where r_t is chosen such that $\lambda(A_t) = t$. Then $G_\Omega\left(\lambda|_{A_t \cup U}\right) \leq G_\Omega \mu_t$, with equality on $\Omega \backslash \overline{B}(x_0, r_t)$. Since

$$G_\Omega\left(\lambda|_{A_t \cup U}\right) \leq G_\Omega\left(\lambda|_{\omega_t}\right) \leq G_\Omega \mu_t,$$

it follows that $\omega_t \subset B(x_0, r_t)$. It is now easy to see that $\lambda(\omega_t) = \mu_t(\Omega) = \lambda(U) + t$, and so $\lambda((\cap_{t>0}\omega_t) \backslash U) = 0$, as required. □

Corollary 11. *If $\nu \ll \lambda$, then $(\mu|_\omega)^{\Omega \backslash \omega} = (\nu|_\omega)^{\Omega \backslash \omega}$ on Ω, where ω is defined by (3), and so*

$$B_\mu^\nu = \nu|_\omega + \mu|_{\Omega \backslash \omega}.$$

Proof. Suppose that $\nu \ll \lambda$. Since $B_\mu^\nu \leq \nu$ on Ω, we see from the above result and (4) that $(\mu|_\omega)^{\Omega \backslash \omega} = (\nu|_\omega)^{\Omega \backslash \omega}$ on Ω. □

4. The Hele-Shaw law

Let U be a smoothly bounded region in the narrow gap between two parallel plates, and suppose that U is filled with a viscous incompressible fluid, and that fluid is being steadily injected at $x_0 \in U$. This results in a growing family of domains $\{U_t : t \geq 0\}$ which represent the extent of the fluid at time t. The evolution of these domains is described by the Hele-Shaw equation, which says that at a given point of ∂U_t the normal velocity is proportional to the density of $\delta_{x_0}^{U_t^c}$ with respect to arc length measure on ∂U_t. From this it can be shown that the partial balayage of the measure $t\delta_{x_0} + \lambda|_U$ with respect to λ is $\lambda|_{U_t}$, and that the 'derivative' of $B^{\lambda}_{t\delta_{x_0}+\lambda|_U}$ with respect to t is $\delta_{x_0}^{U_t^c}$. In this section we will adapt the proof of Theorem 10 to show that, for a reasonably general family of domains U, the 'derivative' of $B^{\lambda}_{t\delta_{x_0}+\lambda|_U}$ at $t = 0$ is $\delta_{x_0}^{U^c}$.

Now let U be a bounded domain in $\mathbb{R}^N$ and let $x_0 \in U$. As before we choose r_0 such that $U \subset B(x_0, r_0)$, and define $\Omega = \mathbb{R}^N$ if $N \geq 3$, or $\Omega = B(x_0, r_0 + 1)$ if $N = 2$. We would like to be able to say that

$$\left(B^{\lambda}_{t\delta_{x_0}+\lambda|_U} - \lambda|_U\right)/t \overset{w^*}{\to} \delta_{x_0}^{U^c} \quad (t \to 0+). \tag{7}$$

However, this is not true in general. For example, if L is the closed upper halfspace and $U = B(0,2)\backslash(L \cap \partial B(0,1))$, then

$$B^{\lambda}_{t\delta_0+\lambda|_U} = B^{\lambda}_{t\delta_0+\lambda|_{B(0,2)}} = \lambda|_{B(0,r_t)} \quad (t > 0),$$

where r_t satisfies $\lambda(B(0,r_t)) = \lambda(B(0,2)) + t$, and thus

$$\left(B^{\lambda}_{t\delta_0+\lambda|_U} - \lambda|_U\right)/t \overset{w^*}{\to} \delta_0^{B(0,2)^c} \neq \delta_0^{U^c} \quad (t \to 0+).$$

The above example partially explains the formulation of the following generalisation of the Hele-Shaw law. We recall that the fine topology of classical potential theory is the coarsest topology which makes all superharmonic functions continuous. A set which is open with respect to this topology will be called finely open.

Theorem 12. *Let U be a bounded domain in $\mathbb{R}^N$ and let $x_0 \in U$. Suppose that, for any finely open set V satisfying $\lambda(V\backslash U) = 0$, the set $V\backslash U$ is actually polar. Then (7) holds.*

Proof. Let μ_t, ν_t and ω_t be as in the proof of Theorem 10. We saw there that $B^{\lambda}_{\mu_t} = B^{\lambda}_{\nu_t} = \lambda|_{\omega_t}$ and $\lambda(\omega_t) = \lambda(U) + t$, that $U \subseteq \omega_t$ and that $\omega_t \subset B(x_0, r_0 + \frac{1}{2})$ for all sufficiently small t. Hence $B^{\lambda}_{\mu_t} - \lambda|_U = \lambda|_{\omega_t\backslash U}$ for all $t > 0$. Let (t_n) be a positive l^1 sequence. By weak* compactness we can replace (t_n) by an appropriate subsequence so that the associated sequence of probability measures $\left(t_n^{-1}\lambda|_{\omega_{t_n}\backslash U}\right)$ is weak* convergent to some measure ν. Since $G_\Omega\mu_t \geq G_\Omega\nu_t \geq G_\Omega\left(\lambda|_{\omega_t}\right)$, with equality outside ω_t, we have

$$G_\Omega\left(t^{-1}\lambda|_{\omega_t\backslash U}\right) \leq G_\Omega\left(t^{-1}\left(\nu_t - \lambda|_U\right)\right) = G_\Omega\delta_{x_0}^{U^c}, \text{ with equality outside } \omega_t.$$

It follows (by Exercise 5.13 in [1]) from weak* convergence that $G_\Omega\nu \leq G_\Omega\delta_{x_0}^{U^c}$ with equality quasi-everywhere on $\Omega\backslash\omega^*$, where $\omega^* = \cap_{m\geq 1}\cup_{n\geq m}\omega_{t_n}$. Clearly $\lambda(\omega^*) = \lambda(U)$. The finely open set $V = \{G_\Omega\nu < G_\Omega\delta_{x_0}^{U^c}\}$ satisfies $\lambda(V\backslash U) \leq \lambda(\omega^*\backslash U) = 0$ and so, by hypothesis, $V\backslash U$ is polar. It follows that $G_\Omega\nu = G_\Omega\delta_{x_0}^{U^c}$ quasi-everywhere on $\Omega\backslash U$.

Since $\delta_{x_0}^{U^c}(U) = 0$, and $G_\Omega \delta_{x_0}^{U^c}$ is finite valued, the Maria-Frostman domination principle shows that $G_\Omega \nu \geq G_\Omega \delta_{x_0}^{U^c}$ and hence $\nu = \delta_{x_0}^{U}$. The result now follows, in view of the arbitrary nature of the sequence (t_n). □

The following example shows the necessity of the hypothesis in the above result.

Example 13. Let (x_k) be a sequence in $(B(0,1)\backslash B(0,\frac{1}{2}))\cap L$ with limit set $E = L\cap \partial B(0,1)$, and let

$$F = \{x \in B(0,1) : U\mu(x) \geq (1-|x|)^{2-N}\},$$

where $\mu = \sum m_k \delta_{x_k}$ and $0 < m_k \leq 2^{-k}$. (If $N = 2$, we replace $(1-|x|)^{2-N}$ by $-\log(1-|x|)$.) By choosing the m_k to be sufficiently small we can ensure that the open set $U = B(0,2)\backslash(E\cup F)$ is connected. The proof of Theorem 12 can be modified to show that

$$t^{-1}\left(B^{\lambda}_{t\delta_0+\lambda|_U} - \lambda|_U\right) = t^{-1}\left(B^{\lambda}_{t\delta_0+\lambda|_{B(0,2)\backslash F}} - \lambda|_U\right) \xrightarrow{w^*} \delta_0^{V^c} \neq \delta_0^{U^c} \quad (t \to 0+),$$

where $V = B(0,2)\backslash F$, and so (7) fails to hold for this choice of U. The crucial point here is that the set V is finely open (compare Example 7.9.3 in [1]) and $\lambda(V\backslash U) = 0$, but $V\backslash U = L\cap \partial B(0,1)$, which is nonpolar.

5. Application to the exterior inverse problem

A bounded domain Ω in $\mathbb{R}^N$ is called *solid* if $\mathbb{R}^N \setminus \overline{\Omega}$ is connected and $\Omega = \overline{\Omega}^{\circ}$. The *exterior inverse problem of potential theory*, which dates back to work of Novikov in the 1930's, is as follows:

If Ω_1 and Ω_2 are solid domains in $\mathbb{R}^N$ such that $\lambda|_{\Omega_1}$ and $\lambda|_{\Omega_2}$ produce the same potential in the complement of $\Omega_1 \cup \Omega_2$, must Ω_1 and Ω_2 coincide?

Novikov himself proved that the answer is yes if both domains are assumed to be convex or, more generally, starshaped with respect to a common point. Although it is nowadays suspected that the answer to the general question may be negative, it has long been conjectured that convexity of *one* of the domains should be enough for a positive answer. That this is, in fact, the case was recently proved in [6]. The proof, which relies on partial balayage and the 'moving plane' method, will be outlined in this section. We begin by formulating the main result.

Theorem 14. *If Ω_1, Ω_2 are bounded solid domains in $\mathbb{R}^N$, where Ω_2 is convex, and if*

$$U^{\chi_{\Omega_1}} = U^{\chi_{\Omega_2}} \text{ in } \mathbb{R}^N \setminus (\Omega_1 \cup \Omega_2),$$

then $\Omega_1 = \Omega_2$.

We will use partial balayage in the special case where $\nu = \lambda$, so $v(x)$ can be taken to be a suitable multiple of $-|x|^2$, and $\Omega = \mathbb{R}^N$. This creates a problem when $N = 2$, since $\mathbb{R}^2$ is not Greenian, but our measure μ will always have compact support and the construction and basic properties of partial balayage can be extended to cover this case (see [8], for instance). For ease of exposition we will assume below that $N \geq 3$.

In order to describe the moving plane method, we need some notation. Points in $\mathbb{R}^N$ will be denoted by $(x', x_N) \in \mathbb{R}^{N-1} \times \mathbb{R}$, and we will write

$$W_+ = \{(x', x_N) : x_N > 0\},\ W_- = \{(x', x_N) : x_N < 0\},\ H = \{(x', x_N) : x_N = 0\}.$$

Given a positive measure μ we denote by $\Omega(\mu)$ the largest open set U for which $(\lambda - B_\mu^\lambda)(U) = 0$. (Thus $\Omega(\mu)$ will contain the set ω in Theorem 7.) By Corollary 11 we know that

$$B_\mu^\lambda = \lambda|_{\Omega(\mu)} + \mu|_{\Omega(\mu)^c}, \tag{8}$$

where $\mu|_{\Omega(\mu)^c} \leq \lambda$.

Lemma 15. *Let μ be a measure with compact support contained in $W_- \cup H$ and let $A = \{x' : (x', 0) \in \Omega(\mu) \cap H\}$. Then there is a continuous function $g : A \to (0, \infty)$, continuously vanishing on ∂A, such that*

$$\Omega(\mu) \cap W_+ = \{(x', x_N) : x' \in A \text{ and } 0 < x_N < g(x')\}. \tag{9}$$

Since it is not immediately obvious why this should be true we will supply a proof below, based on [8]. However, the plausibility of the lemma is suggested by the intuition, using the connection with Hele-Shaw flow, that the measure B_μ^λ will be as 'close' to μ as possible subject to the constraint that it is majorized by λ. The name 'moving plane' comes from the fact that we can apply the above lemma to all hyperplanes disjoint from the convex hull K of the support of μ to see that $\Omega(\mu)$ cannot have any 'holes' outside K (see the following figure).

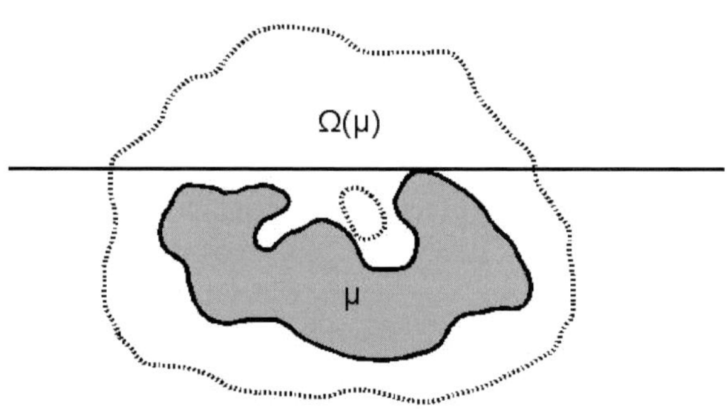

Proof. Let $u = U\mu - UB_\mu^\lambda$. Thus $u \geq 0$. We may assume, by means of a limiting argument, that supp$\mu \subset W_-$, and so u is continuously differentiable on an open set containing $W_+ \cup H$. Let $\overline{u}(x) = u(x', -x_N)$. We note that $U\mu - \overline{u} - v$ is subharmonic on W_+, and $UB_\mu^\lambda - v$ is subharmonic on all of $\mathbb{R}^N$. Since

$$U\mu - \overline{u} - v = U\mu - u - v = UB_\mu^\lambda - v \quad \text{on } H,$$

the function

$$w = \begin{cases} \max\{U\mu - \overline{u} - v, UB_\mu^\lambda - v\} & \text{on } W_+ \\ UB_\mu^\lambda - v & \text{on } W_- \cup H \end{cases}$$

is a subharmonic minorant of $U\mu - v$. Thus $w \le UB_\mu^\lambda - v$ by the definition of UB_μ^λ, whence $U\mu - \overline{u} \le UB_\mu^\lambda$ on W_+ and so $u \le \overline{u}$ there. It follows that $\partial u/\partial x_N \le 0$ on H.

Let $\Omega_+ = \Omega(\mu) \cap W_+$. Since $u = 0$ on ω^c (where ω is given by (3)), and so on $\Omega(\mu)^c$, and since every point of $\partial\Omega(\mu)$ is the limit of some sequence of points of Lebesgue density of $\Omega(\mu)^c$, we see that $|\nabla u| = 0$ on $\partial\Omega_+ \cap W_+$. We note from (8) that Δu is constant in Ω_+, so the function $\partial u/\partial x_N$ is harmonic there, and hence $\partial u/\partial x_N \le 0$ on Ω_+, by the maximum principle. Further, since u is nonconstant in each component of Ω_+, and $u = 0$ on $W_+ \setminus \Omega_+$, we actually have $\partial u/\partial x_N < 0$ on Ω_+. We now define

$$g(x') = \sup\{t > 0 : (x', t) \in \Omega_+\} \quad (x' \in A).$$

Clearly $\Omega(\mu) \cap W_+$ lies under the graph of g. Conversely, if (x', x_N) lies under the graph of g and $x_N > 0$, then $u(x', x_N) > 0$ and so $(x', x_N) \in \omega(\mu) \subseteq \Omega(\mu)$. Thus (9) holds.

It remains to check that g is continuous and vanishes at ∂A. In fact, since $\Omega(\mu)$ is open and

$$\{x' : g(x') > c\} = \left\{x' : (x', c) \in \Omega(\mu)\right\} \quad (c > 0),$$

it is clear that g is lower semicontinuous. On the other hand, if we apply the result of the previous paragraph with hyperplanes of varying orientation, we see that each point of $\partial\Omega_+ \cap W_+$ is the vertex of a vertical cone lying in $\Omega(\mu)^c$, and so g is also upper semicontinuous. In fact g vanishes continuously at ∂A, since we can apply the preceding reasoning with H replaced by a slightly lower hyperplane. □

We now proceed to the proof of Theorem 14. Let η be the Riesz measure associated with $\min\{U^{\chi_{\Omega_1}}, U^{\chi_{\Omega_2}}\}$, and $\Omega_0 = \Omega(\eta)$; thus $a_N\eta = -\Delta \min\{U^{\chi_{\Omega_1}}, U^{\chi_{\Omega_2}}\}$, where a_N is a dimensional constant. The first step is to prove a simple lemma about Ω_0.

Lemma 16. *With the above notation, $B_\eta^\lambda = \lambda|_{\Omega_0}$ and $\Omega_1 \cap \Omega_2 \subset \Omega_0$.*

Proof. An application of Theorem 9 shows that the set of points

$$\{x : U^{\chi_{\Omega_1}}(x) = U^{\chi_{\Omega_2}}(x)\} \cap [(\Omega_1 \setminus \Omega_2) \cup (\Omega_2 \setminus \Omega_1)]$$

has Lebesgue measure zero. Next, we can apply Kato's inequality, also known as Grishin's lemma (see [4, 5]), to $U^{\chi_{\Omega_i}} - U\eta$ $(i = 1,2)$ to see that $\eta \ge \lambda|_{\Omega_1 \cap \Omega_2}$. Hence η is of the form $\eta|_D + \eta_1$, where D is an open set, $\eta \ge \lambda|_D$ and η_1 is singular with respect to Lebesgue measure. Since $B_\eta^\lambda = \lambda|_{\Omega_0} + \eta|_{\Omega_0^c}$ we see that $D \subset \Omega_0$. It follows that $\eta|_{\Omega_0^c} = \eta_1|_{\Omega_0^c} = 0$, since $B_\eta^\lambda \le \lambda$. □

It is easy to see that $\lambda(\Omega_1) = \lambda(\Omega_2) = \lambda(\Omega_0)$ and, by construction, we have $U^{\chi_{\Omega_0}} \le U^{\chi_{\Omega_i}}$ $(i = 1,2)$. Further, $\partial\Omega_i \subset \overline{\Omega_0}$ $(i = 1,2)$, because if there were an open ball B in $\mathbb{R}^N \setminus \overline{\Omega_0}$ with center in $\partial\Omega_i$, then we would arrive at the impossible situation that $U^{\chi_{\Omega_i}} - U^{\chi_{\Omega_0}}$ is non-negative and superharmonic on B and vanishes on $B \setminus \Omega_i$ but not on all of B.

Let $\mu = \lambda|_{\Omega_2} + \lambda|_E$, where $E = \Omega_2 \setminus \Omega_0$. Also let $D = \Omega_0 \cup E = \Omega_0 \cup \Omega_2$. The next step in the proof is to show that $B_\mu^\lambda = \lambda|_D$ and hence $D \subset \Omega(\mu)$. From this we will be able to obtain a contradiction if $\Omega_1 \ne \Omega_2$ by an application of the moving plane method. What follows is a sketch of the remaining proof; full details may be found in [6].

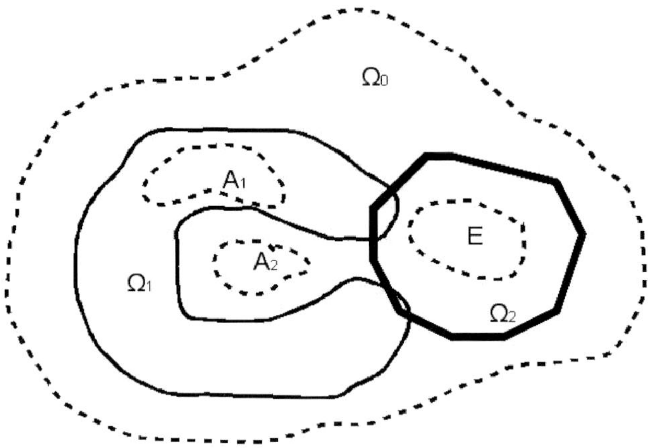

On the set $A_2 = \mathbb{R}^N \setminus \overline{(\Omega_0 \cup \Omega_1 \cup \Omega_2)} = \mathbb{R}^N \setminus \overline{(D \cup \Omega_1)}$ we have by assumption $U\eta = U^{\chi_{\Omega_0}} = U^{\chi_{\Omega_1}} = U^{\chi_{\Omega_2}}$, so $U\mu = U^{\chi_D}$ here. But on the set $A_1 = \Omega_1 \setminus \overline{\Omega}_0 = \Omega_1 \setminus \overline{D}$ it is also clear that $U\eta = U^{\chi_{\Omega_2}}$, since $-\Delta U^{\chi_{\Omega_1}} = a_N \lambda$ on A_1. Hence $U\mu = U^{\chi_D}$ on $\mathbb{R}^N \setminus (\partial\Omega_1 \cup \overline{D})$. It follows by continuity that $U\mu = U^{\chi_D}$ on $\mathbb{R}^N \setminus \overline{D}^{\circ}$, and we also have $U\mu \geq U^{\chi_D}$ everywhere. Hence $\Omega(\mu) \subset \overline{D}^{\circ}$. But $\lambda(\Omega(\mu)) = \lambda(D)$, so $D \subset \Omega(\mu)$, and these sets differ by at most a Lebesgue null set. Thus $B_\mu^\lambda = \lambda|_D$. Since $\lambda(\partial\Omega_2) = 0$ and the set $\Omega(\mu) \setminus (E \cup \partial\Omega_2)$ is open, the latter is a subset of Ω_0 by construction (it differs from $D \setminus E = \Omega_0$ by at most a Lebesgue null set). Therefore $D \subset \Omega(\mu) \subset D \cup \partial\Omega_2$.

According to the moving plane method it is not possible for $\Omega(\mu)$ to have any holes outside the set $\overline{\Omega}_2$ (like A_1 or A_2 in the figure). Hence $\Omega_1 \subset \Omega(\mu)$, since $\partial\Omega_1 \subset \overline{\Omega}_0$, so $\Omega_1 \subset \Omega_0$, and using the fact that Ω_1 and Ω_0 have equal Lebesgue measure, and are both saturated with respect to Lebesgue measure, it follows that they are in fact identical.

But this implies that $U^{\chi_{\Omega_1}} \leq U^{\chi_{\Omega_2}}$ everywhere. Since $U^{\chi_{\Omega_2}} - U^{\chi_{\Omega_1}}$ is superharmonic on $\mathbb{R}^N \setminus \overline{\Omega}_1$ (which is connected) and attains the value 0 it follows that $U^{\chi_{\Omega_1}} = U^{\chi_{\Omega_2}}$ on this set. So $\Omega_2 \subset \Omega_1$, and since they are both solid with equal Lebesgue measure it follows that they are also identical.

Acknowledgments

This research was supported by Science Foundation Ireland under Grant 06/RFP/MAT057, and is also part of the programme of the ESF Network "Harmonic and Complex Analysis and Applications" (HCAA).

References

[1] D. H. Armitage and S. J. Gardiner, *Classical Potential Theory*. Springer, London, 2001.

[2] J. Bliedtner and W. Hansen, *Potential Theory. An Analytic and Probabilistic Approach to Balayage*. Springer, Berlin, 1986.

[3] J. Bourgain, On the Hausdorff dimension of harmonic measure in higher dimension. *Invent. Math.* **87** (1987), 477-483.

[4] H. Brezis and A.C. Ponce, Kato's inequality when Δu is a measure. *C.R. Math. Acad. Sci. Paris,* **338** (2004), 599-604.

[5] B. Fuglede, Some properties of the Riesz charge associated with a δ-subharmonic function. *Potential Analysis* **1** (1992), 355-371.

[6] S. J. Gardiner and T. Sjödin, Convexity and the exterior inverse problem of potential theory. *Proc. Amer. Math. Soc.* **136** (2008), 1699-1703.

[7] D. Gilbarg and N. S. Trudinger, *Elliptic Partial Differential Equations of Second Order.* Springer, Berlin, 2001.

[8] B. Gustafsson and M. Sakai, Properties of some balayage operators, with applications to quadrature domains and moving boundary problems. *Nonlinear Anal.* **22** (1994), 1221-1245.

[9] W. Hansen and H. Hueber, Singularity of harmonic measure for sub-Laplacians. *Bull. Sci. Math.* **112** (1988), 53-64.

[10] B. K. Øksendal, Null sets for measures orthogonal to $R(X)$. *Amer. J. Math.* **94** (1972), 331-342.

[11] M. Sakai, *Quadrature Domains* (Lecture Notes in Math. **934**). Springer, Berlin, 1982.

SCHOOL OF MATHEMATICAL SCIENCES, UNIVERSITY COLLEGE DUBLIN, BELFIELD, DUBLIN 4, IRELAND.
E-mail address: stephen.gardiner@ucd.ie

DEPARTMENT OF MATHEMATICS, LINKÖPING UNIVERSITY, SE-581 83 LINKÖPING, SWEDEN.
E-mail address: tosjo@mai.liu.se

Potential Theory and Stochastics in Albac, 125–137
Aurel Cornea Memorial Volume

THE INFLUENCE OF THE DISTANCE FUNCTION IN SOME SINGULAR ELLIPTIC PROBLEMS

MARIUS GHERGU AND VICENŢIU RĂDULESCU

This paper is dedicated to the memory of Aurel Cornea

ABSTRACT. We present some existence and nonexistence results for classical solutions to singular elliptic problems of the form

$$-\Delta u \pm \delta(x)^{-\alpha} u^{-\beta} = \lambda f(x,u) \quad \text{in } \Omega,$$

subject to homogeneous Dirichlet boundary conditions. Here $\Omega \subset \mathbb{R}^N (N \geq 1)$ is a smooth and bounded domain, $\delta(x) = \text{dist}(x, \partial\Omega)$, α, β and λ are positive real numbers, while f has either a linear or a sublinear growth with respect to the second variable.

1. INTRODUCTION AND THE MAIN RESULTS

Let $\Omega \subset \mathbb{R}^N$ $(N \geq 1)$ be a bounded domain with a smooth boundary. We are concerned in this paper with singular elliptic problems of the following type

$$\begin{cases} -\Delta u \pm \delta(x)^{-\alpha} u^{-\beta} = \lambda f(x,u) & \text{in } \Omega, \\ u > 0 & \text{in } \Omega, \\ u = 0 & \text{on } \partial\Omega, \end{cases} \qquad (P_\lambda)^\pm$$

where $\delta(x) = \text{dist}(x, \partial\Omega)$, $\alpha, \beta, \lambda > 0$. Throughout this paper we suppose that $f : \overline{\Omega} \times [0,\infty) \to [0,\infty)$ is a Hölder continuous function which is nondecreasing with respect to the second variable and such that f is positive on $\overline{\Omega} \times (0,\infty)$. The analysis we develop in this paper concerns the cases where f is either linear or f is sublinear with respect to the second variable. This last case means that f fulfills the hypotheses

$(f1)$ the mapping $(0,\infty) \ni t \longmapsto \dfrac{f(x,t)}{t}$ is nonincreasing for all $x \in \overline{\Omega}$;

$(f2)$ $\displaystyle\lim_{t\to 0^+} \frac{f(x,t)}{t} = +\infty$ and $\displaystyle\lim_{t\to +\infty} \frac{f(x,t)}{t} = 0$, uniformly for $x \in \overline{\Omega}$.

Such singular boundary value problems arise in the context of chemical heterogeneous catalysts and chemical catalyst kinetics (such as the Langmuir–Hinshelwool model), in the theory of heat conduction in electrically conducting materials, singular minimal surfaces, as well as in the study of non-Newtonian fluids or boundary layer phenomena for viscous fluids (we refer for more details to [5, 6, 8, 9, 11, 12] and the

---

2000 *Mathematics Subject Classification.* Primary 35J60. Secondary 35B25, 35B50, 58J55.

Key words and phrases. Nonlinear elliptic equation, singular nonlinearity, bifurcation problem, asymptotic behavior of solutions.

more recent papers [7, 14, 15, 17, 20, 21, 23, 24]). We also point out that, due to the meaning of the unknowns (concentrations, populations, etc.), only the positive solutions are relevant in most cases. For instance, problems of this type characterize some reaction-diffusion processes where $u \geq 0$ is viewed as the density of a reactant and the region where $u = 0$ is called the *dead core*, where no reaction takes place (see Aris [3] for the study of a single, irreversible steady-state reaction). Nonlinear singular elliptic equations are also encountered in glacial advance, in transport of coal slurries down conveyor belts and in several other geophysical and industrial contents (see Callegari and Nachman [6] for the case of the incompressible flow of a uniform stream past a semi-infinite flat plate at zero incidence).

To the best of our knowledge, there does not exist a qualitative theory for the study of singular boundary value problems with nonlinearities in the Kato class $K_N^{\text{loc}}(\mathbb{R}^N)$. This theory was introduced by Aizenman and Simon in [2] to describe wide classes of functions arising in Potential Theory. We refer to the recent paper [22] for existence and bifurcation results on Dirichlet boundary value problems with indefinite nonlinearities.

In this paper we aim to study the influence of the distance function $\delta(x)$ in such singular elliptic problems. We first establish the following result related to problem $(P_\lambda)^+$.

Theorem 1.1. *Assume that f satisfies $(f1), (f2)$.*

(i) *If $\alpha + \beta \leq 1$ then $(P_\lambda)^+$ has no classical solutions;*

(ii) *If $\alpha + \beta > 1$, then there exists $\lambda^* > 0$ such that $(P_\lambda)^+$ has at least one classical solution for all $\lambda > \lambda^*$ and no solution exists if $0 < \lambda < \lambda^*$.*

In contrast to the results in Theorem 1.1, the study of $(P_\lambda)^-$ is quite different. The one dimensional case, namely the problem

$$\begin{cases} -H''(t) = \delta(t)^{-\alpha} H(t)^{-\beta}, H > 0 & \text{in } (0,1), \\ H(0) = H(1) = 0, \end{cases} \tag{1.1}$$

was discussed in [25] and [1, Section 2]. It has been shown that (1.1) has solutions if and only if $\alpha < 2$. We first obtain that condition $\alpha < 2$ is also necessary and sufficient to obtain a solution in higher dimension provided f has a sublinear growth. More precisely we have:

Theorem 1.2. *Assume that f satisfies $(f1), (f2)$.*

(i) *If $\alpha \geq 2$ then $(P_\lambda)^-$ has no classical solutions;*

(ii) *If $0 < \alpha < 2$, then for all $\lambda > 0$ problem $(P_\lambda)^-$ has al least one solution.*
Moreover, there exist $0 < \eta < 1$ and $C_1, C_2 > 0$ such that u_λ satisfies

(ii1) *If $\alpha + \beta > 1$, then*

$$C_1 \delta(x)^{\frac{2-\alpha}{1+\beta}} \leq u_\lambda(x) \leq C_2 \delta(x)^{\frac{2-\alpha}{1+\beta}}, \quad \textit{for all } x \in \Omega; \tag{1.2}$$

(ii2) *If $\alpha + \beta = 1$, then*

$$C_1 d(x)(-\ln \delta(x))^{\frac{1}{2-\alpha}} \leq u_\lambda(x) \leq C_2 d(x)(-\ln \delta(x))^{\frac{1}{2-\alpha}}, \tag{1.3}$$

for all $x \in \Omega$ with $\delta(x) < \eta$;

(ii3) *If* $\alpha+\beta<1$, *then*

$$C_1\delta(x)\le u_\lambda(x)\le C_2\delta(x),\quad \textit{for all } x\in\Omega. \tag{1.4}$$

If $\alpha+\beta<1$, we are able to obtain the uniqueness and to provide the regularity of solution to $(P_\lambda)^-$ by means of the associated Green function.

Theorem 1.3. *Assume that f satisfies $(f1),(f2)$ and $\alpha+\beta<1$. Then, for all $\lambda>0$ problem $(P_\lambda)^-$ has a unique solution u_λ which in addition satisfies*

(i) $(0,\infty)\ni\lambda\longmapsto u_\lambda$ *is increasing in* Ω;
(ii) $u_\lambda\in C^2(\Omega)\cap C^{1,1-\alpha-\beta}(\overline{\Omega})$.

We now consider the case where f is asymptotically linear. More precisely, we assume that f satisfies $(f1)$ and

$$(f3)\qquad m:=\lim_{t\to0^+}\frac{f(x,t)}{t}\in(0,\infty).$$

In case $\alpha+\beta<1$ we have the following result.

Theorem 1.4. *Assume that $\alpha+\beta<1$ and f satisfies $(f1),(f3)$. Then, problem $(P_\lambda)^-$ has solutions if and only if $0<\lambda<\lambda_1/m$. Moreover, for all $0<\lambda<\lambda_1/m$, there exists a unique solution u_λ of $(P_\lambda)^-$ such that*

(i) $u_\lambda\in C^2(\Omega)\cap C^{1,1-\alpha-\beta}(\overline{\Omega})$;
(ii) $\lim_{\lambda\nearrow\lambda_1/m}u_\lambda=\infty$ *uniformly on compact subsets of* Ω.

The next sections contains the proofs of the above results.

2. Proof of Theorem 1.1

Several times in this paper we apply the following comparison result (we refer to [16, Lemma 2.1] for a complete proof).

Lemma 2.1. *Let $\Phi:\Omega\times(0,\infty)\to\mathbb{R}$ be a Hölder continuous function such that the mapping $(0,\infty)\ni t\longmapsto\Phi(x,t)/t$ is strictly decreasing for each $x\in\Omega$. Assume that there exist $v,w\in C^2(\Omega)\cap C(\overline{\Omega})$ such that*

(a) $\Delta w+\Phi(x,w)\le0\le\Delta v+\Phi(x,v)$ *in* Ω;
(b) $v,w>0$ *in* Ω *and* $v\le w$ *on* $\partial\Omega$;
(c) $\Delta v\in L^1(\Omega)$ *or* $\Delta w\in L^1(\Omega)$.

Then $v\le w$ *in* Ω.

Proof of Theorem 1.1. (i) Assume that $\alpha+\beta\ge1$ and there exists $\lambda>0$ such that $(P_\lambda)^+$ has a classical solution u and let $C=\max_{\overline{\Omega}}\lambda f(u)>0$. Let also $v\in C^2(\overline{\Omega})$ be the unique solution of

$$\begin{cases}-\Delta v=C & \text{in }\Omega,\\ v>0 & \text{in }\Omega,\\ v=0 & \text{on }\partial\Omega.\end{cases}\tag{2.1}$$

Moreover, there exist $c_1,c_2>0$ such that

$$c_1d(x)\le v\le c_2d(x),\quad\text{for all }x\in\Omega.\tag{2.2}$$

By maximum principle, it follows that $u \le v$ in Ω. Next we consider the perturbed problem

$$\begin{cases} -\Delta u + (\delta(x)+\varepsilon)^{-\alpha}(u+\varepsilon)^{-\beta} = C & \text{in } \Omega, \\ u > 0 & \text{in } \Omega, \\ u = 0 & \text{on } \partial\Omega. \end{cases} \tag{2.3}$$

Then, u and v are, respectively, sub- and super-solution of (2.3). By standard arguments and elliptic regularity (see [18]), there exists $u_\varepsilon \in C^2(\overline{\Omega})$ a solution of (2.3) such that $u \le u_\varepsilon \le v$ in Ω. Integrating in (2.3) we obtain

$$-\int_\Omega \Delta u_\varepsilon dx + \int_\Omega (\delta(x)+\varepsilon)^{-\alpha}(u_\varepsilon+\varepsilon)^{-\beta} dx = C|\Omega|.$$

Hence

$$-\int_{\partial\Omega} \frac{\partial u_\varepsilon}{\partial n} ds + \int_\Omega (\delta(x)+\varepsilon)^{-\alpha}(u_\varepsilon+\varepsilon)^{-\beta} dx \le M, \tag{2.4}$$

where M is a positive constant. Taking into account that $\partial u_\varepsilon/\partial n \le 0$ on $\partial\Omega$ and $v \le u_\varepsilon$ in Ω we have $\int_\Omega (\delta(x)+\varepsilon)^{-\alpha}(u_\varepsilon+\varepsilon)^{-\beta} dx \le M$. Thus, for any compact subset $\omega \subset\subset \Omega$ we have

$$\int_\omega (\delta(x)+\varepsilon)^{-\alpha}(u_\varepsilon+\varepsilon)^{-\beta} dx \le M.$$

Passing to the limit with $\varepsilon \to 0^+$ we obtain $\int_\omega \delta(x)^{-\alpha} u_\varepsilon^{-\beta} dx \le M$, for all $\omega \subset\subset \Omega$. Therefore

$$\int_\Omega \delta(x)^{-\alpha} u_\varepsilon^{-\beta} dx \le M. \tag{2.5}$$

On the other hand, using (2.2) and the hypothesis $\alpha+\beta \ge 1$, it follows that

$$M \ge \int_\Omega \delta(x)^{-\alpha} u_\varepsilon(x)^{-\beta} dx \ge c \int_\Omega \delta(x)^{-\alpha-\beta} dx = +\infty,$$

which is a contradiction. Hence, problem $(P_\lambda)^+$ has no classical solutions and the proof of Theorem 1.1 is complete.

(ii) We first establish the existence of solutions to $(P_\lambda)^+$ for $\lambda > 0$ large. By virtue of [23, Lemma 2.4] (see also ([24, Theorem 2.2]), the problem

$$\begin{cases} -\Delta U = \lambda f(x,U) & \text{in } \Omega, \\ U > 0 & \text{in } \Omega, \\ U = 0 & \text{on } \partial\Omega, \end{cases} \tag{2.6}$$

has at least one classical solution U_λ, for all $\lambda > 0$ and U_λ is a super-solution of $(P_\lambda)^+$. The main point is to find a sub-solution $\underline{u}_\lambda$ of $(P_\lambda)^+$ such that $\underline{u}_\lambda \le U_\lambda$ in Ω. To this aim, let

$$\Psi : [0,\infty) \to [0,\infty), \quad \Psi(t) = \int_0^t \left[2\left(1 - \frac{1}{1-\alpha-\beta} s^{1-\alpha-\beta}\right)\right]^{-1/2} ds.$$

Since $\alpha+\beta < 1$, Ψ is well defined. Moreover, Ψ is a bijective map. Let $h : [0,\infty) \to [0,\infty)$ be the inverse of Ψ. Then $h \in C^2(0,\infty) \cap C^1[0,\infty)$ and h satisfies

$$\begin{cases} h''(t) = h(t)^{-\alpha-\beta}, h > 0 & \text{in } (0,\infty), \\ h(0) = h'(0) = 0. \end{cases} \tag{2.7}$$

The following technical result provides a suitable sub-solution for our problem.

Lemma 2.2. (see [13]) *There exist two positive constants $c>0$ and $M>0$ such that $\underline{u}_\lambda := Mh(c\varphi_1)$ is a sub-solution of $(P_\lambda)^+$ provided $\lambda>0$ is large enough.*

Using Lemma 2.1, it follows that $\underline{u}_\lambda \le U_\lambda$ in Ω and by standard elliptic arguments (see [18]) we obtain a classical solution u_λ of $(P_\lambda)^+$ such that $\underline{u}_\lambda \le u_\lambda \le U_\lambda$ in Ω.

Next we prove that $(P_\lambda)^+$ has no solutions for $\lambda>0$ small. since f satisfies $(f1)$–$(f2)$, we can find $m>0$ such that

$$f(x,t)-p(d(x))g(t)<mt, \qquad \text{for all } (x,t)\in\Omega\times(0,+\infty). \tag{2.8}$$

Set $\lambda_0=\min\{1,\lambda_1/2m\}$. We claim that problem $(P_\lambda)^+$ has no classical solution for $0<\lambda\le\lambda_0$. Indeed, assume by contradiction that u_0 is a classical solution of $(P_\lambda)^+$ with $\lambda\in(0,\lambda_0]$. Then, according to (2.8), u_0 is a sub-solution of

$$\begin{cases} -\Delta v = \dfrac{\lambda_1}{2}v & \text{in } \Omega,\\ v>0 & \text{in } \Omega,\\ v=0 & \text{on } \partial\Omega. \end{cases} \tag{2.9}$$

By Lemma 2.1 we have $u_0\le U_\lambda$ in Ω which yields $cu_0\le\varphi_1$ in Ω for some positive constant $c>0$. Note that cu_0 is still a sub-solution of (2.9) while φ_1 is a super-solution of (2.9). By standard elliptic arguments, it follows that problem (2.9) has a solution $v\in C^2(\overline{\Omega})$. Multiplying by φ_1 in (2.9) and integrating on Ω we have

$$-\int_\Omega \varphi_1\Delta v\,dx=\frac{\lambda_1}{2}\int_\Omega v\varphi_1\,dx,$$

that is,

$$\lambda_1\int_\Omega v\varphi_1\,dx=-\int_\Omega v\Delta\varphi_1\,dx=\frac{\lambda_1}{2}\int_\Omega v\varphi_1\,dx.$$

The above equality yields $\int_\Omega v\varphi_1\,dx=0$, but this is clearly a contradiction, since v and φ_1 are both positive in Ω. It follows that $(P_\lambda)^+$ has no classical solutions for $0<\lambda\le\lambda_0$.

Next we define

$$A=\left\{\lambda>0;\ \text{problem } (P_\lambda)^+ \text{ has at least one classical solution}\right\}.$$

From the above arguments we deduce that A is nonempty and $\lambda^*:=\inf A$ is positive. We only need to show that if $\lambda\in A$, then $(\lambda,\infty)\subseteq A$ but this follows by the sub and super-solution method. Hence $(\lambda^*,\infty)\subseteq A\subseteq[\lambda^*,\infty)$ and the proof is now complete. □

3. Proof of Theorem 1.2

(i) We proceed similarly as in the proof of Theorem 1.1.

(ii) Remark first that for all $\lambda>0$, the solution U_λ of (2.6) is a sub-solution for $(P_\lambda)^-$. However, we need a suitable sub-solution for $(P_\lambda)^-$ that will provide the asymptotic behavior for u_λ as describet in (1.2)–(1.4).

Lemma 3.1. (see [13]) *Assume* $0 < \alpha < 2$ *and let* H *be the solution of (1.1). Then, for all* $\lambda > 0$ *there exist positive constants* $C, c > 0$ *and* $M, m > 0$ *(depending on* λ*) such that* $\underline{u}_\lambda := mH(c\varphi_1)$, $\overline{u}_\lambda := MH(C\varphi_1)$ *is a sub-solution and a super-solution of* $(P_\lambda)^-$ *such that* $\underline{u}_\lambda \le \overline{u}_\lambda$ *in* Ω..

Hence, for all $\lambda > 0$ there exists an ordered pair of sub- and super-solution of $(P_\lambda)^-$. By standard elliptic arguments, problem $(P_\lambda)^-$ has at least one solution u_λ such that

$$mH(c\varphi_1) \le u_\lambda \le MH(C\varphi_1) \quad \text{in } \Omega. \tag{3.1}$$

(ii1) Remark that

$$H(t) = \left(\frac{(1+\beta)^2}{(2-\alpha)(\alpha+\beta-1)} \right)^{1/(1+\beta)} t^{\frac{2-\alpha}{1+\beta}}, \qquad t > 0,$$

is a solution of (1.1) if $\alpha + \beta > 1$. The conclusion in this case follows now from (3.1).

(ii2) Since H is concave, there exists $H'(0^+) \in (0, \infty]$. Taking $0 < b < 1$ small enough, we can assume that $H' > 0$ on $(0, b]$. Hence H satisfies

$$\begin{cases} H''(t) = -t^{-\alpha} H^{\alpha-1}(t), & \text{for all } 0 < t \le b < 1, \\ H(0) = 0, \\ H, H' > 0 \quad \text{in } (0, b]. \end{cases} \tag{3.2}$$

Since H is concave, it follows that

$$H(t) > tH'(t), \quad \text{for all } 0 < t \le b. \tag{3.3}$$

Relations (3.2) and (3.3) yield

$$-H''(t) < t^{-1}(H'(t))^{\alpha-1}, \quad \text{for all } 0 < t \le b.$$

Hence

$$-H''(t)(H'(t))^{1-\alpha} \le \frac{1}{t}, \quad \text{for all } 0 < t \le b. \tag{3.4}$$

Integrating in (3.4) over $[t, b]$ we get

$$(H')^{2-\alpha}(t) - (H')^{2-\alpha}(b) \le (2-\alpha)(\ln b - \ln t), \quad \text{for all } 0 < t \le b.$$

Hence, there exist $c_1 > 0$ and $\delta_1 \in (0, b)$ such that

$$H'(t) \le c_1(-\ln t)^{\frac{1}{2-\alpha}}, \quad \text{for all } 0 < t \le \delta_1. \tag{3.5}$$

Fix $t \in (0, \delta_1]$. Integrating over $[\varepsilon, t]$, $0 < \varepsilon < t$, in (3.5) we have

$$H(t) - H(\varepsilon) \le c_1 t(-\ln t)^{\frac{1}{2-\alpha}} + \frac{c_1}{2-\alpha} \int_\varepsilon^t (-\ln s)^{\frac{\alpha-1}{2-\alpha}} ds. \tag{3.6}$$

Note that

$$\int_0^t (-\ln s)^{\frac{\alpha-1}{2-\alpha}} ds < +\infty \qquad \text{and} \quad \lim_{t\to 0^+} \frac{\int_0^t (-\ln s)^{\frac{\alpha-1}{2-\alpha}} ds}{t(-\ln t)^{\frac{1}{2-\alpha}}} = 0. \tag{3.7}$$

Therefore, taking $\varepsilon \to 0^+$ in (3.6) we deduce that there exist $c_2 > 0$ and $\delta_2 \in (0, \delta_1)$ such that

$$H(t) \le c_2 t(-\ln t)^{\frac{1}{2-\alpha}}, \quad \text{for all } 0 < t \le \delta_2. \tag{3.8}$$

From (3.2) and (3.8) we obtain

$$-H''(t) \ge c_2^{\alpha-1} t^{-1}(-\ln t)^{\frac{\alpha-1}{2-\alpha}}, \quad \text{for all } 0 < t \le \delta_2.$$

Integrating over $[t,\delta_2]$ in the above inequality we get

$$H'(t) \geq (2-\alpha)c_2^{\alpha-1}\left[(-\ln t)^{\frac{1}{2-\alpha}} - (-\ln\delta_2)^{\frac{1}{2-\alpha}}\right], \quad \text{for all } 0<t\leq\delta_2.$$

Therefore, there exist $c_3>0$ and $\delta_3\in(0,\delta_2)$ such that

$$H'(t) \geq c_3(-\ln t)^{\frac{1}{2-\alpha}}, \quad \text{for all } 0<t\leq\delta_3.$$

With the same arguments as in (3.5)-(3.8) we obtain $c_4>0$ and $\delta_4\in(0,\delta_3)$ such that

$$H(t) \geq c_4 t(-\ln t)^{\frac{1}{2-\alpha}}, \quad \text{for all } 0<t\leq\delta_4. \tag{3.9}$$

The conclusion of (ii2) in Theorem 1.2 follows now from (3.8) and (3.9).

(ii3) Using the fact that $H'(0+)\in(0,\infty]$ and the inequality (3.3), we get the existence of $c>0$ such that

$$H(t) > ct, \quad \text{for all } 0<t\leq b.$$

This yields

$$-H''(t) \leq c^{-\beta}t^{-(\alpha+\beta)}, \quad \text{for all } 0<t\leq b.$$

Since $\alpha+\beta<1$, it follows that $H'(0+)<+\infty$, that is, $H\in C^1[0,b]$. Thus, there exists $c_1,c_2>0$ such that

$$c_1 t \leq H(t) \leq c_2 t, \quad \text{for all } 0<t\leq b. \tag{3.10}$$

The conclusion in Theorem 1.2 (ii3) is now immediately by (3.10). This finishes the proof of Theorem 1.2. □

4. Proof of Theorem 1.3

(i) From Theorem 1.2, for all $\lambda>0$ there exists a solution u_λ of $(P_\lambda)^-$ such that (1.4) holds. Since $\alpha+\beta<1$ this implies

$$0\leq -\Delta u_\lambda = \delta(x)^{-\alpha}u_\lambda^{-\alpha+\beta} + \lambda f(x,u_\lambda) \leq C\delta(x)^{-\alpha-\beta}\in L^1(\Omega).$$

Therefore, by Lemma 2.1 we derive that u_λ is the unique solution of $(P_\lambda)^-$.

(ii) Let $0<\lambda_1<\lambda_2$ and u_{λ_1}, u_{λ_2} be the corresponding solutions to $(P_{\lambda_1})^-$ and $(P_{\lambda_2})^-$ respectively. From the above arguments we have that $\Delta u_{\lambda_i}\in L^1(\Omega)$, $i=1,2$. We only have to apply Lemma 2.1 for the mapping $\Phi(x,t)=\delta(x)^{-\alpha}t^{-\beta}+\lambda_1 f(x,t)$, $(x,t)\in\Omega\times(0,\infty)$ in order to deduce $u_{\lambda_1}\leq u_{\lambda_2}$ in Ω. By maximum principle in follows that $u_{\lambda_1}<u_{\lambda_2}$ in Ω.

(iii) For the proof of regularity we use a similar method to that developed by Gui and Lin [19]. Let $\mathcal{G}$ be the Green function associated with the Laplace operator in Ω. We shall use the following estimates, which are due to Widman [26].

Lemma 4.1. *There exists a positive constant $c>0$ such that for all $x,y\in\Omega$, $x\neq y$ we have*

(i) $|\mathcal{G}_x(x,y)| \leq c\,\dfrac{\min\{|x-y|,d(y)\}}{|x-y|^N}$;

(ii) $|\mathcal{G}_{xx}(x,y)| \leq c\,\dfrac{\min\{|x-y|,d(y)\}}{|x-y|^{N+1}}$.

Then

$$u_\lambda(x) = -\int_\Omega \mathcal{G}(x,y)\Phi(y)\,dy \quad \text{for all } x \in \Omega,$$

where $\Phi(y) = \delta(y)^{-\alpha} u_\lambda(y)^{-\beta} + \lambda f(y, u_\lambda(y))$. Hence

$$\nabla u_\lambda(x) = -\int_\Omega \mathcal{G}_x(x,y)\Phi(y)\,dy \quad \text{for all } x \in \Omega. \tag{4.1}$$

A very useful tool in our approach is the following technical result.

Lemma 4.2. *There exists* $c_0 = c_0(\Omega) > 0$ *and* $\delta_0 = \delta_0(\Omega) > 0$ *such that for all* $x_1, x_2 \in \Omega$, $0 < |x_1 - x_2| < \delta_0$ *there exists a* C^1 *path* $\xi : [0,1] \to \Omega$ *with the properties*

(i) $\xi(0) = x_1$ *and* $\xi(1) = x_2$;
(ii) $|\xi'(t)| \le c_0 |x_1 - x_2|$, *for all* $0 \le t \le 1$.

Fix $x_1, x_2 \in \Omega$, $0 < |x_1 - x_2| < \delta_0$ and let $\xi : [0,1] \to \Omega$ be the corresponding path in Lemma 4.2.

By (4.1) we have

$$\begin{aligned}|\nabla u_\lambda(x_1) - \nabla u_\lambda(x_2)| &\le \int_\Omega |\mathcal{G}_x(x_1,y) - \mathcal{G}_x(x_2,y)|\Phi(y)\,dy \\ &\le \underbrace{\int_{B_r(x_1)} |\mathcal{G}_x(x_1,y) - \mathcal{G}_x(x_2,y)|\Phi(y)\,dy}_{I} \\ &\quad + \underbrace{\int_{\Omega\setminus B_r(x_1)} |\mathcal{G}_x(x_1,y) - \mathcal{G}_x(x_2,y)|\Phi(y)\,dy}_{II},\end{aligned} \tag{4.2}$$

where $r = (c_0 + 1)|x_1 - x_2|$ and c_0 is the constant appearing in Lemma 4.2.

Before evaluating I and II, let us remark that there exists $c_1 > 0$ such that $\Phi(y) \le c_1 \delta(y)^{-\alpha-\beta}$ for all $y \in \Omega$. Then

$$\begin{aligned} I &\le c_1 \int_{B_r(x_1)} |\mathcal{G}_x(x_1,y) - \mathcal{G}_x(x_2,y)|\delta(y)^{-\alpha-\beta}\,dy \\ &\le c_1 \int_{B_r(x_1)} |\mathcal{G}_x(x_1,y)|\delta(y)^{-\alpha-\beta}\,dy + c_1 \int_{B_R(x_2)} |\mathcal{G}_x(x_2,y)|\delta(y)^{-\alpha-\beta}\,dy,\end{aligned}$$

where $R = r + |x_1 - x_2|$. Let $y \in B_r(x_1)$.
If $d(y) \ge |x_1 - y|$ then, by Lemma 4.1 (i), we have

$$|\mathcal{G}_x(x_1,y)|\delta(y)^{-\alpha-\beta} \le c|x_1 - y|^{-N+1}\delta(y)^{-\alpha-\beta} \le c|x_1 - y|^{-N+1-\alpha-\beta}.$$

If $\delta(y) < |x_1 - y|$ then, by Lemma 4.1 (i), we obtain

$$|\mathcal{G}_x(x_1,y)|\delta(y)^{-\alpha-\beta} \le c|x_1 - y|^{-N}\delta(y)^{1-\alpha-\beta} \le c|x_1 - y|^{-N+1-\alpha-\beta}.$$

Therefore, for all $y \in B_r(x_1)$ we have

$$|\mathcal{G}_x(x_1,y)|\delta^{-lpha-\beta}(y) \le c|x_1 - y|^{-N+1-\alpha-\beta},$$

and similarly

$$|\mathcal{G}_x(x_2,y)|\delta(y)^{-\alpha-\beta} \le c|x_2 - y|^{-N+1-\alpha-\beta} \quad \text{for all } y \in B_R(x_2).$$

Hence,

$$
\begin{aligned}
I &\le c_2\int_{B_r(x_1)}|x_1-y|^{-N+1-\alpha-\beta}dy+c_2\int_{B_R(x_2)}|x_2-y|^{-N+1-\alpha-\beta}dy\\
&\le c_3\int_0^R t^{-\alpha-\beta}dt\le c_4|x_1-x_2|^{1-\alpha-\beta}.
\end{aligned}
\tag{4.3}
$$

To evaluate II, we first apply the mean value theorem. We have

$$
\begin{aligned}
II &\le c_1\int_{\Omega\setminus B_r(x_1)}|\mathcal{G}_x(\xi(0),y)-\mathcal{G}_x(\xi(1),y)|\delta(y)^{-\alpha-\beta}dy\\
&\le c_1\int_{\Omega\setminus B_r(x_1)}\int_0^1|\mathcal{G}_{xx}(\xi(t),y)||\xi'(t)|\delta(y)^{-\alpha-\beta}dt\,dy\\
&\le c_5|x_1-x_2|\int_{\Omega\setminus B_r(x_1)}\int_0^1|\mathcal{G}_{xx}(\xi(t),y)|\delta(y)^{-\alpha-\beta}dt\,dy.
\end{aligned}
$$

As shown earlier, by Lemma 4.1 (ii) we obtain

$$
|\mathcal{G}_x(\xi(t),y)|\delta(y)^{-\alpha-\beta}\le c|\xi(t)-y|^{-N-\alpha-\beta}\qquad \text{for all } y\in B_r(x_1) \text{ and } 0\le t\le 1.
$$

Let $c_6=1/(1+c_0)$, where c_0 is the constant from Lemma 4.2 that depends only on Ω and not on x_1,x_2. Then for all $y\in\Omega\setminus B_r(x_1)$,

$$
\begin{aligned}
|\xi(t)-y| &\ge |x_1-y|-|\xi(t)-x_1|=|x_1-y|-|\xi(t)-\xi(0)|\\
&\ge |x_1-y|-t|\xi'(c_t)|\ge|x_1-y|-c_0|x_1-x_2|\\
&\ge c_6|x_1-y|.
\end{aligned}
$$

Combining the last two estimates we obtain

$$
|\mathcal{G}_x(\xi(t),y)|d^{-\alpha-\beta}(y)\le c_7|x_1-y|^{-N-\alpha-\beta}\qquad \text{for all } y\in B_r(x_1) \text{ and } 0\le t\le 1.
$$

Hence, we may write

$$
\begin{aligned}
II &\le c_7|x_1-x_2|\int_{\Omega\setminus B_r(x_1)}|x_1-y|^{-N-\alpha-\beta}dy\\
&\le c_7|x_1-x_2|\int_r^\infty t^{-1-\alpha-\beta}dt\le c_8|x_1-x_2|r^{-\alpha-\beta}\\
&\le c_9|x_1-x_2|^{1-\alpha-\beta}.
\end{aligned}
\tag{4.4}
$$

The conclusion follows now from (4.2), (4.3), and (4.4). The proof of Theorem 1.3 is now complete. □

5. Proof of Theorem 1.4

Let $\lambda^*=\lambda_1/m$. We first prove that $(P_\lambda)^-$ has no classical solutions for $\lambda\ge\lambda^*$. Indeed, assume that there exists $\lambda\ge\lambda^*$ such that $(P_\lambda)^-$ has a solution u and let $c=\max_{\overline{\Omega}}\delta(x)^{-\alpha}u(x)^{-\beta}$. Since the mapping $t\longmapsto f(x,t)/t$ is nonincreasing, it follows that $\lambda f(x,u)\ge\lambda_1 u$ in Ω. Then, u satisfies $-\Delta u\ge c+\lambda_1 u$ in Ω. Multiplying by φ_1 in the last inequality and integrating by parts, we have

$$
\lambda_1\int_\Omega u\varphi_1\,dx\ge\int_\Omega(c+\lambda_1 u)\varphi_1\,dx,
$$

which is clearly a contradiction.

Let $0<\lambda<\lambda^*$. Since $\alpha+\beta<1$, by [7, Theorem 2] there exists $v\in C^2(\Omega)\cap C(\overline{\Omega})$ such that

$$\begin{cases} -\Delta v = v^{-\alpha-\beta}+\lambda f(v) & \text{in } \Omega,\\ v>0 & \text{in } \Omega,\\ v=0 & \text{on } \partial\Omega, \end{cases} \tag{5.1}$$

and $c_1\delta(x)\le v\le c_2\delta(x)$ in Ω, for some $c_1,c_2>0$. Therefore, v satisfies

$$-\Delta v\ge c\delta(x)^{-\alpha}v^{-\beta}+\lambda f(v)\quad \text{in } \Omega, \tag{5.2}$$

where $c>0$. Let now $M>1$ large such that $M^{1+\beta}c>1$. Since $Mf(v)\ge f(Mv)$ in Ω, from (5.2) we deduce that $\overline{u}_\lambda:=Mv$ satisfies

$$-\Delta\overline{u}_\lambda\ge\delta(x)^{-\alpha}\overline{u}_\lambda^{-\beta}+\lambda f(\overline{u}_\lambda)\quad \text{in } \Omega.$$

Hence $\overline{u}_\lambda$ is a super-solution of $(P_\lambda)^-$ and obviously $\overline{u}_\lambda\ge\underline{u}_\lambda$ in Ω. Thus, $(P_\lambda)^-$ has at least one solution u_λ. Furthermore, from the last inequality we also have $u_\lambda\ge c\delta(x)$ in Ω, for some $c>0$. This implies that $\Delta u_\lambda\in L^1(\Omega)$ and by Lemma 2.1 we obtain the uniqueness. The regularity in (i) follows exactly in the same way as in Theorem 1.3. We now prove (ii).

Remark that $\{u_\lambda\}_{0<\lambda<\lambda^*}$ is a sequence of positive increasing super-harmonic functions. By [4, Theorem 3.7.3] it follows that

$$u^*(x):=\lim_{\lambda\nearrow\lambda^*}u_\lambda(x)\quad x\in\Omega,$$

is either identically ∞ or a superharmonic function. Assume by contradiction that $u^*\not\equiv\infty$, so that u^* is superharmonic. Hence, $u^*\in L^1_{\rm loc}(\Omega)$ (see, e.g., Theorem 3.1.3 in [4]).

We claim that $\{u_\lambda\}_{0<\lambda<\lambda^*}$ is bounded in $L^2(\Omega)$. We argue by contradiction. Thus, passing eventually at a subsequence we have $u_\lambda=k(\lambda)w_\lambda$, where

$$k(\lambda)=\|u_\lambda\|_2\to\infty \text{ as } \lambda\nearrow\lambda^* \text{ and } w_\lambda\in L^2(\Omega),\ \|w_\lambda\|_2=1. \tag{5.3}$$

Notice that $f(x,t)\le at+b$ for all $(x,t)\in\overline{\Omega}\times(0,\infty)$, where $a,b>0$. This implies

$$\frac{1}{k(\lambda)}(\delta(x)^{-\alpha}u_\lambda^{-\beta}+\lambda f(x,u_\lambda))\to 0\quad \text{in } L^1_{\rm loc}(\Omega) \text{ as } \lambda\nearrow\lambda^*.$$

That is,

$$-\Delta w_\lambda\to 0\quad \text{in } L^1_{\rm loc}(\Omega) \text{ as } \lambda\nearrow\lambda^*. \tag{5.4}$$

By Green's first identity, for all $\phi\in C_0^\infty(\Omega)$ we have

$$\int_\Omega\nabla w_\lambda\cdot\nabla\phi\,dx=-\int_\Omega\phi\Delta w_\lambda\,dx=-\int_{{\rm Supp}\,\phi}\phi\Delta w_\lambda\,dx. \tag{5.5}$$

Using (5.4) we obtain

$$\begin{aligned}\left|\int_{{\rm Supp}\,\phi}\phi\Delta w_\lambda\,dx\right| &\le\int_{{\rm Supp}\,\phi}|\phi||\Delta w_\lambda|\,dx\\ &\le\|\phi\|_\infty\int_{{\rm Supp}\,\phi}|\Delta w_\lambda|\,dx\to 0\quad \text{as } \lambda\nearrow\lambda^*.\end{aligned} \tag{5.6}$$

Combining (5.5) and (5.6) we derive that for all $\phi\in C_0^\infty(\Omega)$ we have

$$\int_\Omega\nabla w_\lambda\cdot\nabla\phi\,dx\to 0\quad \text{as } \lambda\nearrow\lambda^*. \tag{5.7}$$

By definition, the sequence $(w_\lambda)_{0<\lambda<\lambda^*}$ is bounded in $L^2(\Omega)$. We claim that $(w_\lambda)_{0<\lambda<\lambda^*}$ is bounded in $H_0^1(\Omega)$. Indeed, using Sobolev-Hardy and Hölder inequality, we have

$$\begin{aligned}\int_\Omega |\nabla w_\lambda|^2 dx &= -\int_\Omega w_\lambda \Delta w_\lambda dx\\ &= \frac{1}{k(\lambda)}\int_\Omega w_\lambda(\delta(x)^{-\alpha}u_\lambda^{-\beta} + \lambda f(x,u_\lambda))dx\\ &= \frac{1}{k(\lambda)^{1+\beta}}\int_\Omega \frac{w_\lambda^{1-\beta}}{\delta(x)^\alpha}dx + a\lambda\int_\Omega w_\lambda^2 dx + \frac{\lambda}{k(\lambda)}\int_\Omega w_\lambda dx\\ &\le \frac{1}{k(\lambda)^{1+\beta}}\left(\int_\Omega |\nabla w_\lambda|^2 dx\right)^{(1-\beta)/2} + a\lambda^* + \frac{\lambda^*}{k(\lambda)}|\Omega|^{1/2}.\end{aligned}$$

From the above estimates it follows that $\{w_\lambda\}_{0<\lambda<\lambda^*}$ is bounded in $H_0^1(\Omega)$, so the claim follows. Thus, there exists $w \in H_0^1(\Omega)$ such that up to a subsequence and as $\lambda \nearrow \lambda^*$ we have

$$\begin{aligned} w_\lambda &\rightharpoonup w \quad \text{weakly in } H_0^1(\Omega),\\ w_\lambda &\to w \quad \text{strongly in } L^2(\Omega).\end{aligned} \tag{5.8}$$

On the one hand, by (5.3) and (5.8), we derive that $\|w\|_2 = 1$. Furthermore, using (5.7) and (5.8), we infer that

$$\int_\Omega \nabla w\cdot\nabla\phi\, dx = 0 \quad \text{for all } \phi\in C_0^\infty(\Omega).$$

Because $w\in H_0^1(\Omega)$, using the previous relation and the definition of $H_0^1(\Omega)$, we find $w = 0$. This contradiction shows that $\{u_\lambda\}_{0<\lambda<\lambda^*}$ is bounded in $L^2(\Omega)$. As noted earlier for $\{w_\lambda\}_{0<\lambda<\lambda^*}$, we derive that $\{u_\lambda\}_{0<\lambda<\lambda^*}$ is bounded in $H_0^1(\Omega)$. Hence, there exists $u^*\in H_0^1(\Omega)$ such that, up to a subsequence and as $\lambda\nearrow\lambda^*$, there holds

$$\begin{aligned} u_\lambda &\rightharpoonup u^* \quad \text{weakly in } H_0^1(\Omega),\\ u_\lambda &\to u^* \quad \text{strongly in } L^2(\Omega),\\ u_\lambda &\to u^* \quad \text{almost everywhere in } \Omega.\end{aligned} \tag{5.9}$$

Now we obtain the desired contradiction in the same manner as in the proof of (i)–(ii). By $(f1)$ we have $f(x,u_\lambda)\ge mu_\lambda$ in Ω. We next multiply by φ_1 in $(P_\lambda)^-$ and then we integrate over Ω. We obtain

$$\lambda_1\int_\Omega u_\lambda\varphi_1 dx = \int_\Omega(\delta(x)^{-\alpha}u_\lambda^{-\beta} + \lambda f(u_\lambda))\varphi_1 dx \ge \int_\Omega(\delta(x)^{-\alpha}u_\lambda^{-\beta} + \lambda m u_\lambda)\varphi_1 dx, \tag{5.10}$$

for all $0<\lambda<\lambda^*$. Passing to the limit in (5.10) with $\lambda\nearrow\lambda^*$, by virtue of Lebesgue's theorem on dominated convergence we find

$$\lambda_1\int_\Omega u^*\varphi_1 = \int_\Omega(\delta(x)^{-\alpha}u^{*-\beta} + \lambda_1 u^*)\varphi_1 dx,$$

which is a contradiction. Hence $u^*\equiv\infty$. It is easy now to see, using the monotonicity of $\{u_\lambda\}_{0<\lambda<\lambda^*}$ that $\lim_{\lambda\nearrow\lambda^*} u_\lambda = \infty$ uniformly on compact subsets of Ω, and the proof is now complete. □

Acknowledgments

The authors are partially supported by Grants 2-CEx06-11- 18/2006 and CNCSIS PN II-79/2007 *Procese Neliniare Degenerate şi Singulare.*

References

[1] R. Agarwal and D. O'Regan, Existence theory for single and multiple solutions to singular positone boundary value problems. *J. Differential Equations* **175** (2001), 393-414.

[2] M. Aizenman and B. Simon, Brownian motion and Harnack inequality for Schrödinger operators. *Comm. Pure Appl. Math.* **35** (1982), 209-273.

[3] R. Aris, *The Mathematical Theory of Diffusion and Reaction in Permeable Catalysts.* Clarendon Press, Oxford, 1975.

[4] D.H. Armitage and S.J. Gardiner, *Classical Potential Theory.* Springer-Verlag, London, 2001.

[5] L. Caffarelli, R. Hardt and L. Simon, Minimal surfaces with isolated singularities. *Manuscripta Math.* **48** (1984), 1-18.

[6] A. Callegari and A. Nachman, A nonlinear singular boundary value problem in the theory of pseudoplastic fluids. *SIAM J. Appl. Math.* **38** (1980), 275-281.

[7] F. Cîrstea, M. Ghergu and V. Rădulescu, Combined effects of asymptotically linear and singular nonlinearities in bifurcation problems of Lane-Emden-Fowler type. *J. Math. Pures Appliquées* **84** (2005), 493-508.

[8] M.M. Coclite and G. Palmieri, On a singular nonlinear Dirichlet problem. *Comm. Partial Differential Equations* **14** (1989), 1315-1327.

[9] D.S. Cohen and H.B. Keller, Some positive problems suggested by nonlinear heat generators. *J. Math. Mech.* **16** (1967), 1361-1376.

[10] M.G. Crandall, P.H. Rabinowitz and L. Tartar, On a Dirichlet problem with a singular nonlinearity. *Comm. Partial Differential Equations* **2** (1977), 193-222.

[11] J.I. Díaz, *Nonlinear Partial Differential Equations and Free Boundaries. Vol. I. Elliptic Equations* (Research Notes in Mathematics, Vol. **106**). Pitman (Advanced Publishing Program), Boston, MA, 1985.

[12] J.I. Díaz, J.M. Morel and L. Oswald, An elliptic equation with singular nonlinearity. *Comm. Partial Differential Equations* **12** (1987), 1333-1344.

[13] L. Dupaigne, M. Ghergu and V. Rădulescu, Lane-Emden-Fowler equations with convection and singular potential. *J. Math. Pures Appl.* **87** (2007), 563-581.

[14] M. Ghergu and V. Rădulescu, Sublinear singular elliptic problems with two parameters. *J. Differential Equations* **195** (2003), 520-536.

[15] M. Ghergu and V. Rădulescu, Bifurcation for a class of singular elliptic problems with quadratic convection term. *C. R. Acad. Sci. Paris, Ser. I* **338** (2004), 831-836.

[16] M. Ghergu and V. Rădulescu, On a class of sublinear singular elliptic problems with convection term. *J. Math. Anal. Appl.* **311** (2005), 635-646.

[17] M. Ghergu and V. Rădulescu, *Singular Elliptic Problems: Bifurcation and Asymptotioc Analysis* (Oxford Lecture Notes in Mathematics **37**). Oxford University Press, 2008.

[18] D. Gilbarg and N.S. Trudinger, *Elliptic Partial Differential Equations of Second Order*, 2nd ed.. Springer-Verlag, Berlin Heidelberg New York, 1983.

[19] C. Gui and F.H. Lin, Regularity of an elliptic problem with a singular nonlinearity. *Proc. Royal Soc. Edinburgh Sect. A* **123** (1993), 1021-1029.

[20] J. Hernández, F.J. Mancebo and J.M. Vega, On the linearization of some singular nonlinear elliptic problems and applications. *Ann. Inst. H. Poincaré, Anal. Non Linéaire* **19** (2002), 777-813.

[21] A. Meadows, Stable and singular solutions of the equation $\Delta u = 1/u$. *Indiana Univ. Math. J.* **53** (2004), 1681-1703.

[22] N.B. Rhouma and M. Mosbah, On the existence of positive solutions for semilinear elliptic equations with indefinite nonlinearities. *J. Differential Equations* **215** (2005), 37-51.

[23] J. Shi and M. Yao, On a singular nonlinear semilinear elliptic problem. *Proc. Royal Soc. Edinburgh Sect. A* **128** (1998), 1389-1401.

[24] J. Shi and M. Yao, Positive solutions for elliptic equations with singular nonlinearity. *Electron. J. Differential Equations* **4** (2005), 1-11.
[25] S.D. Taliaferro, A nonlinear singular boundary value problem. *Nonlinear Anal., T.M.A.* **3** (1979), 897-904.
[26] K.O. Widman, Inequalities for the Green function and boundary continuity of the gradient of solutions of elliptic differential equations. *Math. Scand.* **21** (1967), 17-37.

"SIMION STOILOW" INSTITUTE OF MATHEMATICS OF THE ROMANIAN ACADEMY, P.O. BOX 1-764, 014700 BUCHAREST, ROMANIA
E-mail address: marius.ghergu@imar.ro

"SIMION STOILOW" INSTITUTE OF MATHEMATICS OF THE ROMANIAN ACADEMY, P.O. BOX 1-764, 014700 BUCHAREST, ROMANIA and DEPARTMENT OF MATHEMATICS, UNIVERSITY OF CRAIOVA, 200585 CRAIOVA, ROMANIA
E-mail address: vicentiu.radulescu@imar.ro, vicentiu.radulescu@math.cnrs.fr

Potential Theory and Stochastics in Albac, 139–144
Aurel Cornea Memorial Volume

A FATOU TYPE THEOREM FOR MULTIPLY SUPERHARMONIC FUNCTIONS

KOHUR GOWRISANKARAN

ABSTRACT. We consider a multiply harmonic function $u > 0$ on the product of two Brelot spaces. We construct filters corresponding to a set of full measure (representing the function u) for all extremal elements belonging to a compact base of the cone of positive multiply harmonic functions. Utilizing these filters we prove a Fatou type limit theorem for quotients of the form v/u where v is a multiply superharmonic function > 0.

1. INTRODUCTION

The classical result of Fatou on the existence of non-tangential limits for almost every point of the (boundary) circle for a class of harmonic functions defined on the unit disc was generalized by Doob [2] in 1957 to the case of a Green space. This result was extended to the case of axiomatic harmonic functions on a Brelot space by us [3, 4]. We recall this result below.

Let Ω be a basic Brelot space (that is, a locally compact, connected, locally connected Hausdorff space with a structure of harmonic functions satisfying Axioms 1, 2, 3 of Brelot [1].Throughout this article we will assume that the Brelot spaces we consider have (1) a countable base for open sets and (2) a potential > 0.

Let Δ_1 denote the set of minimal (extremal) harmonic functions on Ω belonging to a compact base of positive superharmonic functions. For each $h \in \Delta_1$, we define the 'fine filter' $\mathscr{F}_h$ as the filter on Ω consisting of sets E such that $R_h^{\complement E} \neq h$. It is a simple matter to verify that this is also the filter with the base consisting of sets of form $\{x \in \Omega : \frac{p(x)}{h(x)} < \epsilon\}$ where p is an arbitrary potential and $\epsilon > 0$.

Theorem 1.1 (Fatou Theorem for Brelot spaces). *Let $v > 0$ be superharmonic on Ω and $u > 0$ be a harmonic function with the canonical representing measure μ_u on Δ_1. Then $\lim_{\mathscr{F}_h} \frac{v(x)}{u(x)}$ exists as a real number for μ_u-almost every $h \in \Delta_1$.*

Let Ω_1, Ω_2 be Brelot harmonic spaces as above. Let $u > 0$ be a 2-harmonic function on $\Omega_1 \times \Omega_2$ (refer 2 for definitions) with the canonical measure μ_u on ${}_1\Delta_1 \times_2 \Delta_1$ representing u. Here $j\Delta_1$ is the set of minimal harmonic functions belonging to a convenient compact base of the set of positive superharmonic functions on Ω_j. The objective of this article is to prove a Fatou type theorem for quotients $\frac{v}{u}$, v being a 2-superharmonic function > 0 with filters $\tau^u_{h_1,h_2}$ that depend on u for each $(h_1, h_2) \in$

2000 *Mathematics Subject Classification.* 31D05.

Key words and phrases. Fatou type limit, Brelot space, multiply superharmonic function, strict subdivision, Lusin space.

${}_1\Delta_1 \times {}_2\Delta_1$ (see Theorem 3). We hope that this result will lead to establishing a more general result where the filters are independent of u as described in [7].

These results were presented at a conference held in honour of the memory of Aurel Cornea in September 2007. To have had Aurel as a friend was a pleasure and to have had the opportunity to discuss mathematics with him was a privilege. I would also like to thank Jurgen Bliedtner and Peter Loeb for their comments pointing out possible connection with their earlier work. I believe that relation is more appropriate with '2-harmonic' filters which will be subject of a future forthcoming article.

2. Multiply superharmonic functions and some results concerning Polish spaces

Definition 2.1. Let $v : \Omega_1 \times \Omega_2 \to \mathbb{R} \cup \{\infty\}$ be a lower semi-continuous function. It is called a *2-superharmonic* function if (i) $v \neq +\infty$ and for every fixed x(resp. y), $v(x,\cdot)$ (resp($v(\cdot, y)$)) is hyperharmonic on the other space. A *2-harmonic* function is one which is harmonic in each variable separately.

We recall some of the basic results regarding these functions (see [5, 6]).

1. The cone of all positive 2-harmonic functions is a lattice with the natural order. Provided with the compact-open topology, it is metrizable and has a compact base. The extreme points of a suitable compact base can be identified with the set ${}_1\Delta_1 \times_2 \Delta_1$ of functions of the form $h_1\, h_2$ where $h_j \in {}_j\Delta_1$. Further to each $u > 0$ and 2-harmonic corresponds a unique regular Borel measure μ_u on ${}_1\Delta_1 \times_2 \Delta_1$ such that

$$u = \int h_1 h_2\, \mu_u(dh_1\, dh_2).$$

2. ${}_1\Delta_1 \times_2 \Delta_1$ is a G_g -subset of a compact metric space and is hence a Polish space (for this and other results on strict subdivision etc. see [8]).

3. Every 2-superharmonic function ≥ 0 has a greatest 2-harmonic minorant. We denote by **2-P**, the set of all 2-superharmonic functions > 0 with greatest 2-harmonic minorant 0.

4 Now $X = {}_1\Delta_1 \times_2 \Delta_1$ being a Polish space has a 'strict subdivision'. It can be described as follows: For each $n \in \mathbb{N}$, corresponds a *countable* set C_n (of indices) and mappings $\Pi_n : C_n \to C_{n-1}$ which is a projection (here we designate $C_0 = \{1\}$). Further, $\forall\, c \in C_n$ corresponds a subset $E_c \subset X \quad (E_1 = X \,\text{for}\, 1 \in C_0)$ and verifies the following properties.

(a) $\forall\, c \in C_{n-1}(\forall n),\ E_c = \bigcup\{E_d : d \in C_n \quad \pi_n(d) = c\}$.

(b) Every sequence of sets $\{E_{c_n}\}$ for $c_n \in C_n$ for $n = 1\,\text{to}\,\infty$ that is coherent, verifying $\Pi_n(c_n) = c_{n-1} \quad \forall\, n$, (forms a base of a filter that) converges to an element contained in all of them.

(c) Every point of X is a limit of such a coherent sequence.

(d) $\forall\, n \geq 1, \quad \{E_c\}_{c \in C_n}$ consist of mutually disjoints sequence of sets.

Lemma 2.2. *There is a topology T_1 on X generated by the collections of sets $\{E_c\}_{c\in C_n}$ for all C_n, as a base for open sets. Then T_1 is finer than the original topology on X. Further (X, T_1) is a Lusin space since the given strict subdivision is also a strict subdivision for T_1. Further the Borel subsets are identical for the two topologies.*

Proof. The fact that (X,T) is a Hausdorff topological space is obvious. The rest of the results are easily deduced from properties of Lusin spaces [8]. □

3. Construction of the filters

Now let us start with the fixed 2-harmonic function $u > 0$ and the associated Radon measure μ_u on $X = {}_1\Delta_1 \times_2 \Delta_1$. Consider all the sets $\{E_c : c \in E_n, n \in \mathbb{N}$ with $\mu_u(E_c) = 0\}$. This being a countable collection, the union of this collection has μ_u- measure zero. In what follows we will only consider those E_c such that $\mu_u(E_c) > 0$. Suppose $(h_1,h_2) \in X$ and $\mu_u(E_c) > 0$ for every set in the strict subdivision that contains (h_1,h_2). Note that the set of such $(h_1,h_2)$is of full μ_u-measure and is hence non-trivial.

Definition 3.1. Suppose $E \subset X$, $\mu_u(E) > 0$ and E Borel. Let U_E be the 2-harmonic function with canonical representing measure equal to the restriction of μ_u to E. Corresponding to this U_E define $\mathcal{G}_E$ as follows. For a 2-potential p and $\epsilon > 0$, set

$$A_{p,\epsilon} = \{(x,y) : p(x,y)/u_E(x,y) \ < \epsilon\} \text{ and}$$

$$\mathcal{G}_E = \{A_{p,\epsilon} : p \text{ a 2-potential and} \epsilon > 0\}.$$

Lemma 3.2. *$\mathcal{G}_E$ is a base of a filter on $\Omega_1 \times \Omega_2$.*

Proof. Consider $A_{p_1,\epsilon_1}, A_{p_2,\epsilon_2}$. Let $q = p_1 + p_2$ which is a 2-potential and $\epsilon = \min(\epsilon_1, \epsilon_2)$. clearly, $A_{q,\epsilon} \subset A_{p_1,\epsilon_1}, \cap A_{p_2,\epsilon_2}$. □

Here are some remarks that are no more than simple observations:
(1) $\lim_{\mathcal{G}_E} \frac{p}{U_E} = 0 \quad \forall p \in$ **2-P.**
(2) If $E_1 \subset E_2$ are Borel subsets of X with $\mu_u(E_1) > 0$, then $\mathcal{G}_{E_1}$ is finer than $\mathcal{G}_{E_2}$.

Definition 3.3. The filter $\tau'(h_1,h_2)$.

Let $(h_1,h_2) \in X$ such that $\mu_u(E_c) > 0$ for every set of the subdivision $E_c \ni (h_1,h_2)$. There is a unique coherent sequence of such sets E_c. Clearly, $\{\mathcal{G}_{E_c}\}$ is increasing as E_c decreases to the element (h_1,h_2). We denote by $\tau'(h_1,h_2)$ the union or the limit of these filters.

Lemma 3.4. *Let $A, B \subset X$ be Borel sets such that $\mu_u(A) > 0$, $\mu_u(B) > 0$ and $\mu_u(A \cap B) = 0$. Then, $\lim \frac{U_A}{U_B} = 0$ following $\mathcal{G}_{u_B}$ and a similar result with roles of A and B reversed.*

Proof. Under the given hypothesis, the greatest 2-harmonic minorant of $\min(U_A, U_B)$ is zero since it is $\le U_A$ and U_B and so its canonical measure is $\le \mu_u$ restricted to $A \cap B$. So $\frac{1}{U_B} \min(U_A, U_B)$ has the limit 0 following $\mathcal{G}_{u_B}$. This proves the lemma. □

Let us now take a fixed (h_1,h_2) with the filter $\tau\prime(h_1,h_2)$ as constructed above. Consider any two T_1-compact sets K_1, and K_2 containing (h_1,h_2) such that $\mu_u(K_j) > 0$ for $j = 1,2$ and recall that $\mu_u(E_c) > 0$ for all sets E_c in the subdivision with $(h_1,h_2) \in E_c$. First of all we note that there is a filter finer than $\mathcal{G}_{E_c}$ and each of $\mathcal{G}_{U_{K_j}}$ for $j = 1,2$.. This is seen as follows. If $K_1 \cap E_c$ has μ_u measure > 0, $E_c \ni (h_1,h_2)$, the intersection C_1 is a T_1 compact set (since E_c is T_1 closed too). Clearly, $\mathcal{G}_{U_{C_1}}$, is a filter finer than $\mathcal{G}_{E_c}$ and $\mathcal{G}_{U_{K_1}}$. If however $\mu_u(E_c \cap K_1) = 0$ for some E_c, then using the above lemma, it is simple to deduce that $\mathcal{G}_{U_{K_1}}$ is in fact smaller than $\mathcal{G}_{U_{E_c}}$(details are same as the more general situation we deal with below).

Hence there is a filter $\mathscr{F}_j$ finer than $\tau\prime(h_1,h_2)$ and $\mathscr{G}_{U_{K_1}}$ for $j=1,2$. We claim that there is a filter finer than $\mathscr{F}_1$ and $\mathscr{F}_2$ of the same form. Let A_j be a set in $\mathscr{F}_j$, that is of the form $E_j \cap F$ where $E_j \in \mathscr{G}_{U_{K_j}}$ and $F \in \mathscr{G}_{U_{E_c}}$ for some c (clearly we can take one c and one F in that filter for both $\mathscr{F}_1$ and $\mathscr{F}_2$). The case where $K_1 \cap K_2 \cap E_c$ has a μ_u measure >0 is dealt with easily . If $\mu_u(K_1 \cap K_2)>0$ we are back to the case considered above. The most interesting case is $\mu_u(K_1 \cap K_2)=0$. Suppose A_j is given by $\{\frac{p_j}{U_{K_j}}<\epsilon_j\} \cap \{\frac{p_j}{U_{E_c}}<\epsilon_j\}$ for $j=1,2$. We can assume that it is also the intersection with $\{\frac{U_k}{U_{k_j}}<\frac{\epsilon}{2}\}\,(i \neq j)$. This is due to the Lemma 3.4 whereby $\frac{U_{K_i}}{U_{K_j}}$ tends to zero following $\mathscr{G}_{U_{K_j}}$. We may with simple manipulation assume all p_j's are all equal to a p in **2-P** and ϵ_j are all equal to $\epsilon<1$.

Consider

$$\Big\{\frac{p}{U_{K_1 U K_2}}<\frac{\epsilon}{2}\Big\} \cap \Big\{\frac{p}{U_{E_c}}<\epsilon\Big\}=A$$

Since $\mu_u(K_1 \cap K_2)=0, \quad U_{K_1 U K_2}=U_{K_1}+U_{K_2}$. For every $x \in B$,

$$0<\frac{p(x)}{U_{K_1}(x)}=\Big(\frac{p(x)}{U_{K_1}(x)+U_{K_2}(x)}\Big)\Big(\frac{U_{K_2}(x)}{U_{K_1}(x)}+1\Big)$$

$$<\frac{\epsilon}{2}\Big(\frac{\epsilon}{2}+1\Big)<\epsilon.$$

Similarly we can show that $\forall\, x \in B$, $\frac{p(x)}{U_{K_2}(x)}<\epsilon$. This proves our claim. This leads us to the following definition.

Definition 3.5. For every $(h_1,h_2) \in X$ with $\mu_u(E_c)>0$ for all sets of the subdivision $\ni (h_1,h_2)$ we define the filter $\tau(h_1,h_2)$ to be the smallest one finer than $\tau\prime(h_1,h_2)$ and all the filters $\mathscr{G}_{U_K}$ for all T_1 compact set $K \ni (h_1,h_2)$ with $\mu_u(K)>0$.

4. Case of u-bounded 2-harmonic functions

Lemma 4.1. *Let K be a T_1-compact set with $\mu_u(K)>0$ Then $\lim_{\tau(h_1,h_2)} \frac{U_K}{u}=\chi_K$, μ_u-almost everywhere.*

Proof. Note that $p=\min(U_K,U_{X\setminus K})$ is a potential and hence $\frac{p}{U_k}$ has a limit 0 for $\tau(h_1,h_2)$ for every (h_1,h_2) in K. This implies that $\frac{U_{X-K}}{U_K}$ has a limit of 0 at those points and hence $\frac{U_{X\setminus K}}{u}$. It is immediate that $\lim_{\tau(h_1,h_2)} \frac{U_K}{u}=1$ for all $(h_1,h_2) \in K$. If $\mu_u(X\setminus K)=0$ there is nothing to prove. If not, for every T_1-compact set $C \subset X\setminus K$ with $\mu_u(C)>0$ we can apply the previous result. Since μ_u is also Randon measure for (X,T_1), we conclude that $\lim_{\tau(h_1,h_2)} \frac{U_{X\setminus K}}{u}=1$ for μ_u-almost every (h_1,h_2) in $X\setminus K$. This concludes the proof. □

Theorem 4.2. *Suppose f is a bounded Borel function on X. Then $\lim_{\tau(h_1,h_2)} U_f/u=f$ for μ_u-almost every (h_1,h_2) where $U_f=\int f h_1,h_2\ \mu_u(dh_1,dh_2)$.*

Proof. Note that we may assume that f is T_1-Borel. Consider a T_1-Borel set E with $\mu_u(E)>0$. For every T_1-compact set $K \subset E$ with $\mu_u(K)>0$, we apply the previous lemma. By the regularity of μ_u(on (X,T_1)) we conclude that $\lim_{\tau(h_1,h_2)} \frac{U_E}{u}=1$ for μ_u-almost every element of E. If $\mu_u(X\setminus E)=0$ there is nothing to prove. Otherwise apply the above argument to $X\backslash E$ and clude that $\frac{U_E}{u}$ has the limit $= X_E$ following almost every $\tau_{(h_1,h_2)}$. Obviously the result is true for simple functions and for all bounded and ≥ 0 Borel functions since they are the uniform limit of simple functions. The proof is complete. □

An important consequence is the following minimum principle.

Theorem 4.3. *Let $v \in 2-S$ be such that (i) v/u is lower bounded and (ii) for μ_u-almost every (h_1,h_2), $\lim_{\tau(h_1,h_2)} \sup\ v/u \geq 0$. Then $v \geq 0$.*

Proof. By taking $\min(v,u)$ we may without loss of generality assume v/u is bounded. Hence $v = p + u_1$ where p is a 2-potential and u_1, 2-harmonic and u-bounded. Thus, u_1 is represented by $f\,d\mu_u$ for some bounded Borel function f on X. Now, p/u has limit 0, following $\tau_{(h_1,h_2)}$ and u_1/u tends to f at almost every (h_1,h_2) in X, following $\tau_{(h_1,h_2)}$. The hypothesis then implies that $f \geq 0$, that is $u_1 \geq 0$. Thus $v \geq 0$ completing the proof. □

5. FATOU TYPE THEOREM WITH $\tau_{(h_1,h_2)}$

We shall prove the main result concerning the existence of $\tau_{(h_1,h_2)}$ limits (a.e) for v/u where v is a general $2-S^+$ function. The method is pretty standard and depends on developing a Dirichlet problem. We shall omit some of the details of the proof which follow standard arguments or are found in [4, 6]. Define, for an extended real valued function f on X, the following two classes of functions.

$$\sum\nolimits_{f,u} = \{v : v-2\text{-hyperharmonic and } v/u \text{ lower bounded, for a.e. } (h_1,h_2), \lim_{\tau(h_1,h_2)} \inf v/u \geq f(h_1,h_2)\}.$$

If further $f \geq 0$,

$$\widetilde{\sum}_{f,u} = \{v : v-2\text{-hyperharmonic and } v/u \text{ lower bounded}, \lim_{\tau(h_1,h_2)} \sup v/u \geq f(h_1,h_2) \text{ for } \mu_u \text{ a.e} (h_1,h_2)\}.$$

Let $\overline{\mathcal{H}}_{f,u} = \inf\left\{v : v \in \sum_{f,u}\right\}$. Also $\mathcal{D}_{f,u} = \inf\left\{v : v \in \widetilde{\sum}_{f,u}\right\}$ when $f \geq 0$.

Lemma 5.1. *For an extended real valued function f, $\overline{\mathcal{H}}_{f,u} = \underline{\mathcal{H}}_{f,u}$ (and not $\pm\infty$ identically) if and only if f is μ_u-integrable. In this case both functions coincide with $\int_X h_1 h_2\, f(h_1,h_2)\, \mu_u(dh_1,dh_2)$, which is clearly a 2-harmonic function.*

Lemma 5.2. *For a function $f \geq 0$. $\overline{\mathcal{H}}_{f,u} = 0$ if and only if $f = 0$ expect on a μ_u-measurable set of μ_u-measure 0.*

Lemma 5.3. *For an extended real valued function $f \geq 0$, $\overline{\mathcal{D}}_{f,u} = \overline{\mathcal{H}}_{f,u}$.*

Theorem 5.4. *(Main result). Let $v > 0$ be 2-superharmonic. Then, for μ_u-almost every (h_1,h_2) in X, the limit of v/u following $\tau_{(h_1,h_2)}$ exist. The limit function which is defined μ_u-almost everywhere is a Radon-Nikodym derivative of the measure representing the greatest 2-harmonic minorant of v relative to the measure μ_u.*

Proof. Let $v = p + w$ where w is the greatest 2-harmonic minorant of v. Further $w = w_1 + w_2$ where the measure corresponding to w_1 (resp. w_2) is the absolutely continuous part of that of w relative to μ_u (resp. singular part). We know that $\tau_{(h_1,h_2)}$ limit of p/u is zero and the same is true for $\frac{w_2}{u}$ since $\min(w_2, u)$ is a 2-Potential. Let the measure representing w_1 be $f\, d u_u$ where $f \geq 0$ is a μ_u-integrable function. Utilizing Theorem 4.2, it is easily seen that

(1) $\lim_{\tau_{(h_1,h_2)}} \inf \frac{w_1}{u} \geq f(h_1, h_2)$ μ_u-almost everywhere.

Let $g(h_1, h_2) = \lim_{\tau_{(h_1,h_2)}} \sup \frac{w_1}{u}$. Then clearly $g \geq f$ μ_u-almost everywhere. From Lemma 5.2 and Lemma 5.5 we conclude that

$$\overline{\mathscr{H}}_{f,u} = w_1 \geq \mathscr{D}_{g,u} = \overline{\mathscr{H}}_{g,u} \geq \overline{\mathscr{H}}_{f,u} = \mathscr{D}_{g,u} = \overline{\mathscr{H}}_{f,u}.$$

It is simple to deduce $\overline{\mathscr{H}}_{g-f} = 0$ and from Lemma 5.2, we conclude that $g - f = 0$ μ_u-almost everywhere. This proves that $\tau_{(h_1,h_2)}$ limit of $\frac{w_1}{u}$ equals f μ_u-almost everywhere, completing the proof. □

Acknowledgments

This paper was prepared when tho author was on sabbatical leave at the Centre de Recherches Mathematiques (Montreal). He would like to thank CRM for the support.

References

[1] M. Brelot, *Lectures on Potential Theory.* Tata Institute of Fundamental Research, 1960.
[2] J.L. Doob, A Non-probabilistic proof of a relative Fatou theorem. *Annales Inst. Fourier* **9** (1959), 293-300.
[3] K. Gowrisankaran, Extreme harmonic functions and boundary value problems. *Annales Inst. Fourier* **13** (1963), 307-356.
[4] K. Gowrisankaran, Fatou-Naim-Doob theorem for axiomatic Brelot spaces. *Annales Inst. Fourier* **16** (1966), 455-467.
[5] K. Gowrisankaran, Multiply harmonic functions. *Nagoya Math. J* **28** (1966), 27-48.
[6] K. Gowrisankaran, Polydiscs and non-tangential limits. *Proc. Amer. Math. Soc.* **115** (1992), 977-984.
[7] K. Gowrisankaran, Multiply superharmonic functions, *Annales Inst. Fourier* **25** (1975), 235-244.
[8] L. Schwartz, *Radon Measures on General Topological spaces.* Oxford University Press, 1972.

McGill University, Department of Mathematics and Statistics, 805 Sherbrooke Street, West Montreal, Quebec H3A 2K6, Canada

E-mail address: gowri@math.mcgill.ca

Potential Theory and Stochastics in Albac, 145–156
Aurel Cornea Memorial Volume

SOME REMARKS ON REPRESENTATIONS OF NON-SYMMETRIC LOCAL DIRICIILET FORMS

ZE-CHUN HU, ZHI-MING MA, AND WEI SUN

ABSTRACT. In this note, we give some remarks on the structure of non-symmetric local Dirichlet forms. We first present LeJan's transformation rule for their co-symmetric parts. Combined with the classical LeJan's formula, it shows that essentially any local Dirichlet form must be given by differential operators (up to a killing part). Then, as applications, we use this formula to give explicit representations for local Dirichlet forms on both infinite and finite dimensional state spaces.

1. INTRODUCTION

Based on the celebrated Beurling-Deny formula [1] and LeJan's formula [13], the structure of symmetric Dirichlet forms is well understood (cf. also Fukushima et al. [7], Dong et al. [6], Kuwae [12] and the references therein). However, the structure of non-symmetric Dirichlet forms is still not completely known until now. In [8, 9], we obtained the Beurling-Deny formula for quasi-regular semi-Dirichlet forms. Furthermore, we gave in [10] the LeJan's transformation rule for non-symmetric Dirichlet forms. In this note, we will give some further results and remarks in the case that the underlying non-symmetric Dirichlet form is local. Since the SPV integrable condition (cf. [9, 10]) is automatically satisfied if there is no jumping part, we can obtain neater results for local Dirichlet forms. The further results and remarks in this note are novel and supplemental to [10]. But for the convenience of the reader we will restate some of the results obtained in [10], in particular we will repeat some relevant material of [10, Section 4] when it makes the presentation more clear.

This short note is organized as follows. In Section 2, we present LeJan's transformation rule for the co-symmetric parts (cf. (2.4) below) of local Dirichlet forms. Combined with the classical LeJan's formula, it shows that essentially any local Dirichlet form must be given by differential operators (up to a killing part). Based on the works on quasi-regular topologies for Dirichlet forms (cf. Dong [5] and Beznea et al. [2]), one finds that the obtained transformation rule depends essentially only on the measurable structure of Dirichlet forms. In Section 3, we use both infinite and finite dimensional examples to exhibit applications of the transformation rule obtained in Section 2. Some explicit representations of non-symmetric local Dirichlet forms are given. For the notions and notations used in this note, we refer to Fukushima et al. [7] and Ma and Röckner [14]. We hope this note can stimulate the applications of non-symmetric

2000 *Mathematics Subject Classification.* 31C25, 60J45.
Key words and phrases. Non-symmetric local Dirichlet form, LeJan's formula.

local Dirichlet forms to investigate Markov processes on more general state spaces like configuration spaces, loop spaces or measure-valued processes.

2. LeJan's Transformation Rule

2.1. **Quasi-regular Case.** Let E be a Hausdorff topological space and $\mathscr{B}(E)$ be the Borel σ-algebra of E. We assume that $\mathscr{B}(E) = \sigma(C(E))$ where $C(E)$ denotes the set of all continuous functions on E. Let m be a σ-finite measure on $(E, \mathscr{B}(E))$ and $(\mathscr{E}, D(\mathscr{E}))$ be a quasi-regular Dirichlet form on $L^2(E;m)$. We assume that $(\mathscr{E}, D(\mathscr{E}))$ has local property, i.e., for any $u, v \in D(\mathscr{E})$ with $\mathrm{supp}[u] \cap \mathrm{supp}[v] = \emptyset$, we have $\mathscr{E}(u, v) = 0$. Hereafter, we use $\mathrm{supp}[\cdot]$ to denote the support of a measure or a function on E. Note that when dealing with quasi-regular Dirichlet forms one could assume without loss of generality that E is a topological Lusin space (cf. Ma and Röckner [14, Page 101]). Furthermore, a Dirichlet form is quasi-regular if and only if it is quasi-homeomorphic to a regular Dirichlet form on a locally compact separable metric space (cf. Chen et al. [4]). Hence most results originally stated and proved for regular Dirichlet forms also hold for quasi-regular Dirichlet forms. In what follows, we will freely use the above facts without explicitly mentioning them. We denote by $(\tilde{\mathscr{E}}, D(\mathscr{E}))$ and $(\hat{\mathscr{E}}, D(\mathscr{E}))$ the symmetric part and the dual form of $(\mathscr{E}, D(\mathscr{E}))$, respectively. That is, for $u, v \in D(\mathscr{E})$,

$$\tilde{\mathscr{E}}(u, v) := \frac{1}{2}(\mathscr{E}(u, v) + \mathscr{E}(v, u)),\ \hat{\mathscr{E}}(u, v) := \mathscr{E}(v, u).$$

By [9], noting that the jumping measure $J = 0$, for any $u, v \in D(\mathscr{E})$, we have the following unique decompositions:

$$\mathscr{E}(u, v) = \mathscr{E}^c(u, v) + \int_E \tilde{u}(x)\tilde{v}(x)K(dx), \tag{2.1}$$

$$\hat{\mathscr{E}}(u, v) = \hat{\mathscr{E}}^c(u, v) + \int_E \tilde{u}(x)\tilde{v}(x)\hat{K}(dx), \tag{2.2}$$

$$\tilde{\mathscr{E}}(u, v) = \tilde{\mathscr{E}}^c(u, v) + \int_E \tilde{u}(x)\tilde{v}(x)\tilde{K}(dx),$$

where $\tilde{u}$ and $\tilde{v}$ are quasi-continuous versions of u and v, $\mathscr{E}^c$ and K (resp. $\hat{\mathscr{E}}^c$ and $\hat{K}$) are the diffusion part and killing measure of $(\mathscr{E}, D(\mathscr{E}))$ (resp. $(\hat{\mathscr{E}}, D(\mathscr{E}))$), respectively. $\mathscr{E}^c$ and $\hat{\mathscr{E}}^c$ satisfy the *left-strongly local property* in the sense that $\mathscr{E}^c(u, v) = \hat{\mathscr{E}}^c(u, v) = 0$ whenever $v \in D(\mathscr{E}), u \in I_q(v) := \{u \in D(\mathscr{E}) : \tilde{u}$ is constant $\mathscr{E} - q.e.$ on a quasi-open set containing $\mathrm{supp}_q[v]\}$. Hereafter, we denote by $\mathrm{supp}_q[v]$ the *quasi-support* of v, which is the smallest (up to an $\mathscr{E}$-exceptional set) quasi-closed set F such that $\int_{E\setminus F} |v(x)|m(dx) = 0$. $\tilde{\mathscr{E}}^c$ and $\tilde{K}$ are respectively the diffusion part and killing measure of $(\tilde{\mathscr{E}}, D(\mathscr{E}))$ with

$$\tilde{\mathscr{E}}^c(u, v) = \frac{1}{2}(\mathscr{E}^c(u, v) + \hat{\mathscr{E}}^c(u, v)),\ \ \forall u, v \in D(\mathscr{E}),$$

$$\tilde{K} = \frac{1}{2}(K + \hat{K}).$$

Note that for $\mu = K, \hat{K}$ or $\tilde{K}$,

$$\int_E |\tilde{u}(x)\tilde{v}(x)|\mu(dx) \le 2\int_E |\tilde{u}(x)\tilde{v}(x)|\tilde{K}(dx) \le 2(\mathscr{E}(u, u)\mathscr{E}(v, v))^{\frac{1}{2}}. \tag{2.3}$$

Denote $D(\mathscr{E})_b := D(\mathscr{E}) \cap L^\infty(E;m)$. Let $u, v \in D(\mathscr{E})_b$. Then we obtain by (2.1) and (2.2) that

$$\begin{aligned}
\check{\mathscr{E}}(u,v) &:= \frac{1}{2}(\mathscr{E}(u,v) - \mathscr{E}(v,u)) \\
&= \frac{1}{2}(\mathscr{E}^c(u,v) - \hat{\mathscr{E}}^c(u,v)) + \int_E \tilde{u}(x)\tilde{v}(x)\frac{K - \hat{K}}{2}(dx) \\
&:= \check{\mathscr{E}}^c(u,v) + \int_E \tilde{u}(x)\tilde{v}(x)\check{K}(dx).
\end{aligned}$$

Note that $\check{\mathscr{E}}^c(u,v)$ is in general not anti-symmetric. In fact, one finds that $\check{\mathscr{E}}^c(u,v) = -\check{\mathscr{E}}^c(v,u)$ for any $u, v \in D(\mathscr{E})_b$ if and only if $\check{K} = 0$, i.e., $K = \hat{K}$. Hence we refer $\check{\mathscr{E}}^c(u,v)$ as the *co-symmetric* diffusion part. Then $(\mathscr{E}^c, D(\mathscr{E})_b)$ is uniquely decomposed into the symmetric part and the co-symmetric part as follows.

$$\mathscr{E}^c(u,v) = \tilde{\mathscr{E}}^c(u,v) + \check{\mathscr{E}}^c(u,v). \tag{2.4}$$

Note that $\tilde{\mathscr{E}}^c$ obeys LeJan's transformation rule (cf. LeJan [13]) derived from the mutual energy measure $\tilde{\mu}^c_{<u,v>}$, where $\tilde{\mu}^c_{<u,v>}$ is defined as (cf. Fukushima et al. [7, §3.2]):

$$\int_E \tilde{f}\, d\tilde{\mu}^c_{<u,v>} = \tilde{\mathscr{E}}^c(uf,v) + \tilde{\mathscr{E}}^c(vf,u) - \tilde{\mathscr{E}}^c(uv,f) \text{ for } f \in D(\mathscr{E})_b. \tag{2.5}$$

Therefore, to understand the structure of $\mathscr{E}$, we need only to concentrate on $\check{\mathscr{E}}^c$. The difficulty is that the mutual energy measure defined as in (2.5) is not available for the co-symmetric part. For $u, v \in D(\mathscr{E})_b$, we introduce a new linear functional $< L(u,v), \cdot >$ by

$$\begin{aligned}
< L(u,v), f > &:= \check{\mathscr{E}}^c(u, vf) \\
&= \frac{1}{2}\left(\mathscr{E}^c(u,vf) - \hat{\mathscr{E}}^c(u,vf)\right) \text{ for } f \in D(\mathscr{E})_b.
\end{aligned} \tag{2.6}$$

We can see from the following theorem that $< L(u,v), \cdot >$ is a good substitution of the mutual energy measure.

Theorem 2.1. *Let $u_1, \ldots, u_m, v, f \in D(\mathscr{E})_b$. Denote $u := (u_1, \ldots, u_m)$. If $\phi \in C^2(\mathbf{R}^m)$ with $\phi(0) = 0$, then $\phi(u), \phi_{x_i}(u) \in D(\mathscr{E})_b, 1 \le i \le m$, and*

$$< L(\phi(u), v), f >= \sum_{i=1}^m < L(u_i, v), \phi_{x_i}(u) f > . \tag{2.7}$$

The proof of Theorem 2.1 follows from [10] by quasi-homeomorphism, (2.3) and approximation. We omit the details here. Note that any Dirichlet form given by differential operators must be local (cf. Schmuland [16, Section 2]). Combined with the classical LeJan's formula, Theorem 2.1 shows that the converse is true. That is, any local Dirichlet form must be given by differential operators (up to a killing part).

Remark 2.2. Suppose that $(\mathscr{E}, D(\mathscr{E}))$ is a regular (non-symmetric) local Dirichlet form and E is a locally compact separable metric space. One finds that $< L(u_1, v_1), f >=< L(u_2, v_2), f >$ if $u_1 = u_2$, $v_1 = v_2$ m-a.e. on a neighborhood of supp$[f]$. Define

$$\begin{aligned}
D(\mathscr{E})_{b,loc} := \{u| &\text{ for any relatively compact open subset } G \text{ of } E, \text{there exists } v \in D(\mathscr{E})_b \\
&\text{such that } u = v \text{ on } G\}.
\end{aligned}$$

Then we may extend the definition of the linear functional $< L(u,v),\cdot >$ from $u,v\in D(\mathcal{E})_b$ to $u,v\in D(\mathcal{E})_{b,loc}$. We use $C_0(E)$ to denote the set of all continuous functions on E with compact supports. By Theorem 2.1, following the proof of [10, Theorem 3.8], we obtain

Theorem 2.3. *Let $u_1,\dots,u_m,v\in D(\mathcal{E})_{b,loc}$. Denote $u:=(u_1,\dots,u_m)$. If $\phi\in C^2(\mathbf{R}^m)$, then for any $f\in C_0(E)\cap D(\mathcal{E})$, we have*

$$< L(\phi(u),v),f >=\sum_{i=1}^m < L(u_i,v),\phi_{x_i}(u)f > .$$

2.2. **Arbitrary Case on Lusin Measurable Spaces.** We would like to point out that Theorem 2.1 applies to local Dirichlet forms on arbitrary Lusin measurable state spaces. Let $(E,\mathcal{B})$ be a Lusin measurable space (i.e. the image of a Polish space under a measurable isomorphism) and $(\mathcal{E},D(\mathcal{E}))$ be a (non-symmetric) Dirichlet form on $L^2(E;m)$ with m being a σ-finite measure. We suppose that $(\mathcal{E},D(\mathcal{E}))$ has local property in the sense of Bouleau and Hirsch [3, Chapter I, Corollary 5.1.4], i.e., for any $u,v\in D(\mathcal{E})$ with $uv=0$ m-a.e., we have

$$\mathcal{E}(u,v)=0.$$

By Beznea et al. [2, Corollary 3.4], there exists a (larger) Lusin topological space E_1 such that $E\subset E_1$, E belongs to $\mathcal{B}_1$ (the σ-algebra of all Borel subsets of E_1), $\mathcal{B}=\mathcal{B}_1|_E$, and $(\mathcal{E},D(\mathcal{E}))$ regarded as a Dirichlet form on $L^2(E_1;\bar{m})$ is quasi-regular, where $\bar{m}$ is the measure on $(E_1,\mathcal{B}_1)$ extending m by zero on $E_1\backslash E$. Since every Lusin space is Suslin hence strongly Lindelöf, we obtain by Schmuland [16, Proposition 1.4] that $(\mathcal{E},D(\mathcal{E}))$, regarded as a Dirichlet form on $L^2(E_1;\bar{m})$, is local in the usual sense.

In addition, we assume that one of the following conditions is satisfied:
(i) $1\in D(\mathcal{E})$.
(ii) $(\mathcal{E},D(\mathcal{E}))$ satisfies the strong sector condition (cf. Ma and Röckner [14, Page 16] and Oshima [15, Page 1 and §1.6]) and $1\in D(\mathcal{E})_e$, the extended Dirichlet space.

For $u,v\in D(\mathcal{E})_b$, define

$$\mathcal{E}^c(u,v):=\mathcal{E}(u,v)-\mathcal{E}(1,uv). \tag{2.8}$$

One can check that definition (2.8) is equivalent to the usual definition under assumption (i) or (ii). Therefore, (2.7) holds for local Dirichlet forms on arbitrary Lusin measurable state spaces.

3. Applications

3.1. **Infinite Dimensional Space Case.** Let E be a locally convex topological vector space and H be a separable Hilbert space such that H is densely and continuously embedded into E. Identify the dual H^* of H with H and regard the dual E^* of E as a dense subset of H. Let $\{e_k\}_{k\in\mathbb{N}}$ be an ONB of H such that $e_k\in E^*$ for all $k\in\mathbb{N}$. Let m be a finite Borel measure on $(E,\mathcal{B}(E))$ with $\mathrm{supp}[m]=E$. We assume that

$$\sum_{k=1}^N\langle l,e_k\rangle e_k\longrightarrow l \text{ in } m \text{ as } N\to\infty \text{ for all } l\in E^*, \tag{3.1}$$

where $\langle\cdot\rangle$ denotes the inner product of H. Note that assumption (3.1) is fairly weak. It is satisfied, e.g. if E is a Banach space and its norm is measurable (cf. Kuo [11]). As usual we use $\mathscr{F}C_b^\infty(E)$ to denote the family of all smooth cylindrical functions on E, i.e.

$$\mathscr{F}C_b^\infty(E) := \{f(l_1,\cdots,l_n) | n \in \mathbb{N}, f \in C_b^\infty(\mathbf{R}^n), l_1,\cdots,l_n \in E^*\},$$

where $C_b^\infty(\mathbf{R}^n)$ denotes the set of all infinitely differentiable (real-valued) functions on $\mathbf{R}^n$ with all partial derivatives bounded. For $u \in \mathscr{F}C_b^\infty(E)$, we write $\partial u/\partial e_k(x) := \lim_{t\to 0}(1/t)[u(x+te_k)-u(x)]$ for $x \in E$.

Let $(\mathscr{E}, D(\mathscr{E}))$ be a quasi-regular (non-symmetric) local Dirichlet form on $L^2(E;m)$ satisfying the strong sector condition. Then we have the following representation for $(\tilde{\mathscr{E}}, D(\mathscr{E}))$.

Theorem 3.1. *(Dong et al.* [6, Theorem 1.4]*) Suppose that the following two conditions hold:*

(i) $e_k \in D(\mathscr{E})_e$ for all $k \in \mathbb{N}$ and $1 \in D(\mathscr{E})$.

(ii) $\{\sum_{k=1}^N \langle l, e_k\rangle e_k\}_{N\in\mathbb{N}}$ forms an $\mathscr{E}$-Cauchy sequence for all $l \in E^*$.

Then $\mathscr{F}C_b^\infty(E) \subset D(\mathscr{E})$ and there exists a unique family of finite signed Borel measures $\{\nu_{k,k'}\}_{k,k'\in\mathbb{N}}$ on E such that

$$\tilde{\mathscr{E}}(u,v) = \sum_{k,k'=1}^{\infty} \int_E \frac{\partial u}{\partial e_k}\frac{\partial v}{\partial e'_k} d\nu_{k,k'} + \int_E uv d\tilde{K}, \ \forall u,v \in \mathscr{F}C_b^\infty(E), \tag{3.2}$$

where $\tilde{K}$ is the killing measure of $(\tilde{\mathscr{E}}, D(\mathscr{E}))$ and $\sum_{k,k'=1}^\infty$ is understood to be $\lim_{N\to\infty}\sum_{k,k'=1}^N$. Conversely, if the killing part of $(\tilde{\mathscr{E}}, D(\mathscr{E}))$ is controlled by its diffusion part, then conditions (i) and (ii) are also necessary for the existence of $\{\nu_{k,k'}\}_{k,k'\in\mathbb{N}}$ satisfying (3.2).

In what follows we assume that conditions (i) and (ii) of Theorem 3.1 hold. We will investigate the structure of the co-symmetric diffusion part of $(\mathscr{E}, D(\mathscr{E}))$. Let $u = f(l_1,\ldots,l_n)$, $v \in \mathscr{F}C_b^\infty(E)$. Define $\{\varphi_m\}_{m\in\mathbb{N}}$ as in Dong et al. [6, Page 359]. By (2.6), Theorem 2.1 and the assumption that $1 \in D(\mathscr{E})$, approximating l_i by $\varphi_m(l_i)$, $1 \le i \le n$, we obtain that (cf. the proofs of Dong et al. [6, Theorem 1.4] and Fukushima et al. [7, Theorem 3.2.2])

$$\begin{aligned}
\check{\mathscr{E}}^c(u,v) &= \ < L(u,1), v > \\
&= \sum_{i=1}^n \check{\mathscr{E}}^c(l_i, \partial_i f(l_1,\ldots,l_n)v) \\
&= \sum_{i=1}^n \lim_{N\to\infty}\sum_{k=1}^N \check{\mathscr{E}}^c(\langle l_i, e_k\rangle e_k, \partial_i f(l_1,\cdots,l_n)v) \\
&= \lim_{N\to\infty}\sum_{k=1}^N\sum_{i=1}^n \check{\mathscr{E}}^c(\langle l_i, e_k\rangle e_k, \partial_i f(l_1,\cdots,l_n)v) \\
&= \sum_{k=1}^\infty \check{\mathscr{E}}^c\left(e_k, \frac{\partial u}{\partial e_k}v\right).
\end{aligned} \tag{3.3}$$

By (2.3), $u \to \check{\mathscr{E}}^c(u,v)$, $v \in D(\mathscr{E}^g)_e$ is a continuous linear functional on $D(\mathscr{E}^g)_e$ for any $g \in L^1(E;m)$ with $0 < g \le 1$ m-a.e. Here $D(\mathscr{E}^g)_e$ is the domain of extended Dirichlet space of $(\mathscr{E}^g, D(\mathscr{E}))$, where $\mathscr{E}^g(u,v) = \mathscr{E}(u,v) + \int_E uvgdm$, $u,v \in D(\mathscr{E})$. By way of the quasi-homeomorphism, (3.3) and [7, Lemma 1.6.7] together imply that there exist $\{F_k\}_{k\in\mathbb{N}} \subset D(\mathscr{E})_e$ and a family of finite signed Borel measures $\{\nu_k\}_{k\in\mathbb{N}}$ on E such that

$$\begin{aligned}\check{\mathscr{E}}^c(u,v) &= \sum_{k=1}^{\infty}\left[\tilde{\mathscr{E}}\left(F_k,\frac{\partial u}{\partial e_k}v\right)+\int_E \frac{\partial u}{\partial e_k}v d\nu_k\right]\\ &= \frac{1}{2}\sum_{k=1}^{\infty}\int_E \tilde{\mu}^c_{<F_k,\frac{\partial u}{\partial e_k}v>}+\sum_{k=1}^{\infty}\left[\int_E \tilde{F}_k\frac{\partial u}{\partial e_k}v d\tilde{K}+\int_E\frac{\partial u}{\partial e_k}v d\nu_k\right]. \qquad (3.4)\end{aligned}$$

Let $F \in D(\mathscr{E})_e$ and $w = g(h_1,\ldots,h_m) \in \mathscr{F}C_b^\infty(E)$. Then we obtain by Dong et al. [6, Lemma 2.4] that

$$\begin{aligned}\int_E \tilde{\mu}^c_{<F,w>} &= \sum_{i=1}^{m}\int_E \partial_i g(h_1,\ldots,h_m)\tilde{\mu}^c_{<F,h_i>}\\ &= \sum_{i=1}^{m}\int_E \lim_{N\to\infty}\sum_{k'=1}^{N}\partial_i g(h_1,\ldots,h_m)\tilde{\mu}^c_{<F,\langle h_i,e_{k'}\rangle e_{k'}>}\\ &= \lim_{N\to\infty}\sum_{k'=1}^{N}\int_E\sum_{i=1}^{m}\partial_i g(h_1,\ldots,h_m)\tilde{\mu}^c_{<F,\langle h_i,e_{k'}\rangle e_{k'}>}\\ &= \sum_{k'=1}^{\infty}\int_E\frac{\partial g}{\partial e_{k'}}\tilde{\mu}^c_{<F,e_{k'}>}. \qquad (3.5)\end{aligned}$$

Hence, we obtain by (3.4) and (3.5) that

$$\begin{aligned}\check{\mathscr{E}}^c(u,v) &= \frac{1}{2}\sum_{k=1}^{\infty}\sum_{k'=1}^{\infty}\int_E\frac{\partial}{\partial e_{k'}}\left(\frac{\partial u}{\partial e_k}v\right)d\tilde{\mu}^c_{<F_k,e_{k'}>}\\ &\quad+\sum_{k=1}^{\infty}\left[\int_E\tilde{F}_k\frac{\partial u}{\partial e_k}v d\tilde{K}+\int_E\frac{\partial u}{\partial e_k}v d\nu_k\right].\end{aligned}$$

Define

$$\mu_{k,k'} := \frac{1}{2}\tilde{\mu}^c_{<F_k,e_{k'}>} \text{ and } \mu_k := \tilde{F}_k\tilde{K}+\nu_k.$$

Therefore

$$\check{\mathscr{E}}^c(u,v) = \sum_{k=1}^{\infty}\sum_{k'=1}^{\infty}\int_E\frac{\partial}{\partial e_{k'}}\left(\frac{\partial u}{\partial e_k}v\right)d\mu_{k,k'}+\sum_{k=1}^{\infty}\int_E\frac{\partial u}{\partial e_k}v d\mu_k.$$

Now we obtain the following representation.

Theorem 3.2. *Suppose conditions (i) and (ii) of Theorem 3.1 hold. Then $\mathscr{F}C_b^\infty(E) \subset D(\mathscr{E})$ and there exist finite signed Borel measures $\{\nu_{k,k'}\}_{k,k'\in\mathbb{N}}$, $\{\mu_{k,k'}\}_{k,k'\in\mathbb{N}}$ and $\{\mu_k\}_{k\in\mathbb{N}}$ on E such that for $u,v \in \mathscr{F}C_b^\infty(E)$, $\mathscr{E}(u,v)$ has the following decomposition:*

$$\begin{aligned}\mathscr{E}(u,v) &= \sum_{k,k'=1}^{\infty}\int_E\frac{\partial u}{\partial e_k}\frac{\partial v}{\partial e'_k}d\nu_{k,k'}+\sum_{k=1}^{\infty}\sum_{k'=1}^{\infty}\int_E\frac{\partial}{\partial e_{k'}}\left(\frac{\partial u}{\partial e_k}v\right)d\mu_{k,k'}\\ &\quad+\sum_{k=1}^{\infty}\int_E\frac{\partial u}{\partial e_k}v d\mu_k+\int_E uv dK,\end{aligned}$$

where K is the killing measure of $(\mathscr{E}, D(\mathscr{E}))$, while $\sum_{k,k'=1}^{\infty}$ is understood to be $\lim_{N\to\infty}\sum_{k,k'=1}^{N}$.

3.2. **Finite Dimensional Space Case.** This subsection follows from [10, Section 4] by replacing $\mathbf{R}^d$ with U, which is an open subset of $\mathbf{R}^d$. Throughout this subsection we assume that $E = U$ and $(\mathscr{E}, D(\mathscr{E}))$ is a regular (non-symmetric) local Dirichlet form on $L^2(U;m)$ satisfying $C_0^\infty(U) \subset D(\mathscr{E})$. Denote by $C^\infty(U)$ the set of all infinitely differentiable functions on U. We extend the definition of the linear functional $< L(u,v), \cdot >$ to include $u, v \in C^\infty(U)$ (cf. Remark 2.2). Similar to Theorem 2.3, we have

Theorem 3.3. *(cf. [10, Theorem 4.4]) Let $u_1, \ldots, u_m, v \in C^\infty(U)$, $\phi \in C^\infty(\mathbf{R}^m)$ and $f \in C_0^\infty(U)$. Denote $u := (u_1, \ldots, u_m)$. Then*

$$< L(\phi(u), v), f >= \sum_{i=1}^{m} < L(u_i, v), \phi_{x_i}(u) f > .$$

Let $u, v, f \in C_0^\infty(U)$. By Theorem 3.3, we get

$$\check{\mathscr{E}}^c(u, vf) =< L(u,v), f >=< L(u,1), vf >= \sum_{i=1}^{d} < L(x_i, 1), \frac{\partial u}{\partial x_i} vf > .$$

Taking an $f \in C_0^\infty(U)$ satisfying $f|_{\mathrm{supp}[v]} = 1$, we get

$$\check{\mathscr{E}}^c(u, v) = \sum_{i=1}^{d} < L(x_i, 1), \frac{\partial u}{\partial x_i} v > . \tag{3.6}$$

Let G be an open subset of U. We use $(\tilde{\mathscr{E}}_G, D(\tilde{\mathscr{E}}_G))$ to denote the part of $\tilde{\mathscr{E}}$ on G. Note that $(\tilde{\mathscr{E}}_G, D(\tilde{\mathscr{E}}_G))$ is a regular Dirichlet form on $L^2(G;m)$ (cf. Fukushima et al. [7, Theorem 4.4.3]).

Lemma 3.4. [10, Lemma 4.5] *Let $u \in C_0^\infty(U)$ and G be an open subset of U. Then, there exists an $F(u) \in D(\tilde{\mathscr{E}}_G)$ such that*

$$\check{\mathscr{E}}^c(u, v) \quad = \quad \tilde{\mathscr{E}}_1(F(u), v) \ \textit{ for all } v \in C_0^\infty(G).$$

Lemma 3.5. [10, Lemma 4.6] *Let $v \in D(\mathscr{E})$, $u_1, \ldots, u_m \in D(\mathscr{E})_b$ and $\phi \in C^1(\mathbf{R}^m)$. Denote $u := (u_1, \ldots, u_m)$. Then*

$$d\tilde{\mu}^c_{<\phi(u), v>} = \sum_{i=1}^{m} \phi_{x_i}(\tilde{u}) d\tilde{\mu}^c_{<u_i, v>},$$

where $\tilde{\mu}^c_{<u_i,v>}$ is the mutual energy measure for each $1 \le i \le m$ (cf. (2.5)).

Take a sequence $\{G_k\}_{k\in\mathbf{N}}$ of relatively compact open subsets of U such that $\bar{G}_n \subset G_{n+1}$, $\forall n \ge 1$, and $\cup_{n=1}^\infty G_n = U$. Let $\{w_k^{(x_i)}\}_{k\in\mathbf{N}}, 1 \le i \le d$, (resp. $\{w_k^{(1)}\}_{k\in\mathbf{N}}$) be approximating sequences of $x_i, 1 \le i \le d$, (resp. 1), respectively. That is, for any $1 \le i \le d$ and any $k \in \mathbf{N}$, $w_k^{(x_i)} \in C_0^\infty(U)$ (resp. $w_k^{(1)} \in C_0^\infty(U)$) and $w_k^{(x_i)} = x_i$ (resp. $w_k^{(1)} = 1$) on G_k. Then there exists a $k_0 \in \mathbf{N}$ such that $\cup_{i=1}^d \mathrm{supp}[\frac{\partial u}{\partial x_i}] \cup \mathrm{supp}[v] \subset G_{k_0}$. Thus, for any

$k \geq k_0$,

$$\begin{aligned}\langle L(x_i,1),\frac{\partial u}{\partial x_i}v\rangle &= \langle L(w_k^{(x_i)},w_k^{(1)}),\frac{\partial u}{\partial x_i}v\rangle\\ &= \check{\mathscr{E}}^c\left(w_k^{(x_i)},w_k^{(1)}\frac{\partial u}{\partial x_i}v\right)\\ &= \check{\mathscr{E}}^c\left(w_k^{(x_i)},\frac{\partial u}{\partial x_i}v\right).\end{aligned}\tag{3.7}$$

By Lemma 3.4, for each $k \geq k_0$ and $1 \leq i \leq d$, there exists an $F(w_k^{(x_i)}) \in D(\tilde{\mathscr{E}}_{G_k})$ such that

$$\check{\mathscr{E}}^c\left(w_k^{(x_i)},\frac{\partial u}{\partial x_i}v\right) = \tilde{\mathscr{E}}_1\left(F(w_k^{(x_i)}),\frac{\partial u}{\partial x_i}v\right).\tag{3.8}$$

By the classical Beurling-Deny formula (cf. Fukushima et al. [7, Lemma 4.5.4]), we find that

$$\begin{aligned}\tilde{\mathscr{E}}_1\left(F(w_k^{(x_i)}),\frac{\partial u}{\partial x_i}v\right) &= \int_{G_k}\frac{\partial u}{\partial x_i}vF(w_k^{(x_i)})dm+\tilde{\mathscr{E}}^c\left(\frac{\partial u}{\partial x_i}v,F(w_k^{(x_i)})\right)\\ &\quad+\int_{G_k}\widetilde{F(w_k^{(x_i)})}(x)\frac{\partial u}{\partial x_i}(x)v(x)\tilde{K}(dx).\end{aligned}\tag{3.9}$$

Further, we obtain by Lemma 3.5 that

$$\tilde{\mathscr{E}}^c\left(\frac{\partial u}{\partial x_i}v,F(w_k^{(x_i)})\right) = \frac{1}{2}\sum_{j=1}^d\int_{G_k}\frac{\partial}{\partial x_j}\left(\frac{\partial u}{\partial x_i}v\right)d\tilde{\mu}^c_{<x_j,F(w_k^{(x_i)})>}.\tag{3.10}$$

For $k \geq k_0$ and $1 \leq i,j \leq d$, define

$$\mu_i^{(k)}(dx) := F(w_k^{(x_i)})(x)m(dx)+\widetilde{F(w_k^{(x_i)})}(x)\tilde{K}(dx)\tag{3.11}$$

and

$$\mu_{ij}^{(k)}(dx) := \frac{1}{2}\tilde{\mu}^c_{<x_j,F(w_k^{(x_i)})>}(dx).\tag{3.12}$$

Therefore, it follows from (3.8)-(3.12) that

$$\check{\mathscr{E}}^c\left(w_k^{(x_i)},\frac{\partial u}{\partial x_i}v\right) = \int_{G_k}\frac{\partial u}{\partial x_i}v\,d\mu_i^{(k)}+\sum_{j=1}^d\int_{G_k}\frac{\partial}{\partial x_j}\left(\frac{\partial u}{\partial x_i}v\right)d\mu_{ij}^{(k)}.\tag{3.13}$$

Now we obtain the following representation.

Theorem 3.6. *Let $(\mathscr{E},D(\mathscr{E}))$ be a regular (non-symmetric) local Dirichlet form on an open subset U of $\mathbf{R}^d$ satisfying $C_0^\infty(U)\subset D(\mathscr{E})$. Then, for $u,v\in C_0^\infty(U)$, $\mathscr{E}(u,v)$ has the following decomposition:*

$$\mathscr{E}(u,v) = \sum_{i,j=1}^d\int_U\frac{\partial u}{\partial x_i}\frac{\partial v}{\partial x_j}d\nu_{ij}+\sum_{i=1}^d < F_i,\frac{\partial u}{\partial x_i}v > +\int_U u(x)v(x)K(dx),$$

where $\{\nu_{ij}\}_{1\leq i,j\leq d}$ are signed Radon measures on U such that for every compact subset $F\subset U$, $\nu_{ij}(F)=\nu_{ji}(F)$ and $\sum_{i,j=1}^d\xi_i\xi_j\nu_{ij}(F)\geq 0$ for all $\xi_1,\dots,\xi_d\in\mathbf{R}^d$, $\{F_i\}_{1\leq i\leq d}$ are generalized functions on U, and K is the killing measure.

Moreover, for any relatively compact open subset G of U satisfying $G \subset \bar{G} \subset U$, there exist signed Radon measures $\{\mu_i^G\}_{1\le i\le d}$, $\{\mu_{ij}^G\}_{1\le i,j\le d}$ defined on G such that

$$< F_i, v > \;=\; \int_G v\, d\mu_i^G + \sum_{j=1}^d \int_G \frac{\partial v}{\partial x_j} d\mu_{ij}^G \tag{3.14}$$

for all $v \in C_0^\infty(G)$.

Proof. By (2.1) and (2.4), it is enough to show that there exist unique generalized functions $\{F_i\}_{1\le i\le d}$ on U such that for any $u, v \in C_0^\infty(U)$,

$$\check{\mathscr{E}}^c(u,v) = \sum_{i=1}^d < F_i, \frac{\partial u}{\partial x_i} v >, \tag{3.15}$$

and (3.14) holds.

For $1 \le i \le d$, we define the generalized function F_i by

$$< F_i, v > := \lim_{k\to\infty} \left[\int_U v\, d\mu_i^{(k)} + \sum_{j=1}^d \int_U \frac{\partial v}{\partial x_j} d\mu_{ij}^{(k)} \right], \quad v \in C_0^\infty(U). \tag{3.16}$$

where $\mu_i^{(k)}$ and $\mu_{ij}^{(k)}$ are given as above. First, we show that F_i is well defined. Suppose that $\mathrm{supp}[v] \subset G_{k_0}$ for some $k_0 \in \mathbf{N}$. Let $w \in C_0^\infty(U)$ be a function satisfying $w|_{\mathrm{supp}[v]} = x_i$. Then for any $k \ge k_0$, we obtain by (3.13) and (3.7) that

$$\begin{aligned} \int_{G_k} v\, d\mu_i^{(k)} + \sum_{j=1}^d \int_{G_k} \frac{\partial v}{\partial x_j} d\mu_{ij}^{(k)} &= \int_{G_k} \frac{\partial w}{\partial x_i} v\, d\mu_i^{(k)} + \sum_{j=1}^d \int_{G_k} \frac{\partial}{\partial x_j}\left(\frac{\partial w}{\partial x_i} v\right) d\mu_{ij}^{(k)} \\ &= < L(x_i, 1), v > . \end{aligned} \tag{3.17}$$

Thus, F_i is well defined and (3.15) follows from (3.6), (3.17) and (3.16). The uniqueness of F_i is obvious by noting that $< F_i, v >=< L(x_i,1), v >$ for any $v \in C_0^\infty(U)$. Furthermore, we fix a $k \in \mathbf{N}$ such that $\overline{G} \subset G_k$ and define $\{\mu_i^G\}_{1\le i\le d}$, $\{\mu_{ij}^G\}_{1\le i,j\le d}$ to be $\{\mu_i^{(k)}\}_{1\le i\le d}$, $\{\mu_{ij}^{(k)}\}_{1\le i,j\le d}$, respectively. Then (3.14) holds. □

Remark 3.7. The second term on the right hand side of (3.14) is essential and cannot be eliminated in general. Let us consider the following example.

Let $(\mathscr{E}, D(\mathscr{E}))$ be a regular non-symmetric local Dirichlet form on $L^2(\mathbf{R}^d; m)$ with the following expression (cf. Ma and Röckner [14, II.2. d)]).

$$\mathscr{E}(u,v) = \int_{\mathbf{R}^d} \langle \tilde{A}\nabla u, \nabla v\rangle dx + \int_{\mathbf{R}^d} \langle \check{A}\nabla u, \nabla v\rangle dx + \int_{\mathbf{R}^d} \langle \nabla u, b\rangle v dx \quad \text{for } u, v \in C_0^\infty(\mathbf{R}^d),$$

where $\tilde{A}(x) = (\tilde{a}_{ij}(x))$ is a symmetric, strictly positive definite matrix, $\check{A}(x) = (\check{a}_{ij}(x))$ is an anti-symmetric matrix with $\sup_{i,j} \|\check{a}_{ij}\|_\infty < \infty$, and $b(x) = (b_i(x), 1 \le i \le d)$ is a smooth bounded vector on $\mathbf{R}^d$. Then, we have

$$\begin{aligned}\check{\mathscr{E}}^c(u,v) &= \int_{\mathbf{R}^d}\langle \check{A}\nabla u,\nabla v\rangle dx+\int_{\mathbf{R}^d}\langle\nabla u,b\rangle v dx\\ &= \sum_{i=1}^d\int_{\mathbf{R}^d}\frac{\partial u}{\partial x_i}v b_i dx+\sum_{i=1}^d\int_{\mathbf{R}^d}\frac{\partial u}{\partial x_i}\left(\sum_{j=1}^d\frac{\partial v}{\partial x_j}(-\check{a}_{ij})\right)dx\\ &:= \sum_{i=1}^d\left\langle\frac{\partial u}{\partial x_i}v,F_i\right\rangle\end{aligned}$$

with

$$\langle w,F_i\rangle:=\int_{\mathbf{R}^d}wb_i dx+\sum_{j=1}^d\int_{\mathbf{R}^d}\frac{\partial w}{\partial x_j}(-\check{a}_{ij})dx.$$

Hence if $(\check{a}_{ij}(x))$ are not smooth, the second term on the right hand side of (3.14) cannot be eliminated. However, in some special cases it is possible to eliminate it. In the above example if $(\check{a}_{ij}(x))$ are all differentiable, then we obtain

$$\begin{aligned}\check{\mathscr{E}}^c(u,v) &= \int_{\mathbf{R}^d}\langle \check{A}\nabla u,\nabla v\rangle dx+\int_{\mathbf{R}^d}\langle\nabla u,b\rangle v dx\\ &= \sum_{i=1}^d\int_{\mathbf{R}^d}\frac{\partial u}{\partial x_i}v\,\overline{b}_i dx\end{aligned}$$

with

$$\overline{b}_i=b_i-\sum_{j=1}^d\frac{\partial\check{a}_{ij}}{\partial x_j}.$$

In view of the above observation we provide a practically useful result in Theorem 3.8 below.

Let $(\mathscr{E},D(\mathscr{E}))$ be a regular Dirichlet form on U satisfying $C_0^\infty(U)\subset D(\mathscr{E})$. Let G be a relatively compact open subset of U satisfying $G\subset\bar{G}\subset U$. We consider the following condition on $D(L)\cap D(\hat{L})$, where $(L,D(L))$ and $(\hat{L},D(\hat{L}))$ are the generators of $(\mathscr{E},D(\mathscr{E}))$ and $(\hat{\mathscr{E}},D(\mathscr{E}))$, respectively.

Condition (L^G): There exist $y_1,y_2,\dots,y_d$ in $C_0^\infty(U)\cap D(L)\cap D(\hat{L})$ such that the Jacobian determinant $|(\partial_j y_i)_{i,j=1}^d|$ is strictly positive on $\bar{G}$.

Similar to [10, Theorem 4.10], we obtain the following representation.

Theorem 3.8. *Let $(\mathscr{E},D(\mathscr{E}))$ be a regular local Dirichlet form on U satisfying $C_0^\infty(U)\subset D(\mathscr{E})$. Suppose that Condition (L^G) is satisfied for any relatively compact open subset G of U satisfying $G\subset\bar{G}\subset U$. Then, for $u,v\in C_0^\infty(U)$, $\mathscr{E}(u,v)$ has the following decomposition:*

$$\mathscr{E}(u,v) = \sum_{i,j=1}^d\int_U\frac{\partial u}{\partial x_i}\frac{\partial v}{\partial x_j}d\nu_{ij}+\sum_{i=1}^d\int_U\frac{\partial u}{\partial x_i}v d\nu_i+\int_U u(x)v(x)K(dx),$$

where $\{\nu_{ij}\}_{1\le i,j\le d}$ are signed Radon measures on U such that for every compact subset $F\subset U$, $\nu_{ij}(F)=\nu_{ji}(F)$ and $\sum_{i,j=1}^d\xi_i\xi_j\nu_{ij}(F)\ge 0$ for all $\xi_1,\dots,\xi_d\in\mathbf{R}^d$, $\{\nu_i\}_{1\le i\le d}$ are signed Radon measures on U, and K is the killing measure.

We emphasize that in the one dimensional case, a complete characterization of $(\mathscr{E}, D(\mathscr{E}))$ can be obtained based on Theorem 3.6, without using Condition (L^G). Moreover, the measure of the co-symmetric part is always absolutely continuous. See the following theorem.

Theorem 3.9. *Let $(\mathscr{E}, D(\mathscr{E}))$ be a regular (non-symmetric) local Dirichlet form on an open subset U of $\mathbf{R}^1$. We assume that $C_0^\infty(U) \subset D(\mathscr{E})$. Then, for $u, v \in C_0^\infty(U)$, $\mathscr{E}(u,v)$ has the following decomposition:*

$$\mathscr{E}(u,v) \quad = \quad \int_U u'(x)v'(x)\mu(dx) + \int_U u'(x)v(x)b(x)dx + \int_U u(x)v(x)K(dx),$$

where μ and K are positive Radon measures on U, and b is a locally integrable function on U.

We refer the reader to [10, Section 4.3] for a detailed proof of the above theorem.

Acknowledgments

The authors are grateful to the support of 973 project, NNSFC and NSERC. The authors also thank the anonymous referee for valuable suggestions that helped to improve the presentation of the first version of the paper.

References

[1] A. Beurling and J. Deny, Dirichlet spaces. *Proc. Nat. Acad. Sci. U.S.A.* **45** (1959), 208-215.

[2] L. Beznea, N. Boboc, and M. Röckner, Quasi-regular Dirichlet forms and L^p-resolvents on measurable spaces. *Potential Anal.* **25** (2006), 269-282.

[3] N. Bouleau and F. Hirsch, *Dirichlet Forms and Analysis on Wiener Space.* Walter de Gruyter, Berlin-New York, 1991.

[4] Z. Chen, Z.M. Ma, and M. Röckner, Quasi-homeomorphisms of Dirichlet forms. *Nagoya Math. J.* **136** (1994), 1-15.

[5] Z. Dong, Quasi-regular topologies for Dirichlet forms on arbitrary Hausdorff spaces. *Acta Math. Sinica* **14** (1998), 683-690.

[6] Z. Dong, Z.M. Ma, and W. Sun, A note on Beurling-Deny formulae in infinite dimensional spaces. *Acta Math. Appl. Sinica* **13** (1997), 353-361.

[7] M. Fukushima, Y. Oshima, and M. Takeda, *Dirichlet Forms and Symmetric Markov Processes.* Walter de Gruyter, Berlin-New York, 1994.

[8] Z.C. Hu and Z.M. Ma, Beurling-Deny formula of semi-Dirichlet forms. *C. R. Math. Acad. Sci. Paris* **338** (2004), 521-526.

[9] Z.C. Hu, Z.M. Ma, and W. Sun, Extensions of Lévy-Khintchine formula and Beurling-Deny formula in semi-Dirichlet forms setting. *J. Funct. Anal.* **239** (2006), 179-213.

[10] Z.C. Hu, Z.M. Ma, and W. Sun, On representations of non-symmetric Dirichlet forms. Submitted, 2008.

[11] H.H. Kuo, *Gaussian Measures in Banach Spaces.* Springer-Verlag, Berlin-Heidelberg-New York, 1975.

[12] K. Kuwae, Functional calculus for Dirichlet forms. *Osaka J. Math.* **35** (1998), 683-715.

[13] Y. LeJan, Mesures associées à une forme de Dirichlet, applications. *Bull. Soc. Math. France* **106** (1978), 61-112.

[14] Z.M. Ma and M. Röckner, *Introduction to the Theory of (Non-Symmetric) Dirichlet Forms.* Springer-Verlag, Berlin-Heidelberg-New York, 1992.

[15] Y. Oshima, *Lecture on Dirichlet Spaces.* Universität Erlangen-Nürnberg, 1988.

[16] B. Schmuland, On the local property for positivity preserving coercive forms. In: Z.M. Ma, M. Röckner and J.A. Yan (Eds.), *Dirichlet Forms and Stochastic Processes* (Proceedings of the International Conference held in Beijing, China, October 25-31, 1993), pp. 345-354. Walter de Gruyter, 1995.

DEPARTMENT OF MATHEMATICS, NANJING UNIVERSITY, NANJING 210093, CHINA
E-mail address: huzc@nju.edu.cn

INSTITUTE OF APPLIED MATHEMATICS, ACADEMY OF MATHEMATICS AND SYSTEMS SCIENCE, CHINESE ACADEMY OF SCIENCES, BEIJING 100190,CHINA
E-mail address: mazm@amt.ac.cn

DEPARTMENT OF MATHEMATICS AND STATISTICS, CONCORDIA UNIVERSITY, MONTREAL, H3G 1M8, CANADA
E-mail address: wsun@mathstat.concordia.ca

Potential Theory and Stochastics in Albac, 157–172
Aurel Cornea Memorial Volume

OPERATORS ASSOCIATED WITH MULTI-PARAMETER FAMILIES OF PROBABILITY MEASURES

NIELS JACOB, ALEXANDER POTRYKUS, AND MARKUS SCHICKS

Dedicated to the Memory of Professor Aurel Cornea

ABSTRACT. In this paper we outline that large classes of partial differential operators and pseudo-differential operators of higher order (and with multiple characteristics) are naturally associated with certain multi-parameter families of probability measures. In certain cases these families are related to multi-parameter Markov processes. In other cases a change of variables in the parameter domain may lead to multi-parameter processes. Various examples are discussed.

1. INTRODUCTION

In his paper [3] E. B. Dynkin succeeded to solve a boundary value problem for the operator $(-1)^N\Delta_1\dots\Delta_N$ acting on functions defined on $G_1\times\dots\times G_N$, $G_j\subset\mathbb{R}^{n_j}$ bounded, and Δ_j acts on the variables in G_j only. Boundary data of Dirichlet type were prescribed on a distinguished boundary $\partial G_1\times\dots\times\partial G_N$. Dynkin proved his result by probabilistic means, i.e. by using N Brownian motions. Already for $N=2$ the operator does not fit into the "classical" classes:

$$\Delta_1\Delta_2=\left(\frac{\partial^2}{\partial x_1^2}+\dots+\frac{\partial^2}{\partial x_n^2}\right)\left(\frac{\partial^2}{\partial y_1^2}+\dots+\frac{\partial^2}{\partial y_m^2}\right)=\sum_{l,k}\left(\frac{\partial^2}{\partial x_l\partial y_k}\right)^2. \tag{1.1}$$

Dynkin's proof relies much on the product structure of his problem. In [2] Doppel and the first named author gave a functional analytic, generalised solution to this problem in general bounded domains based on the work [6]. Due to the anisotropy of the underlying function spaces no trace theorems can be proved, i.e. it is open to which extent these generalised solutions are classical solutions (as are those of Dynkin). Dynkin's approach was taken up by G. Mazziotto [10] for elliptic operators of second order and the associated diffusions.

In this paper we want to introduce a wider audience to first considerations about "strange" differential or even pseudo-differential operators which relate to some (families of) stochastic processes. Eventually our starting point should be such an operator and the question is whether or not by solving certain (types of initial-value) problems we can construct probability measures which give rise to stochastic processes suitable

---

2000 *Mathematics Subject Classification.* 35E99, 35G10, 35S99, 47D99, 60G60.

Key words and phrases. Initial value problems for higher order pseudo-differential operators, random fields, multi-parameter operator semi-groups, multi-parameter Markov processes.

to represent the solution operators to the given problem, and hence allow us to investigate these operators by probabilistic means. Guided by Dynkin's example we long for multi-parameter processes, i.e. we are looking for random fields.

An apology is in order: there is a vast literature on random fields, a good recent introduction is given by the monograph [1] of R. J. Adler and J. E. Taylor. However we will almost ignore this existing theory. It seems that our "purely analytic" point of view is new and relates only partly to a special type of multi-parameter processes or random fields. For these we refer to D. Khoshnevisan [9].

After having fixed some notation in Section 2, we discuss in Section 3 how certain multi-parameter families of probability measures naturally lead to (pseudo-) differential operators and we provide a lot of concrete examples. In Section 4 we pay attention to operators associated with such a family of probability measures. We will consider multi-parameter operators $\mathrm{T}_{(s_1,\ldots,s_N)}$ giving solutions to the equation

$$\frac{\partial^N}{\partial s_1 \ldots \partial s_N} u + A(s_1,\ldots,s_n;D_x)u = 0 \tag{1.2}$$

when applied to certain "initial" data. Here $A(s_1,\ldots,s_N;D_x)$ should be seen as a type of generator of these operators as well as of these measures. In this paper we confine ourselves to the spatially invariant case which makes the Fourier transform a powerful tool. Parts of our considerations are devoted to the question whether we can transform the problem by a change of variables $(s_1,\ldots,s_N)$ into a problem with product structure. In Section 5 we give some indication how certain processes can be associated with the given family of probability measures. The basic idea is to find N one-parameter processes which relate to the structure imposed by $A(s_1,\ldots,s_N;D_x)$.

2. Notation and Definitions

Our notation is standard and can be found in [7]. By $\mathbb{R}_+^N$ we will denote the set $\{(s_1,\ldots,s_N)\in\mathbb{R}^N | s_j>0\}$. The Fourier transform is defined according to

$$\widehat{u}(\xi) = (2\pi)^{-\frac{n}{2}} \int_{\mathbb{R}^n} e^{-ix\cdot\xi} u(x)\,\mathrm{d}x, \tag{2.1}$$

which yields for the inverse Fourier transform

$$\left(\mathrm{F}^{-1}v\right)(y) = (2\pi)^{-\frac{n}{2}} \int_{\mathbb{R}^n} e^{iy\cdot\eta} v(\eta)\,\mathrm{d}\eta. \tag{2.2}$$

This normalization leads to the following form of the convolution theorem and Plancherel's theorem, respectively,

$$(\mu * \nu)^\wedge = (2\pi)^{\frac{n}{2}} \widehat{\mu}\cdot\widehat{\nu} \tag{2.3}$$

and

$$\|u\|_0 = \|\widehat{u}\|_0, \tag{2.4}$$

where $\|u\|_0$ is the L^2-norm of u. A function $\psi:\mathbb{R}^n\to\mathbb{C}$ is called continuous negative function if it is continuous and if

$$\psi(0)\geq 0 \tag{2.5}$$

and

$$\xi\mapsto e^{-t\psi(\xi)} \text{ is positive definite for } t>0. \tag{2.6}$$

Equivalently, a function is a continuous negative definite function if it has a Lévy-Khinchin representation.

3. Multi-parameter Families of Probability measures and Related Operators

Let $b : \mathbb{R}_+^N \times \mathbb{R}^n \to \mathbb{C}$ be a continuous function such that $b(s_1, \dots, s_N; \cdot) : \mathbb{R}^n \to \mathbb{C}$ is a continuous negative definite function satisfying $b(s_1, \dots, s_N; 0) = 0$ for all $(s_1, \dots, s_N) \in \mathbb{R}_+^N$. Further assume that $(s_1, \dots, s_N) \mapsto b(s_1, \dots, s_N; \xi)$ is for each $\xi \in \mathbb{R}^n$ at least N-times continuously differentiable and

$$|\partial_s^\sigma b(s_1, \dots, s_N; \xi)| \leq c_\sigma (1 + |\xi|^2)^{\frac{r}{2}} \tag{3.1}$$

holds for all $\sigma \in \mathbb{N}_0^N$, $|\sigma| \leq N$. It follows that for $(s_1, \dots, s_N) \in \mathbb{R}_+^N$ fixed, by

$$\widehat{\mu}_{(s_1,\dots,s_N)}(\xi) := (2\pi)^{-\frac{n}{2}} e^{-b(s_1,\dots,s_N;\xi)} \tag{3.2}$$

a continuous positive definite function is given. Hence we have by Bochner's theorem that $(\mu_{(s_1,\dots,s_N)})_{(s_1,\dots,s_N) \in \mathbb{R}_+^N}$ is a family of probability measures.

On $\mathscr{S}(\mathbb{R}^n)$ we define the operators

$$T_{(s_1,\dots,s_N)} u(x) := (2\pi)^{-\frac{n}{2}} \int_{\mathbb{R}^n} e^{ix\cdot\xi} e^{-b(s_1,\dots,s_N;\xi)} \widehat{u}(\xi) \, \mathrm{d}\xi. \tag{3.3}$$

Since

$$T_{(s_1,\dots,s_N)} u(x) = \int_{\mathbb{R}^n} u(x-y) \, \mu_{(s_1,\dots,s_N)}(\mathrm{d}y), \tag{3.4}$$

each operator extends obviously to a contraction on $B_b(\mathbb{R}^n)$, and by Fubini's Theorem it also extends to a contraction on $L^1(\mathbb{R}^n)$. Hence, by interpolation it gives a contraction on each space $L^p(\mathbb{R}^n)$, $1 \leq p \leq \infty$, and as in Example 4.1.3 in [7] it is easy to see that each operator gives also a contraction on $C_\infty(\mathbb{R}^n)$.

Define the symbol

$$a(s_1, \dots, s_N; \xi) := e^{b(s_1,\dots,s_N;\xi)} \frac{\partial^N}{\partial s_1 \dots \partial s_N} e^{-b(s_1,\dots,s_N;\xi)}, \tag{3.5}$$

and consider on $\mathscr{S}(\mathbb{R}^n)$ the pseudo-differential operator

$$a(s_1, \dots, s_N; D_x) u(x) := (2\pi)^{-\frac{n}{2}} \int_{\mathbb{R}^n} e^{ix\cdot\xi} a(s_1, \dots, s_N; \xi) \widehat{u}(\xi) \, \mathrm{d}\xi. \tag{3.6}$$

For $v(s_1, \dots, s_N; x) := T_{(s_1,\dots,s_N)} u(x)$, $u \in \mathscr{S}(\mathbb{R}^n)$, it follows that

$$\begin{aligned}
\frac{\partial^N}{\partial s_1 \dots \partial s_N} v(s_1, \dots, s_N; x) &= \frac{\partial^N}{\partial s_1 \dots \partial s_N} T_{(s_1,\dots,s_N)} u(x) \\
&= \frac{\partial^N}{\partial s_1 \dots \partial s_N} (2\pi)^{-\frac{n}{2}} \int_{\mathbb{R}^n} e^{ix\cdot\xi} e^{-b(s_1,\dots,s_N;\xi)} \widehat{u}(\xi) \, \mathrm{d}\xi \\
&= (2\pi)^{-\frac{n}{2}} \int_{\mathbb{R}^n} e^{ix\cdot\xi} \left(\frac{\partial^N}{\partial s_1 \dots \partial s_N} e^{-b(s_1,\dots,s_N;\xi)} \right) \widehat{u}(\xi) \, \mathrm{d}\xi \\
&= (2\pi)^{-\frac{n}{2}} \int_{\mathbb{R}^n} e^{ix\cdot\xi} a(s_1, \dots, s_N; \xi) e^{-b(s_1,\dots,s_N;\xi)} \widehat{u}(\xi) \, \mathrm{d}\xi \\
&= a(s_1, \dots, s_N; D_x) T_{(s_1,\dots,s_N)} u(x) \\
&= a(s_1, \dots, s_N; D_x) v(s_1, \dots, s_N; x).
\end{aligned}$$

Thus for $u \in \mathscr{S}(\mathbb{R}^n)$ and $v(s_1,\dots,s_N;x) := T_{(s_1,\dots,s_N)}\,u(x)$ we find that v solves the equation

$$\frac{\partial^N}{\partial s_1 \dots \partial s_N} v(s_1,\dots,s_N;x) - a(s_1,\dots,s_N;D_x)\, v(s_1,\dots,s_N;x) = 0. \tag{3.7}$$

We will later on study the behaviour of $v(s_1,\dots,s_N;x)$ as $s_j \to 0$, $1 \le j \le N$, but first of all we want to collect some examples.

Example 3.1. For continuous negative definite functions $\psi_j : \mathbb{R}^n \to \mathbb{C}$, $1 \le j \le N$, consider

$$b(s_1,\dots,s_N;\xi) := \sum_{j=1}^{N} s_j \psi_j(\xi). \tag{3.8}$$

A straightforward calculation yields

$$a(s_1,\dots,s_N;\xi) = (-1)^N \prod_{j=1}^{N} \psi_j(\xi). \tag{3.9}$$

Note that this includes the case when $\mathbb{R}^n$ is decomposed according to $\mathbb{R}^n = R^{n_1} \times \dots \times \mathbb{R}^{n_N}$, $N \le n$, and ψ_j depends only on ξ_l for $n_1 + \dots + n_{j-1} < l \le n_1 + \dots + n_{j+1}$. If we write $\psi_j(\xi) = \psi_j(\xi^{(j)})$, $\xi^{(j)} \in \mathbb{R}^{n_j}$, we find

$$a(s_1,\dots,s_N;\xi) = (-1)^N \prod_{j=1}^{N} \psi_j(\xi^{(j)}). \tag{3.10}$$

We want to list some rather concrete cases:

(1) If we take $\psi_j(\xi) = |\xi|^2$ then $a(s_1,\dots,s_N;D_x) = (\Delta_x)^N$.

(2) For $\psi_j(\xi^{(j)}) = |\xi^{(j)}|^2$ we obtain $a(s_1,\dots,s_N;D_x) = \Delta_{x^{(1)}} \dots \Delta_{x^{(N)}}$ where $\Delta_{x^{(j)}}$ is the Laplacian acting only on the variable $x^{(j)}$. This operator was investigated by E. B. Dynkin [3]. Note that for $N = 2$, $n_1 = n_2 = 1$, it follows that

$$a(s_1,s_2;D_x) = \frac{\partial^4}{\partial x_1^2 \partial x_2^2} = \left(\frac{\partial^2}{\partial x_1 \partial x_2}\right)^2, \tag{3.11}$$

i.e. it is the square of the wave operator.

(3) For $n = 1$ consider drifts $\xi \mapsto id\xi$, $d \in \mathbb{R}$, and diffusions $\xi \mapsto c\xi^2$, $c > 0$. We can obtain any operator of the type

$$a(s_1,\dots,s_N;D_x) = \lambda \frac{\partial^k}{\partial x^k} \quad ,0 < k \le 2N,\ \lambda \in \mathbb{R}. \tag{3.12}$$

Example 3.2. Let us switch for a moment our point of view. Given the two independent Lévy processes $(X_s^{(1)})_{s\ge 0}$ and $(X_t^{(2)})_{t\ge 0}$ each starting at $0 \in \mathbb{R}^n$. Let ψ_1 and ψ_2 denote the two corresponding continuous negative definite functions. We may consider the family

$$\begin{aligned} T_{s,t}\,u(x) &= T_s^{(1)}\,u(x) + T_t^{(2)}\,u(x) &(3.13)\\ &= \mathbb{E}^x\left(u(X_s^{(1)})\right) + \mathbb{E}^x\left(u(X_t^{(2)})\right) &(3.14)\\ &= (2\pi)^{-\frac{n}{2}} \int_{\mathbb{R}^n} e^{ix\cdot\xi}\left(e^{-s\psi_1(\xi)} + e^{-t\psi_2(\xi)}\right)\hat{u}(\xi)\,\mathrm{d}\xi, &(3.15) \end{aligned}$$

where (3.13)-(3.14) hold for $u \in B_b(\mathbb{R}^n)$ and (3.15) is at least for $u \in \mathcal{S}(\mathbb{R}^n)$ correct. For all $u \in B_b(\mathbb{R}^n)$ we find

$$\frac{\partial^2}{\partial t \partial s} T_{s,t}\, u = 0. \tag{3.16}$$

It is of some interest to relate the result to the standard theory of the simple hyperbolic equation $\frac{\partial^2}{\partial t \partial s} v(s,t) = 0$. The general solution is given by $v(s,t) = h(s) + g(t)$ for two (arbitrary) functions $h, g : \mathbb{R} \to \mathbb{R}$. For $s = 0$ we find $v(0,t) = h(0) + g(t)$ and for $t = 0$ we have $v(s,0) = h(s) + g(0)$. The lines $s = 0$ and $t = 0$ form the characteristic curves of the operator $\frac{\partial^2}{\partial t \partial s}$, hence the problem

$$\frac{\partial^2 v}{\partial s \partial t} = 0, \tag{3.17}$$

$$v|_{\{s=0\}\cup\{t=0\}} = f, \tag{3.18}$$

where $f(s,t) := \begin{cases} h(s) + g(0), & t = 0; \\ g(t) + h(0), & s = 0; \end{cases}$ is a "characteristic initial value problem" which is solved by $v(s,t) = h(s) + g(t)$. Since in our case

$$v(s,t) = T_s^{(1)}\, u(x) + T_t^{(2)}\, u(x)$$

we find of course $v(0,t) = u(x) + T_t^{(2)}\, u(x)$ and $v(s,0) = u(x) + T_s^{(1)}\, u(x)$. This is not an isolated situation. In the following we will often have to handle "non-standard" problems for the (pseudo-differential) operators under consideration.

Example 3.3. Let $\psi_j : \mathbb{R}^n \to \mathbb{C}$, $j = 1,2$, be two continuous negative definite functions and let $d : \mathbb{R}_+^2 \times \mathbb{R}^n \to \mathbb{C}$, $(s,t;\xi) \to d(s,t;\xi)$ be a function smooth in s and t and continuous negative definite with respect to ξ such that $d(0,t;\xi) = d(s,0;\xi) = d(s,t;0) = 0$.

Now consider

$$b(s,t;\xi) := s\psi_1(\xi) + t\psi_2(\xi) + d(s,t;\xi). \tag{3.19}$$

A straightforward calculation gives

$$\begin{aligned} a(s,t;\xi) =& \psi_1(\xi)\,\psi_2(\xi) + \psi_1(\xi)\frac{\partial}{\partial t} d(s,t;\xi) + \psi_2(\xi)\frac{\partial}{\partial s} d(s,t;\xi) \\ &+ \frac{\partial}{\partial s} d(s,t;\xi)\,\frac{\partial}{\partial t} d(s,t;\xi) - \frac{\partial^2}{\partial s \partial t} d(s,t;\xi). \end{aligned} \tag{3.20}$$

We want to use (3.19) and (3.20) to get more insight into the equation

$$\frac{\partial^2}{\partial s \partial t} v(s,t;x) - a(s,t;D_x)\, v(s,t;x) = 0. \tag{3.21}$$

We start with a simple observation using (3.9). For $N = 2$ we may take $\psi_1(\xi) = |\xi|^\alpha$, $0 \le \alpha \le 2$, and $\psi_2(\xi) = |\xi|^\beta$, $0 \le \beta \le 2$, which results in $a(s,t;\xi) = |\xi|^{\alpha+\beta}$. Obviously there are many decompositions of $|\xi|^{\alpha+\beta} = |\xi|^{\alpha'+\beta'}$, $\alpha' + \beta' = \alpha + \beta$. Hence, given $a(s,t;\xi)$ is in general not sufficient to fix a certain "structure" of the corresponding measures. However, appropriate initial data may, but need not, remedy the situation.
In the case of (3.19) and (3.20) we find that

$$v(s,t;\xi) := T_{(s,t)}\, u(x) = (2\pi)^{-\frac{n}{2}} \int_{\mathbb{R}^n} e^{i x\cdot\xi}\, e^{-s\psi_1(\xi) - t\psi_2(\xi) - d(s,t;\xi)}\, \widehat{u}(\xi)\, \mathrm{d}\xi \tag{3.22}$$

solves for $u \in \mathscr{S}(\mathbb{R}^n)$ the equation

$$\frac{\partial^2}{\partial s \partial t} v(s,t;x) - a(s,t;D_x)\, v(s,t;x) = 0. \tag{3.23}$$

Moreover we find

$$\begin{aligned}\lim_{s\to 0} v(s,t;x) &= \lim_{s\to 0} T_{(s,t)}\, u(x)\\ &= (2\pi)^{-\frac{n}{2}} \int_{\mathbb{R}^n} e^{i x\cdot\xi}\, e^{-t\psi_2(\xi)}\, \widehat{u}(\xi)\, \mathrm{d}\xi = T_t^{(2)}\, u(x),\end{aligned}$$

and

$$\begin{aligned}\lim_{t\to 0} v(s,t;x) &= \lim_{t\to 0} T_{(s,t)}\, u(x)\\ &= (2\pi)^{-\frac{n}{2}} \int_{\mathbb{R}^n} e^{i x\cdot\xi}\, e^{-s\psi_1(\xi)}\, \widehat{u}(\xi)\, \mathrm{d}\xi = T_s^{(1)}\, u(x),\end{aligned}$$

where $\left(T_s^{(1)}\right)_{s\in\mathbb{R}_+}$ and $\left(T_t^{(2)}\right)_{t\in\mathbb{R}_+}$ denote the operator semigroups associated with ψ_1 and ψ_2, respectively. Note that they are Feller semigroups as well as L^p-sub-Markovian semigroups. Thus we find that $v(s,t;x) := T_{(s,t)}\, u(x)$ solves (3.23) under the initial conditions

$$v(0,t;x) = T_t^{(2)}\, u(x), \tag{3.24}$$

and

$$v(s,0;x) = T_s^{(1)}\, u(x). \tag{3.25}$$

It is clear that for $u \in \mathscr{S}(\mathbb{R}^n)$ the solution to (3.23)-(3.25) is unique. Note further that the case $d(s,t;x) \equiv 0$ is included and (3.24) together with (3.25) prevent the non-uniqueness due to different decompositions of $a(s,t;\xi)$ as mentioned above. Let us point out that the given data is $u \in \mathscr{S}(\mathbb{R}^n)$, not two arbitrary functions on $\mathbb{R}_+ \times \mathbb{R}^n$, or more precisely one function defined on $\{0\} \times \mathbb{R}_+ \times \mathbb{R}^n \cup \mathbb{R}_+ \times \{0\} \times \mathbb{R}^n$. Thus data are given on $\{0\} \times \{0\} \times \mathbb{R}^n$ which we can consider in analogy with the theory of functions of several complex variables as a distinguished boundary. However, looking at (3.23) as an operator-valued equation with independent variables s and t, we again encounter the fact that we are dealing with a type of "characteristic initial value problem".

Example 3.4. In the situation described in Example 3.3 let us assume for a moment that $d(s,t;\xi) \equiv 0$ and that ψ_1 and ψ_2 are real-valued. We may consider the two unitary one-parameter groups given for $u \in \mathscr{S}(\mathbb{R}^n)$ by

$$U_s^{(1)}\, u(x) := (2\pi)^{-\frac{n}{2}} \int_{\mathbb{R}^n} e^{i x\cdot\xi}\, e^{i s\psi_1(\xi)}\, \widehat{u}(\xi)\, \mathrm{d}\xi \tag{3.26}$$

and

$$U_t^{(2)}\, u(x) := (2\pi)^{-\frac{n}{2}} \int_{\mathbb{R}^n} e^{i x\cdot\xi}\, e^{-i t\psi_2(\xi)}\, \widehat{u}(\xi)\, \mathrm{d}\xi \tag{3.27}$$

as well as the operators

$$W_{(s,t)}\, u(x) := (2\pi)^{-\frac{n}{2}} \int_{\mathbb{R}^n} e^{i x\cdot\xi}\, e^{i s\psi_1(\xi) - i t\psi_2(\xi)}\, \widehat{u}(\xi)\, \mathrm{d}\xi. \tag{3.28}$$

We find that

$$\begin{aligned}
&\frac{\partial^2}{\partial s \partial t} W_{(s,t)}\, u(x)\\
=&(2\pi)^{-\frac{n}{2}} \int_{\mathbb{R}^n} e^{i x\cdot\xi} \big(i\psi_1(\xi)\big)\big(-i\psi_2(\xi)\big)\, e^{is\psi_1(\xi)-it\psi_2(\xi)}\, \widehat{u}(\xi)\, \mathrm{d}\xi\\
=&(\psi_1\cdot\psi_2)\,(D_x)\, W_{(s,t)}\, u(x),
\end{aligned}$$

i.e. $w(s,t;\xi) := W_{(s,t)}\, u(x)$ solves

$$\frac{\partial^2}{\partial s \partial t} w(s,t;\xi) - (\psi_1\cdot\psi_2)\,(D_x)\, w(s,t;x) = 0, \tag{3.29}$$

and

$$\lim_{s\to 0} w(s,t;x) = U_t^{(2)}\, u(x), \tag{3.30}$$

as well as

$$\lim_{t\to 0} w(s,t;x) = U_t^{(1)}\, u(x). \tag{3.31}$$

Comparing (3.23)-(3.25) and (3.29)-(3.31) for $d(s,t;\xi)\equiv 0$ and ψ_1, ψ_2 real-valued shows that we cannot expect that a (unique) solution to

$$\frac{\partial^N}{\partial s_1\ldots\partial s_N} v(s_1,\ldots,s_N;x) - a(s_1,\ldots,s_N;x) = 0 \tag{3.32}$$

and

$$v(s_1,\ldots,s_{j-1},\ldots,s_N;x) = g(s_1,\ldots,s_{j-1},s_{j+1},\ldots,s_N;x),\ 1\le j\le N, \tag{3.33}$$

will in general give rise to a family of probability measures.

Note that the unitary groups $U_s^{(1)}$, $U_t^{(2)}$ reflect again the partially hyperbolic nature of our problem.

Example 3.5. In Example 3.3 the operator $a(s,t;D_x)$ may have any order between 0 and 4. It is noteworthy that $(\psi_1\cdot\psi_2)(D)$ need not be the dominating term in $a(s,t;D_x)$. To see this take $\psi_1(\xi)=|\xi|^{\frac14}$, $\psi_2(\xi)=|\xi|^{\frac14}$ and $d(s,t;\xi)=st|\xi|^{\frac34}$, which gives

$$a(s,t;\xi) = |\xi|^{\frac12} + (s+t)\,|\xi| + st\,|\xi|^{\frac32} - |\xi|^{\frac34}, \tag{3.34}$$

and this observation clearly extends to the general case, i.e. the case

$$b(s_1,\ldots,s_N;\xi) = \sum_{j=1}^{N} s_j\,\psi_j(\xi) + d(s_1,\ldots,s_N;\xi). \tag{3.35}$$

The following example is a special case of Example 3.3 but gives a further interesting feature.

Example 3.6. Let $g,h:[0,\infty)\to\mathbb{R}$ be two smooth functions such that $g|_{[0,\tau_1]}=0$ and $g|_{[\tau_2,\infty)}=\mathrm{id}$ for $0<\tau_1<\tau_2$, as well as $h|_{[0,\sigma_1]}=0$ and $h|_{[\sigma_2,\infty)}=\mathrm{id}$ for $0<\sigma_1<\sigma_2$. Further let $\psi_j:\mathbb{R}^n\to\mathbb{C}$, $j=1,2$, be two continuous negative definite functions. We consider the two-parameter family of measures given by

$$\widehat{\mu}_{(s,t)}(\xi) = (2\pi)^{-\frac{n}{2}}\, e^{-(s+t)\,\psi_1(\xi)-(h(s)+g(t))\psi_2(\xi)}. \tag{3.36}$$

The corresponding operator $a(s,t;D_x)$ has now the symbol

$$a(s,t;\xi) = \big(\psi_1(\xi)+h'(s)\,\psi_2(\xi)\big)\big(\psi_1(\xi)+g'(t)\psi_2(\xi)\big). \tag{3.37}$$

In particular for $0 < s < \sigma_1$ and $0 < t < \tau_1$ we have

$$a(s,t;\xi) = \psi_1(\xi)^2 \tag{3.38}$$

and for $\sigma_2 < s$ and $\tau_2 < t$ we find

$$a(s,t;\xi) = \big(\psi_1(\xi) + \psi_2(\xi)\big)^2. \tag{3.39}$$

Moreover, introducing the semigroups $(T_t^{\psi_1})_{t\geq 0}$, $(T_t^{\psi_2})_{t\geq 0}$ and $(T_t^{\psi_1+\psi_2})_{t\geq 0}$ we find for the family of operators $(T_{(s,t)})_{(s,t)\in\mathbb{R}_+^2}$ associated with the measures (3.36) in the usual manner that for $0 < s < \sigma_1$ and $0 < t < \tau_1$ it holds

$$T_{(s,t)}\, u(x) = T_s^{\psi_1} \circ T_t^{\psi_2}\, u = T_{s+t}^{\psi_1}\, u \tag{3.40}$$

where as for $\sigma_2 < s$ and $\tau_2 < t$ it follows that

$$T_{(s,t)}\, u = T_s^{\psi_1+\psi_2} \circ T_t^{\psi_1+\psi_2} = T_{s+t}^{\psi_1} \circ T_{s+t}^{\psi_2}. \tag{3.41}$$

So far we may summarise our considerations as follows: Given for each $s \in \mathbb{R}_+^N$ a probability measure μ_s on $\mathbb{R}^n$ by (3.2) we can introduce operators T_s, $s \in \mathbb{R}_+^N$, by (3.3) or (3.4). These operators applied to suitable functions defined on $\mathbb{R}^n$ give solutions to certain initial value problems for operators (mainly pseudo-differential operators) which in general do not belong to any standard class. Moreover, in order to obtain uniqueness or positivity preserving solution operators we may have to consider characteristic initial value problems or we have to introduce distinguished boundaries. Initial data might be obtained by extending data given on a distinguished boundary to a (sometimes characteristic) manifold, typical extension operators will be operator semigroups. With these remarks in mind it is easy to transform Example 3.3 into a general theorem (with an obvious proof).

Theorem 3.7. *Let $\psi_j : \mathbb{R}^n \to \mathbb{C}$, $1 \leq j \leq N$, be continuous negative definite functions such that $\psi_j(0) = 0$. Furthermore let $d : \mathbb{R}_+^N \times \mathbb{R}^n \to \mathbb{C}$, $(s_1,\dots,s_N;\xi) \mapsto d(s_1,\dots,s_N;\xi)$ be a function such that $d(\cdot;\xi)$ is of class C^N and $\xi \mapsto d(s_1,\dots,s_N;\xi)$ is a continuous negative definite function satisfying $d(s_1,\dots,s_N;\xi) = 0$ whenever at least one of the "time" coordinates s_j, $1 \leq j \leq N$, is equal to 0, or when $\xi = 0$. Set*

$$b(s_1,\dots,s_N;\xi) := \sum_{j=1}^N s_j\psi_j(\xi) + d(s_1,\dots,s_N;\xi). \tag{3.42}$$

Then we find

$$\begin{aligned} a(s_1,\dots,s_N;\xi) &= e^{b(s_1,\dots,s_N;\xi)} \frac{\partial^N}{\partial s_1 \dots \partial s_N} e^{-b(s_1,\dots,s_N;\xi)} \\ &= (-1)^N \left(\prod_{j=1}^N \psi_j(\xi) + h(s_1,\dots,s_N;\xi) \right). \end{aligned} \tag{3.43}$$

Denoting by $(T_{s_j}^{(j)})_{s_j\in\mathbb{R}_+}$ the semigroup associated with the continuous negative definite function ψ_j, $1 \leq j \leq N$, we find that for every $u \in \mathscr{S}(\mathbb{R}^n)$ a solution to

$$\frac{\partial^N}{\partial s_1 \dots \partial s_N} v(s_1,\dots,s_N;x) - a(s_1,\dots,s_N;D_x)\, v(s_1,\dots,s_N;x) = 0 \tag{3.44}$$

and

$$\lim_{s_j\to 0} v(s_1,\dots,s_N;x) = \tilde{T}^{(j)}_{(s_1,\dots,s_N)}\, u(x) \tag{3.45}$$

is given by

$$v(s_1,\dots,s_N;x) := T_{(s_1,\dots,s_N)}\, u(x) := (2\pi)^{-\frac{n}{2}} \int_{\mathbb{R}^n} e^{i\,x\cdot\xi}\, e^{-b(s_1,\dots,s_N;\xi)}\, \widehat{u}(\xi)\,\mathrm{d}\xi, \tag{3.46}$$

where we have used the definition

$$\tilde{T}^{(j)}_{(s_1,\dots,s_N)}\, u(x) := (2\pi)^{-\frac{n}{2}} \int_{\mathbb{R}^n} e^{i\,x\cdot\xi}\, e^{-\sum_{l=1,l\neq j}^{N} s_l\psi_l(\xi)}\, \widehat{u}(\xi)\,\mathrm{d}\xi. \tag{3.47}$$

Remark 3.8. Using more detailed knowledge about the operators $\tilde{T}^{(j)}_{(s_1,\dots,s_N)}$, i.e. the semigroups $(T^{(j)}_{s_j})_{s_j\in\mathbb{R}_+}$ will allow us to handle easily the case when initial data u are taken from $L^p(\mathbb{R}^n)$, $1<p<\infty$, or $u\in C_\infty(\mathbb{R}^n)$.

Although Theorem 3.7 gives some insight into the problem on how to construct certain random fields by starting with analytic data (operators, symbols, semigroups), from the analytical point of view it is not very deep since in some sense we started with the solution. However, we can now formulate a more challenging problem:

Given a pseudo-differential operator $q(s_1,\dots,s_N;D_x)$ acting on $\mathscr{S}(\mathbb{R}^n)$. Find conditions on the symbol $q(s_1,\dots,s_N;\xi)$ and appropriate initial conditions such that the solution to

$$\frac{\partial^N}{\partial s_1\dots\partial s_N}\, v(s_1,\dots,s_N;x) - q(s_1,\dots,s_N;D_x)\, v(s_1,\dots,s_N;x) = 0 \tag{3.48}$$

and

$$\lim_{s_j\to 0} v(s_1,\dots,s_N;x) = \tilde{T}^{(j)}_{(s_1,\dots,s_N)}\, u(x) \tag{3.49}$$

for certain families of operators $\tilde{T}^{(j)}_{(s_1,\dots,s_N)}$ not depending on s_j, can be expressed by a family of operators $(T_{(s_1,\dots,s_N)})_{(s_1,\dots,s_N)\in\mathbb{R}^N_+}$ such that each operator $T_{(s_1,\dots,s_N)}\, u(x)$ extends to $B_b(\mathbb{R}^n)$ as a Markov operator, i.e.

$$T_{(s_1,\dots,s_N)}\, u(x) = \int_{\mathbb{R}^n} u(y)\, p_{s_1,\dots,s_N}(x,\mathrm{d}y) \tag{3.50}$$

with a suitable Markov kernel $p_{s_1,\dots,s_N}(x,\mathrm{d}y)$.

4. Considerations on Multi-parameter Families of Operators

Let us return to Theorem 3.7 but assume $d(s_1,\dots,s_N;\xi)\equiv 0$. Then the solution to (3.44) and (3.45) is given by

$$\begin{aligned} v(s_1,\dots,s_N;x) &= T_{(s_1,\dots,s_N)}\, u(x) \\ &= (2\pi)^{-\frac{n}{2}} \int_{\mathbb{R}^n} e^{ix\xi} e^{-\sum_{j=1}^{N} s_j\psi_j(\xi)}\, \widehat{u}(\xi)\,\mathrm{d}\xi \\ &= T^{(1)}_{s_1}\circ\dots\circ T^{(N)}_{s_N}\, u(x) \end{aligned} \tag{4.1}$$

where

$$T^{(j)}_{s_j}\, u(x) = (2\pi)^{-\frac{n}{2}} \int_{\mathbb{R}^n} e^{ix\xi} e^{-s_j\psi_j(\xi)}\, \widehat{u}(\xi)\,\mathrm{d}\xi. \tag{4.2}$$

Since each of the operators $T^{(j)}_{s_j}$, $s_j \geq 0$, $1 \leq j \leq N$, is translation invariant we have for all $1 \leq j, l \leq N$, and $s_j \geq 0$, $s_l \geq 0$ that

$$[T^{(j)}_{s_j}, T^{(l)}_{s_l}] = 0, \tag{4.3}$$

i.e. the operators commute. This property follows also from a different consideration which has more far reaching consequences. With its natural addition $\mathbb{R}^N_+$ forms a commutative semigroup and this is inherited by the family $\left(T_{(s_1,\dots,s_N)}\right)_{(s_1,\dots,s_N)\in\mathbb{R}^N}$ i.e. we have

$$T_{(s_1,\dots,s_N)} \circ T_{(t_1,\dots,t_N)} = T_{(s_1+t_1,\dots,s_N+t_N)} \tag{4.4}$$

and

$$T_0 = T_{(0,\dots,0)} = \text{id}. \tag{4.5}$$

In the discussion of multi-parameter Markov processes, compare for example D. Khoshnevisan [9], such multi-parameter operator semigroups play a central role. Although many results are obtained in a straightforward way from the theory of one parameter operator semigroups, there are some non-trivial considerations needed when dealing with generators. In M. Schicks [12] a systematic treatment of multi-parameter operator semigroups is given, here we quote some results and start with

Definition 4.1. *A. An N-parameter family $(T_t)_{t\in\mathbb{R}^N_+}$, $t = (t_1,\dots,t_N)$, of linear operators $T_t : X \to X$ in a Banach space X is called an* N-parameter semigroup of operators, *if*

$$T_0 = \text{id} \tag{4.6}$$

and

$$T_{s+t} = T_s \circ T_t \quad \textit{for all } s,t \in \mathbb{R}^N_+. \tag{4.7}$$

B. We call $(T_t)_{t\geq 0}$ strongly continuous *if*

$$\lim_{t\to 0} \| T_t u - u \|_X = 0 \tag{4.8}$$

for all $u \in X$.
C. If

$$\| T_t \| \leq 1 \quad \textit{for all } t \in \mathbb{R}^N_+ \tag{4.9}$$

we call $(T_t)_{t\geq\mathbb{R}^N_+}$ an N-parameter contraction semigroup.

The concepts of a Feller semigroup and of an L^p-sub-Markovian semigroup extend in an obvious manner. It is important to observe that (4.7) implies $[T_s, T_t] = 0$ for all $s,t \in \mathbb{R}^N_+$. In particular, introducing the N *marginal semigroups*

$$T^{(j)}_{t_j} u := T_{t_j e_j} u, \tag{4.10}$$

where $t_j \in \mathbb{R}_+$ and e_j is the jth unit vector in $\mathbb{R}^n$, we find that

$$T_t = T^{(1)}_{t_1} \circ \dots \circ T^{(N)}_{t_N} = T^{(\pi(1))}_{t_{\pi(1)}} \circ \dots \circ T^{(\pi(N))}_{t_{\pi(N)}} \tag{4.11}$$

for any permutation $\pi : \mathbb{N}^N \to \mathbb{N}^N$. It is exactly this observation which makes part of the theory of multi-parameter operator semigroups trivial.

However more care is needed to investigate an appropriate concept of a generator for $(T_t)_{t\in\mathbb{R}_+^N}$. Denoting by $(A^{(j)}, D(A))$ the generator of $(T_{t_j}^{(j)})_{t_j\geq 0}$ and recalling the fundamental equation

$$\frac{\partial}{\partial t_j} T_{t_j}^{(j)} u = A^{(j)} u, \quad u \in D(A^{(j)}), \tag{4.12}$$

one is led to

$$\frac{\partial^N}{\partial t_1 \ldots \partial t_N} T_t u = Au := A^{(1)} \circ \ldots \circ A^{(N)} u \tag{4.13}$$

for suitable u. We refer to M. Schicks [12] where the domain of A is determined and related to those of $A^{(j)}$ and where other properties of $\big(A, D(A)\big)$ were discussed. Furthermore the case of multi-parameter operator semigroups related to multi-parameter convolution semigroups is treated as well as multi-parameter subordination in the sense of Bochner.

In this note we are interested in a situation beyond the semigroup case. Let us return to the operator family (3.3), i.e.

$$T_{(s_1,\ldots,s_N)} u(x) = (2\pi)^{-\frac{n}{2}} \int_{\mathbb{R}^n} e^{ix\xi} e^{-b(s_1,\ldots,s_N;\xi)} \widehat{u}(\xi)\, \mathrm{d}\xi. \tag{4.14}$$

Of course we are interested in properties of the mapping(s) $(s_1,\ldots,s_N) \longmapsto T_{(s_1,\ldots,s_N)}$ (or $(s_1,\ldots,s_N) \longmapsto T_{(s_1,\ldots,s_N)} u$, or $(s_1,\ldots,s_N) \longmapsto T_{(s_1,\ldots,s_N)} u(x)$). In case $(s_1,\ldots,s_N) \longmapsto T_{(s_1,\ldots,s_N)} u$ is differentiable we may approximate this mapping by a linear operator. However we may ask whether we can find a change of variable in $\mathbb{R}_+^N$ which leads to a family of operators having a simple (algebraic) structure. Most interesting is of cause the situation where we can find a bijective mapping

$$\begin{aligned} G : \mathbb{R}_+^N &\to \mathbb{R}_+^N, \\ (s_1,\ldots,s_N) &\longmapsto (\sigma_1,\ldots,\sigma_N) \end{aligned}$$

i.e. $\sigma_j = g_j(s_1,\ldots,s_N)$, $g_j : \mathbb{R}_+^N \to \mathbb{R}_+$, such that the family

$$\big(S_{(\sigma_1,\ldots,\sigma_N)}\big)_{(\sigma_1,\ldots,\sigma_N)\in\mathbb{R}_+^N}, \tag{4.15}$$

where $S_{(\sigma_1,\ldots,s_N)} = T_{(s_1,\ldots,s_N)}$ is an N-parameter semigroup of operators. Here is a simple example adopted from [12].

Example 4.2. Let $\psi : \mathbb{R}^n \to \mathbb{C}$ be a fixed continuous negative definite function and define the family of measures $(\mu_{(s,t)})_{s,t\in\mathbb{R}_+^2}$ by

$$\widehat{\mu}_{(s,t)}(\xi) = (2\pi)^{-\frac{n}{2}} e^{-(s+t+t\arctan(s)\psi(\xi))}. \tag{4.16}$$

We denote the corresponding two-parameter family of operators (say on $\mathscr{S}(\mathbb{R}^n)$) by $(T_{s,t})_{(s,t)\in\mathbb{R}_+^2}$. The change of variable

$$\begin{aligned} \sigma &= \sigma(s,t) := s, \\ \tau &= \tau(s,t) := \big(1 + \arctan(s)\big)t, \end{aligned} \tag{4.17}$$

with inverse

$$\begin{aligned} s &= \sigma, \\ t &= \frac{\tau}{1+\arctan(\sigma)}, \end{aligned} \tag{4.18}$$

is globally (in $\mathbb{R}^2_+$) smooth since

$$\det\begin{pmatrix} \frac{\partial\sigma}{\partial s} & \frac{\partial\sigma}{\partial t} \\ \frac{\partial\tau}{\partial s} & \frac{\partial\tau}{\partial t} \end{pmatrix} = 1+\arctan(s) \neq 0 \text{ for } s \geq 0. \tag{4.19}$$

We may introduce the family of measures

$$\eta_{(\sigma,\tau)} := \mu_{(s,t)} \tag{4.20}$$

and we find

$$\widehat{\eta}_{(\sigma,\tau)}(\xi) = (2\pi)^{-\frac{n}{2}} e^{-\sigma\psi(\xi)-\tau\psi(\xi)}. \tag{4.21}$$

Hence the operator family $(S_{(\sigma,\tau)})_{(\sigma,\tau)\in\mathbb{R}^2_+}$ defined by

$$\begin{aligned} S_{(\sigma,\tau)} u(x) &= (2\pi)^{-\frac{n}{2}} \int_{\mathbb{R}^n} e^{ix\xi} e^{-\sigma\psi(\xi)-\tau\psi(\xi)} \widehat{u}(\xi)\,\mathrm{d}\xi \\ &\left(= (2\pi)^{-\frac{n}{2}} \int_{\mathbb{R}^n} e^{ix\xi} e^{-(s+t+t\arctan(s))\psi(\xi)} \widehat{u}(\xi)\,\mathrm{d}\xi = T_{(s,t)} u(x)\right) \end{aligned} \tag{4.22}$$

is a two-parameter operator semigroup.

In general it will not be possible to find a global change of coordinates in $\mathbb{R}^N_+$ which will transform $\left(T_{s_1,\ldots,s_N)}\right)_{(s_1,\ldots,s_N)\in\mathbb{R}^N_+}$ into an N-parameter operator semigroup. However it could be possible that $\mathbb{R}^N_+$ is decomposed into sets $R_l \subset \mathbb{R}^N_+$, $l \in I$, such that $\bigcup_{l\in I} R_l \cup P = \mathbb{R}^N_+$ and for each set R_l, $l \in I$, we can find a transformation $G_l : R_l \to \mathbb{R}^N_+$, $(s_1,\ldots,s_N) \mapsto (\sigma_1^{(l)},\ldots,\sigma_N^{(l)})$ such that

$$S^{(l)}_{(\sigma_1^{(l)},\ldots,\sigma_N^{(l)})} u = T_{(s_1,\ldots,s_N)} u,\ (s_1,\ldots,s_N) \in R_l, \tag{4.23}$$

and $(S^{(l)}_{(\sigma_1^{(1)},\ldots,\sigma_N^{(l)})})$ is an N-parameter operator semigroup. Already in the case $N = 1$ we encounter such situations of some interest. Consider $\widehat{\mu}_t(\xi) := (2\pi)^{-\frac{n}{2}} e^{-|\sin(t)|\psi(\xi)}$ with a given continuous negative definite function $\psi : \mathbb{R}^n \to \mathbb{C}$. The corresponding family of operators

$$T_t u(x) = (2\pi)^{-\frac{n}{2}} \int_{\mathbb{R}^n} e^{ix\cdot\xi} e^{-|\sin(t)|\psi(\xi)} \widehat{u}(\xi)\,\mathrm{d}\xi \tag{4.24}$$

is with respect to t a π-periodic function. Hence to study the mapping $t \mapsto T_t$ we need only to consider its restriction to $[-\frac{\pi}{2},\frac{\pi}{2})$. Since $\sin(t) = -\sin(-t)$ the function $t \mapsto |\sin(t)|$, i.e. $t \mapsto T_t$, is not injective on $[-\frac{\pi}{2},\frac{\pi}{2})$, but the restrictions to $[-\frac{\pi}{2},0]$ and $[0,\frac{\pi}{2}]$ are. By symmetry we can restrict ourselves to the case $[0,\frac{\pi}{2}]$. On $[0,\frac{\pi}{2}]$ the change of coordinate $\sigma = \sin(t)$, i.e. $t = \arcsin(\sigma)$, $0 \leq \sigma \leq 1$, yields

$$\begin{aligned} S_\sigma u(x) &= (2\pi)^{-\frac{n}{2}} \int_{\mathbb{R}^n} e^{ix\xi} e^{-\sigma\psi(\xi)} \widehat{u}(\xi)\,\mathrm{d}\xi \\ &= (2\pi)^{-\frac{n}{2}} \int_{\mathbb{R}^n} e^{ix\xi} e^{-\sin(t)\psi(\xi)} \widehat{u}(\xi)\,\mathrm{d}\xi = T_t u(x). \end{aligned}$$

Note that S_σ is defined for all $\sigma \geq 0$, and indeed $(S_\sigma)_{\sigma\geq 0}$ is an operator semigroup. However (modulo periodicity) the relation $S_\sigma = T_s$ holds only on $[0, \frac{\pi}{2}]$. It is also noteworthy that $t \mapsto T_t$ degenerates to the identity for all points $t = k\pi$, $k \in \mathbb{N}$, hence the theory of time inhomogeneous Markov processes, compare E.B. Dynkin [3] or S.D. Eidelman et al. [4] is not applicable in a straightforward manner.

In order to understand the significance of these considerations let us return to the case $N = 2$, and let us assume that we can find a global change of variable turning a family $(T_{(s,t)})_{(s,t)\in\mathbb{R}^2_+}$ into a two-parameter semigroup $(S_{(\sigma,\tau)})_{(\sigma,\tau)\in\mathbb{R}^2_+}$. The coordinate change is related to two families of curves in $\mathbb{R}^2_+$ which form essentially the level surfaces s =const. and t =const. Each of these families of curves gives a foliation of $\mathbb{R}^2_+$. Restricted to a leaf of such a foliation the operators of $(T_{(s,t)})_{(s,t)\in\mathbb{R}^2_+}$ are the pre-images of a marginal semigroup of a two-parameter semigroup. In the next section we will put these observations into the context of associated stochastic processes.

5. On Processes Associated with Multi-parameter Families of Probability Measures

Assume that a multi-parameter family of probability measures $\mu_{(s_1,\dots,s_N)}$ on $\mathbb{R}^n$ is given, where $(s_1,\dots,s_N) \in \mathbb{R}^N_+$. To our best knowledge for $N \geq 2$ there exists no general result which assures the existence of a stochastic process which relates to the measures $\mu_{(s_1,\dots,s_N)}$ as in the one-parameter case. In fact in the existing theories of multiparameter processes one constructs out of a given family of $M \leq N$ processes an N-parameter process, compare D. Khoshnevisan [9], recall however our remark in the introduction. We may pose our problem differently: Consider the set Ω of functions $\omega : \mathbb{R}^N_+ \to \mathbb{R}^n$. Which are the σ-fields and probability measures on Ω we can construct such that the distributions of the projections $X_{(s_1,\dots,s_N)} : \Omega \to \mathbb{R}^n$, $\omega \mapsto X_{(s_1,\dots,s_N)}(\omega) = \omega(s_1,\dots,s_N)$ have some interesting relations to the operators associated with the family of measures $(\mu_{(s_1,\dots,s_N)})_{(s_1,\dots,s_N)\in\mathbb{R}^N_+}$? We are not so much interested in the trivial case where all random variables $X_{(s_1,\dots,s_N)}$ are independent and $\mu_{(s_1,\dots,s_N)}$ is the distribution of $X_{(s_1,\dots,s_N)}$.

Let us restrict ourselves to the two-parameter case for simplicity. Instead of all mappings $\omega : \mathbb{R}^2_+ \to \mathbb{R}^n$ we may first consider subsets of $\mathbb{R}^2_+$ as domain, i.e. for a subset $\Gamma \subset \mathbb{R}^2_+$ we define $\Omega_\Gamma := \{\omega : \Gamma \to \mathbb{R}^2\}$ and on Ω_Γ we want to find a σ-field and a probability measure such that the distribution of $X_{(s,t)} : \Omega_\Gamma \to \mathbb{R}^n$ relates in some natural way to $\mu_{(s,t)}$. Although we are still unprecise on what is meant by "in a natural way", eventually (some of) the operators associated with $(\mu_{(s,t)})_{(s,t)\in\Gamma}$ should relate to the process $(X_{(s,t)})_{(s,t)\in\Gamma}$.

Let us give a simple, non-trivial example. Take as $\mu_{(s,t)}$ the measures given by

$$\widehat{\mu}_{(s,t)}(\xi) = (2\pi)^{-\frac{n}{2}} e^{-s\psi_1(\xi)-g(t,\xi)} \tag{5.1}$$

where ψ_1 as well as $g(t,\cdot)$ for every $t \geq 0$ are continuous negative definite functions. We define

$$\widehat{\nu}_t(\xi) = (2\pi)^{-\frac{n}{2}} e^{-g(t,\xi)} \text{ and } \widehat{\mu}^{(1)}_s(\xi) = (2\pi)^{-\frac{n}{2}} e^{-s\psi(\xi)} \tag{5.2}$$

and find

$$\mu_{(s,t)} = (\mu_s^{(1)} * \nu_t)(\xi). \tag{5.3}$$

Denote the Lévy process associated with ψ_1 by $(Y_s^{\psi_1})_{s\geq 0}$ and by $(T_s^{\psi_1})_{s\geq 0}$ the corresponding operator semigroup, i.e.

$$\begin{aligned} E^x\big(u(Y_s^{\psi_1})\big) = T_s^{\psi_1} u(x) &= \int_{\mathbb{R}^n} u(x-y)\mu_s^{(1)}(\mathrm{d}y) \\ &= (2\pi)^{-\frac{n}{2}} \int_{\mathbb{R}^n} e^{ix\xi} e^{-s\psi_1(\xi)} \widehat{u}(\xi)\mathrm{d}\xi \end{aligned} \tag{5.4}$$

where $u \in B_b(\mathbb{R}^n)$, but in the last term in (5.4) we require $u \in \mathscr{S}(\mathbb{R}^n)$. If $T_{t_0}^g$ denotes the operator

$$T_{t_0}^g u(x) = \int_{\mathbb{R}^n} u(x-y)\nu_{t_0}(\mathrm{d}y) = (2\pi)^{-\frac{n}{2}} \int_{\mathbb{R}^n} e^{ix\xi} e^{-g(t_0,\xi)} \widehat{u}(\xi)\,\mathrm{d}\xi \tag{5.5}$$

we find for $T_{(s,t_0)}$ associated with $(\mu_{(s,t_0)})_{s\geq 0}$

$$T_{(s,t_0)} u(x) = E^x\big((T_{t_0}^g u)(Y_s^{\psi_1})\big). \tag{5.6}$$

Moreover we have for u in the domain of $\psi(D)$

$$\frac{\partial T_{(s,t_0)} u(x)}{\partial s} + \psi(D_x) T_{(s,t_0)} u(x) = 0 \tag{5.7}$$

and

$$T_{(0,t_0)} u(x) = T_{t_0}^g u(x). \tag{5.8}$$

This is of course not surprising, but it is worth observing that we have now a very simple foliation of $\mathbb{R}_+^2$ given by the leafs $\Gamma_{t_0} = \{(s, t_0) | s > 0\}$ and on each leaf we can construct a corresponding, non-trivial process. In case that $g(t,\xi) = t\psi_2(\xi)$, where ψ_2 is a further continuous negative definite function, we can do the same on the leafs $\Gamma_{s_0} = \{(s_0, t) | t > 0\}$ and get

$$T_{t_0}^g u(x) = E^x\big(u(Y_{t_0}^{\psi_2})\big) \tag{5.9}$$

where $(Y_t^{\psi_2})_{t\geq 0}$ is the Lévy process associated with ψ_2 which we may take independent of $(Y_s^{\psi_1})_{s\geq 0}$. Thus we find

$$T_{(s,t)} u(x) = E^x\big(E^x(u(Y_t^{\psi_2}))(Y_s^{\psi_1})\big). \tag{5.10}$$

We are indeed in a very special case of the by now classical theory of two-parameter processes: We have on $\mathbb{R}_+^2$ two foliations, on each leaf of each foliation we can find a reasonable process and they can be merged to a two-parameter process.

As in the case of operator families as considered in Section 4, especially Example 4.2, we may now ask for a (global) change of coordinates which will reduce a more complicated situation to the one discussed above. If this is possible we can use Lévy processes as pre-images to associate with the given family of measures processes which relate to the corresponding operators. In particular, in many cases where $(\mu_{(s,t_0)})_{s\geq 0}$ and (or $(\mu_{(s_0,t)})_{t\geq 0}$ correspond to additive processes (processes with independent increments), compare F. F. Gihman and A. V. Skorohod [5], K. Sato [11] or A. V. Skorohod [13], we can find such transformations. Thus a multi-parameter version of the theory of processes with independent (but not necessarily stationary) increments should be possible. (Thanks are due to R. Schilling for pointing out some of these connections.) We refer to M. Schicks [12] for some further discussion.

Let us add another interesting case, the family of measures given by

$$\widehat{\mu}_{(s,t)}(\xi) = (2\pi)^{-\frac{n}{2}} e^{-st\psi(\xi)} \tag{5.11}$$

with ψ being a continuous negative definite function. For $s = 0$ or $t = 0$ we have a degeneration, hence our picture of extending marginal operators (processes) breaks down. Still, when considering the curves (or leafs) $\gamma_\alpha : \mathbb{R}_+ \to \mathbb{R}^2_+$, $r \mapsto (r^\alpha, r^{1-\alpha})$, $0 < \alpha < 1$, we find that the family of measures

$$\widehat{\mu}_{r,\alpha}(\xi) = \widehat{\mu}_{\gamma_\alpha(r)}(\xi) = (2\pi)^{-\frac{n}{2}} e^{-r\psi(\xi)} \tag{5.12}$$

give rise to Lévy processes.

In some sense we are just at the beginning of our investigations and several open problems are apparent, some were already mentioned before, most noticeable

- the construction of processes in the general case
- the connection of processes along leafs
- a multi-parameter theory for additive processes
- what will happen if only local coordinate changes exist.

Of course there is no need to restrict the problem to spatially homogeneous operators (or measures). However, as in the case of one parameter, the starting point should not be a family of measures since they can not easily be constructed. An appropriate starting point would be to solve the equation

$$\frac{\partial^2}{\partial s \partial t} u(s,t;x) + A(s,t;x,D_x)u(s,t;x) = 0 \tag{5.13}$$

under suitable initial conditions and to prove that certain solution operators are positivity preserving and satisfy a variant of the Chapman-Kolmogorov equations. In [8] A. Potrykus and the first named author made a start to handle this problem. Finally, in a forthcoming paper we will see how we can construct certain martingales using the operator A.

Acknowledgment

The first named author would like to thank Professor Lucian Beznea for his support and hospitality during the conference.

References

[1] R.J. Adler and J.E. Taylor, *Random Fields and Geometry* (Springer Monographs in Mathematics). Springer Verlag, Berlin 2007.

[2] K. Doppel and N. Jacob, A non-hypoelliptic Dirichlet problem from stochastics. *Ann. Acad. Sci. Fenn. Ser. A. I. Math.* **8** (1983), 375-389.

[3] E.B. Dynkin, Harmonic functions associated with several Markov processes. *Adv. in Appl. Math.* **2** (1981), 260-283.

[4] S.D. Eidelman, S.D. Ivasyshen and A.N. Kochubei, *Analytic Methods in the Theory of Differential and Pseudo-Differential Equations of Parabolic Type* (Operator Theory. Advances and Applications **152**). Birkhäuser Verlag, Basel, 2004.

[5] F.F. Gihman and A.V. Skorohod, *The Theory of Stochastic Processes II.* Springer-Verlag, Berlin, 1975.

[6] N. Jacob, On generalised Dirichlet problems. *Math. Scand.* **55** (1984), 245-252.
[7] N. Jacob, *Pseudo-Differential Operators and Markov Processes. Vol. 1: Fourier Analysis and Semigroups.* 2nd printing. Imperial College Press, London, 2007.
[8] N. Jacob and A. Potrykus, *(in preparation)*
[9] D. Khoshnevisan, *Multiparameter Processes. An Introduction to Random Fields.* (Springer Monographs in Mathematics). Springer-Verlag, Berlin, 2002.
[10] G. Mazziotto, Two-parameter Hunt processes and a potential theory. *Ann. Probab.* **16** (1988), 600-619.
[11] K. Sato, *Lévy Processes and Infinitely Divisible Distributions* (Cambridge Studies in Advanced Mathematics **68**). Cambridge University Press, Cambridge, 1999.
[12] M. Schicks, *Investigations on Families of Probability Measures Depending on Several Parameters.* PhD-Thesis, Swansea University, Swansea, 2007.
[13] A.V. Skorohod, *Random Processes with Independent Increments* (Mathematics and Its Applications - Soviet Series **47**). Kluwer Academic Publishers, Dordrecht, 1991.

DEPARTMENT OF MATHEMATICS, SWANSEA UNIVERSITY, SINGLETON PARK, SA2 8PP, UNITED KINGDOM
E-mail address: `N.Jacob@swansea.ac.uk`

DEPARTMENT OF MATHEMATICS, SWANSEA UNIVERSITY, SINGLETON PARK, SA2 8PP, UNITED KINGDOM
E-mail address: `A.K.K.Potrykus@swansea.ac.uk`

INSTITUT FÜR MATHEMATISCHE STOCHASTIK, TU BRAUNSCHWEIG, POCKELSSTRASSE 14, 38106 BRAUNSCHWEIG, GERMANY
E-mail address: `M.Schicks@tu-bs.de`

Potential Theory and Stochastics in Albac, 173–182
Aurel Cornea Memorial Volume

JENSEN'S INEQUALITY OVER CAT(κ)-SPACE WITH SMALL DIAMETER

KAZUHIRO KUWAE

ABSTRACT. In this note, we investigate barycenters of probability measures having a first moment over CAT(κ)-spaces with diameters strictly less than half of diameter of the space form of constant curvature κ. We prove that such spaces admit Jensen's inequality, which is an errata for [8].

1. CAT(κ)-SPACE

Throughout this note, κ is a real number. Let $\mathbb{M}^2(\kappa)$ be the two dimensional space form of constant curvature κ, precisely $\mathbb{M}^2(\kappa) := \mathbb{S}^2(\kappa)$ if $\kappa > 0$, $\mathbb{M}^2(\kappa) := \mathbb{R}^2$ if $\kappa = 0$ and $\mathbb{M}^2(\kappa) := \mathbb{H}^2(\kappa)$ if $\kappa < 0$. Here $\mathbb{S}^2(\kappa)$ (resp. $\mathbb{H}^2(\kappa)$) is the two-dimensional sphere (resp. two-dimensional hyperbolic space) with the Riemannian distance $\sqrt{\kappa}d$ (resp. $\sqrt{-\kappa}d$), where d is the canonical Riemannian distance over $\mathbb{S}^2$ (resp $\mathbb{H}^2$). Let R_κ be the diameter of $\mathbb{M}^2(\kappa)$, that is, $R_\kappa := \pi/\sqrt{\kappa}$ if $\kappa > 0$, and $R_\kappa := \infty$ if $\kappa \leq 0$.

Definition 1.1 (Geodesics). Let (Y,d) be a metric space. A map $\gamma : I \to Y$ is said to be a *curve* if it is continuous, where $I = [a,b] \subset \mathbb{R}$ is a closed interval. The length $L(\gamma)$ of a curve $\gamma : I \to Y$ is defined to be

$$L(\gamma) := \sup\left\{ \sum_{i=1}^{n} d(\gamma(t_{i-1}),\gamma(t_i)) \,\middle|\, a = t_0 < t_1 < \cdots < t_{n-1} < t_n = b \right\}.$$

A curve $\gamma : I \to Y$ is said to be a *minimal geodesic* if $L(\gamma|_{[s,t]}) = d(\gamma_s,\gamma_t)$ holds for any $s,t \in I$, $s < t$, equivalently $d(\gamma_r,\gamma_t) = d(\gamma_r,\gamma_s) + d(\gamma_s,\gamma_t)$ for any $r < s < t$. A curve $\gamma : I \to Y$ is said to be a *geodesic* if for any $s,t \in I$, $s < t$ with sufficiently small $|t-s|$, $L(\gamma|_{[s,t]}) = d(\gamma_s,\gamma_t)$ holds. A metric space (Y,d) is called an R_κ*-geodesic space* if any two points in Y whose distance is strictly less than R_κ can be joined by a minimal geodesic.

Definition 1.2 (CAT(κ)-Inequality, see [1]). Let (Y,d) be a metric space and $\triangle$ a geodesic triangle in Y with perimeter strictly less than $2R_\kappa$. Let $\widetilde{\triangle}$ be a comparison triangle of $\triangle$ in $\mathbb{M}^2(\kappa)$. We say that $\triangle$ satisfies *CAT(κ)-inequality* if all $p,q \in \triangle$ and its corresponding points $\tilde{p},\tilde{q} \in \widetilde{\triangle}$ satisfy

$$d(p,q) \leq d(\tilde{p},\tilde{q}).$$

Definition 1.3 (CAT(κ)-Space, see [1]). A metric space (Y,d) is said to be a *CAT(κ)-space* if (Y,d) is an R_κ-geodesic space and all geodesic triangles in Y with perimeter strictly less than $2R_\kappa$ satisfy CAT(κ)-inequality.

2000 *Mathematics Subject Classification.* Primary 52-06, 52A55. Secondary 53C22, 53C43, 58E10, 58E20, 60G99.

Key words and phrases. Geodesic space, CAT(κ)-space, k-convex space, barycenter, Jensen's inequality.

Hereafter, we shall focus only on the case $\kappa > 0$ by the end of this section. Note that $(Y, \sqrt{\kappa}d)$ is a CAT(1)-space if (Y,d) is a CAT(κ)-space. The following proposition is easily confirmed by way of the cosine formula in $\mathbb{S}^2(\kappa)$:

Proposition 1.1. *If a metric space (Y,d) is a CAT(1)-space, then it is an R_1-geodesic space, and for any points $z,\gamma_0,\gamma_1 \in Y$ satisfying $d(z,\gamma_0)$, $d(z,\gamma_1)$, $d(\gamma_0,\gamma_1) < \pi(= R_1)$, $d(z,\gamma_0)+d(z,\gamma_1)+d(\gamma_0,\gamma_1) < 2\pi$ and any minimal geodesic $\gamma : [0,1] \to Y$ joining γ_0 and γ_1, the following inequality holds:*

$$\cos d(z,\gamma_t) \ge \frac{\sin(1-t)d(\gamma_0,\gamma_1)}{\sin d(\gamma_0,\gamma_1)}\cos d(z,\gamma_0) + \frac{\sin td(\gamma_0,\gamma_1)}{\sin d(\gamma_0,\gamma_1)}\cos d(z,\gamma_1). \tag{1.1}$$

Definition 1.4 (Convex Set and Convex Function)**.** Let (Y,d) be a geodesic space. For a subset K of Y, K is called a *convex set* if for any given points $\gamma_0,\gamma_1 \in K$ there exists a minimal geodesic $\gamma : [0,1] \to K$ joining them. A function $\varphi : Y \to \mathbb{R}$ is said to be *convex* if for any minimal geodesic $\gamma : [0,1] \to Y$, $\varphi \circ \gamma : [0,1] \to \mathbb{R}$ is convex in the following sense: for any $t \in [0,1]$

$$\varphi(\gamma_t) \le (1-t)\varphi(\gamma_0) + t\varphi(\gamma_1)$$

holds.

Let (Y,d) be a CAT(1) space with $\mathrm{diam}(Y) < \pi$. Take three points $z,\gamma_0,\gamma_1 \in Y$ and consider a minimal geodesic $\gamma : [0,1] \to Y$ joining γ_0 and γ_1. By (1.1), we have

$$\begin{aligned} 2\sin^2\frac{d(z,\gamma_t)}{2} \le{}& \frac{\sin(1-t)d(\gamma_0,\gamma_1)}{\sin d(\gamma_0,\gamma_1)} 2\sin^2\frac{d(z,\gamma_0)}{2} \\ &+ \frac{\sin td(\gamma_0,\gamma_1)}{\sin d(\gamma_0,\gamma_1)} 2\sin^2\frac{d(z,\gamma_1)}{2} \\ &- \left(\frac{\sin(1-t)d(\gamma_0,\gamma_1)+\sin td(\gamma_0,\gamma_1)}{\sin d(\gamma_0,\gamma_1)} - 1\right). \end{aligned} \tag{1.2}$$

Proposition 1.2 (Projection to Closed Convex Set)**.** *Let (Y,d) be a complete CAT(κ)-space with* $\mathrm{diam}(Y) < R_\kappa$ *and K a closed convex set in (Y,d).*

(1) *For each $z \in Y$ with $d(z,K) < R_\kappa/2$, there exists a unique $\pi_K z \in K$ such that*

$$d(z,\pi_K z) = d(z,K).$$

(2) *π_K is orthogonal in the following sense: for $z \in Y$ with $d(z,K) < R_\kappa/2$ and $w \in K$,*

$$\cos\sqrt{\kappa}d(w,z) \le \cos\sqrt{\kappa}d(\pi_K z,w)\cdot\cos\sqrt{\kappa}d(\pi_K z,z),$$

in particular, if further $\mathrm{diam}(K) < R_\kappa/2$, *then $d(w,\pi_K z) \le d(w,z)$.*

Proof. We may assume $\kappa = 1$, hence $R_\kappa = \pi$.

(1): The assertion of (1) is written in Chapter II 2.6 Exercises (1) in [1] as an exercise. We shall write its proof for the reader's convenience. Fix $z \in Y$. We set $L := \inf_{w\in K} 2\sin^2\frac{d(z,w)}{2}$. Since $d(z,K) < \pi/2$, we see $L < 1$. Let $\{w_n\} \subset K$ be the minimizing sequence of L and $w_{n,m}$ the midpoint of w_n and w_m. From (1.2)

$$\begin{aligned}2\sin^2\frac{d(z,w_{n,m})}{2} &\le \frac{\sin\frac{d(w_n,w_m)}{2}}{\sin d(w_n,w_m)}\left(2\sin^2\frac{d(z,w_n)}{2}+2\sin^2\frac{d(z,w_m)}{2}\right)\\ &\qquad -\left(2\frac{\sin\frac{d(w_n,w_m)}{2}}{\sin d(w_n,w_m)}-1\right)\\ &= \frac{1}{2\cos\frac{d(w_n,w_m)}{2}}\left(2\sin^2\frac{d(z,w_n)}{2}+2\sin^2\frac{d(z,w_m)}{2}\right)\\ &\qquad -\left(\frac{1}{\cos\frac{d(w_n,w_m)}{2}}-1\right).\end{aligned}$$

Thus,

$$(L-1)\cos\frac{d(w_n,w_m)}{2}\le \sin^2\frac{d(z,w_n)}{2}+\sin^2\frac{d(z,w_m)}{2}-1.$$

Hence

$$\cos\frac{d(w_n,w_m)}{2}\ge\frac{1-\sin^2\frac{d(z,w_n)}{2}-\sin^2\frac{d(z,w_m)}{2}}{1-L},$$

which implies $\underline{\lim}_{n,m\to\infty}\cos\frac{d(w_n,w_m)}{2}=1$, that is, $\{w_n\}$ is a Cauchy sequence. Denote by $\pi_K z\in K$ its limit. Then $L=2\sin^2\frac{d(z,\pi_K z)}{2}$, consequently, $d(z,K)=d(z,\pi_K z)$. The proof of the uniqueness is similar.

(2): Consider a minimal geodesic $(\gamma_t)_{t\in[0,1]}$ joining $\gamma_0:=\pi_K z$ and $\gamma_1:=w$. Then $\gamma_t\in K$ for all $t\in[0,1]$, so $d(z,\gamma_t)\ge d(z,\pi_K z)$. We may assume $d(\pi_K z,w)\neq 0$. Thus

$$\begin{aligned}\cos d(z,\pi_K z)\ge&\frac{\sin(1-t)d(\pi_K z,w)}{\sin d(\pi_K z,w)}\cos d(z,\pi_K z)\\ &+\frac{\sin td(\pi_K z,w)}{\sin d(\pi_K z,w)}\cos d(z,w).\end{aligned}$$

Equivalently

$$\begin{aligned}&\{\sin(1-t)d(\pi_K z,w)(\cos td(\pi_K z,w)-1)\\ &\qquad+\cos(1-t)d(\pi_K z,w)\sin td(\pi_K z,w)\}\cos d(z,\pi_K z)\\ &\qquad\qquad\ge\sin td(\pi_K z,w)\cos d(z,w).\end{aligned}$$

Dividing by $td(\pi_K z,w)$ and letting $t\to 0$, we obtain the desired assertion. □

2. Barycenter over k-convex spaces

In this section, we present the notion of k-convex space investigated by Ohta [9],[10] and the properties of barycenter over such spaces.

Definition 2.1 (k-Convex Space; cf. Ohta [9],[10]). A complete separable metric space (Y,d) is called a *k-convex space* or *2-uniformly convex space* if (Y,d) is a geodesic space and for any three points $x, y, z \in Y$, any geodesic $\gamma := (\gamma_t)_{t\in[0,1]}$ in Y with $\gamma_0 = x, \gamma_1 = y$, and all $t \in [0,1]$,

$$d^2(z,\gamma_t) \le (1-t)d^2(z,x) + td^2(z,y) - \frac{k}{2}t(1-t)d^2(x,y). \tag{2.3}$$

By definition, putting $z = \gamma_t$, we see $k \in]0,2]$. The inequality (2.3) yields the (strict) convexity of $Y \ni x \mapsto d^2(z,x)$ for a fixed $z \in Y$. Any closed convex subset of a k-convex space is again k-convex. Every complete CAT(0)-space is a 2-convex space. Ohta [9] proved that for $\kappa > 0$ any complete CAT(κ)-space Y with $\mathrm{diam}(Y) \le (R_\kappa/2) - \varepsilon$, $\varepsilon \in]0, R_\kappa/2[$ is a $\{(\pi - 2\sqrt{\kappa}\varepsilon)\tan\sqrt{\kappa}\varepsilon\}$-convex space, and any Banach space L^p with $p \in]1,2]$ over a measurable space is a $2(p-1)$-convex space (Propositions 3.1 and 3.4 in [9]). He proved that any two points in a k-convex space can be connected by a unique minimal geodesic (Lemma 2.2 in [9]) and contractible (Lemma 2.3 and Corollary 2.4 in [9]).

Let (Y,d) be a complete k-convex space and $\mathscr{P}(Y)$ a family of Borel probability measures on Y. For $p \ge 1$, we set

$$\mathscr{P}^p(Y) := \left\{\mu \in \mathscr{P}(Y) \,\Big|\, \int_Y d^p(x,y)\mu(dy) < \infty \text{ for } \forall/\exists x \in Y\right\}.$$

Definition 2.2 (Barycenter [14]). For $\mu \in \mathscr{P}^2(Y)$,

$$\mathrm{Var}(\mu) := \inf_{z\in Y}\int_Y d^2(z,y)\mu(dy) (< \infty)$$

is called a *variance* of μ. For $\mu \in \mathscr{P}^2(Y)$, if $z \mapsto \int_Y d^2(z,x)\mu(dx)$ has a unique minimizer $b(\mu) \in Y$ then we call $b(\mu)$ the *barycenter of* $\mu \in \mathscr{P}^2(Y)$. For $\mu \in \mathscr{P}^1(Y)$ and $w \in Y$, we consider the following function F_w:

$$\begin{aligned} F_w(z) :&= \int_Y (d^2(z,x) - d^2(w,x))\mu(dx) \\ &\le d(z,w)\int_Y (d(z,x) + d(w,x))\mu(dx) < \infty. \end{aligned} \tag{2.4}$$

If $Y \ni z \mapsto F_w(z)$ admits a unique minimizer $b(\mu)$ independent of w, we call it *barycenter of* $\mu \in \mathscr{P}^1(Y)$. The barycenter of $\mu \in \mathscr{P}^2(Y)$ is always the barycenter of $\mu \in \mathscr{P}^1(Y)$.

For a subset F of Y, denote by $C(F)$ the closed convex hull of F. That is, $C(F)$ is the smallest closed convex subset of Y containing F.

Definition 2.3 (Pure Barycenter, [8]). For $\mu \in \mathscr{P}^1(Y)$ and $w \in Y$, if the function $C(\mathrm{supp}[\mu]) \ni z \mapsto F_w(z)$ admits a unique minimizer $\bar{b}(\mu) \in C(\mathrm{supp}[\mu])$ independent of w, we call it the *pure barycenter of* $\mu \in \mathscr{P}^1(Y)$.

Remark 2.1. Let (Y,d) be complete k-convex space.

(1) The pure barycenter of (Y,d) is the barycenter of $(C(\mathrm{supp}[\mu]),d)$. If $b(\mu) \in C(\mathrm{supp}[\mu])$, in particular, if μ has full topological support, then $\bar{b}(\mu) = b(\mu)$.
(2) We set $K = C(\mathrm{supp}[\mu])$ for $\mu \in \mathscr{P}^1(Y)$ and assume $d(\pi_K z, y) \le d(z,y)$ for all $z \in Y$, $y \in K$. This condition is satisfied for any CAT(κ)-space with diameter strictly less than $R_\kappa/2$ by Proposition 1.2(2). Then $\bar{b}(\mu) = b(\mu)$. Indeed, for

$w, z \in Y$, we have

$$\begin{aligned}\int_Y (d^2(\bar{b}(\mu), y) - d^2(w, y))\mu(dy) &= \int_K (d^2(\bar{b}(\mu), y) - d^2(w, y))\mu(dy) \\ &\le \int_K (d^2(\pi_K z, y) - d^2(w, y))\mu(dy) \\ &\le \int_Y (d^2(z, y) - d^2(w, y))\mu(dy),\end{aligned}$$

which implies that $\bar{b}(\mu)$ is the unique minimizer of $Y \ni z \mapsto F_w(z)$.

(3) For any k-convex space (Y, d), Ohta [10] proved that the barycenter $b(\mu)$ of $\mu \in \mathscr{P}^1(Y)$ is in $B_{2r}(z)$ if $\mathrm{supp}[\mu] \subset B_r(z)$, $z \in Y$.

The following lemma is shown in [8],[10]:

Lemma 2.1 ([8],[10]). *Let (Y, d) be complete k-convex space. Then $\mu \in \mathscr{P}^1(Y)$ admits the both of barycenter and pure barycenter.*

The following proposition is a variant of Proposition 5.5 in [14].

Proposition 2.1 (Characterization of Barycenter). *Let (Y, d) be complete k-convex space. For $\mu \in \mathscr{P}^1(Y)$, there exists a unique element $z_\mu \in Y$:*

$$F_{z_\mu}(z) \ge \frac{k}{2} d^2(z, z_\mu) \text{ for any } z \in Y. \tag{2.5}$$

Moreover, z_μ coincides with $b(\mu)$.

Corollary 2.1 (Characterization of Pure Barycenter). *Let (Y, d) be complete k-convex space. For $\mu \in \mathscr{P}^1(Y)$, there exists a unique element $\bar{z}_\mu \in C(\mathrm{supp}[\mu])$:*

$$F_{\bar{z}_\mu}(z) \ge \frac{k}{2} d^2(z, \bar{z}_\mu) \text{ for any } z \in C(\mathrm{supp}[\mu]).$$

Moreover, $\bar{z}_\mu$ coincides with $\bar{b}(\mu)$.

Proof of Proposition 2.1. Uniqueness of z_μ in (2.5): Suppose that $z_1, z_2 \in Y$ satisfy

$$F_{z_1}(z) \ge \frac{k}{2} d^2(z, z_1) \text{ for any } z \in Y$$

and

$$F_{z_2}(z) \ge \frac{k}{2} d^2(z, z_2) \text{ for any } z \in Y.$$

After substituting $z = z_2$ in the first one and $z = z_1$ in the second one, we obtain $kd^2(z_1, z_2) \le 0$.

Existence of z_μ in (2.5): It suffices to show that $b(\mu)$ satisfies

$$F_{b(\mu)}(z) \ge \frac{k}{2} d^2(z, b(\mu)) \text{ for any } z \in Y.$$

Let $z = (z_t)_{t \in [0,1]}$ be a minimal geodesic joining $z_0 := b(\mu)$ and $z_1 := z$, and set $w = b(\mu)$. Then we have

$$\begin{aligned}0 = F_w(b(\mu)) \le F_w(z_t) &\le (1-t)F_w(z_0) + tF_w(z_1) - \frac{k}{2} t(1-t) d^2(z_0, z_1) \\ &= tF_w(z) - \frac{k}{2} t(1-t) d^2(z, b(\mu)).\end{aligned}$$

Dividing $t>0$ and $t\to 0$, we obtain the desired assertion. □

Proposition 2.2 (Characterization of Barycenter). *Let $\{(Y_i,d_i)\}_{i\in I}$ be a countable family of k-convex spaces. We set*

$$Y := \left\{ x\in\prod_{i\in I} Y_i \;\middle|\; \sum_{i\in I} m_i d_i^2(x_i,o_i)<\infty \right\}$$

and define the distance d by $d^2(x,y) := \sum_{i\in I} m_i d_i^2(x_i,y_i)$ for $x,y\in Y$. Here $m_i\in]0,\infty[$ and $o=(o_i)_{i\in I}$ is a point in $\prod_{i\in I} Y_i$. Then (Y,d) is a complete k-convex space. Let μ_i (resp. μ_i) be the marginal of $\mu\in\mathscr{P}^1(Y)$ (resp. $\mu\in\mathscr{P}^2(Y)$) on each Y_i, then $\mu_i\in\mathscr{P}(Y_i)$ (resp. $\mu_i\in\mathscr{P}^2(Y_i)$). If $\mu\in\mathscr{P}^2(Y)$, then $b(\mu)=(b(\mu_i))_{i\in I}$. If further I is a finite set, then the same conclusion holds for $\mu\in\mathscr{P}^1(Y)$.

Proof. It is easy to see that (Y,d) is a complete metric space and a k-convex space. Since the marginal μ_i on Y_i of $\mu\in\mathscr{P}(Y)$ is the push-forward measure $\mu_i=(p_i)_*\mu$ by projection $p_i : Y\to Y_i$, we have from $\sqrt{m_i}d(x_i,y_i)\le d(x,y)$ that $\mu\in\mathscr{P}^1(Y)$ (resp. $\mu\in\mathscr{P}^2(Y)$) implies $\mu_i\in\mathscr{P}^1(Y_i)$ (resp. $\mu_i\in\mathscr{P}^2(Y_i)$).

Put $y:=(b(\mu_i))_{i\in I}$. If I is finite, we easily get $y\in Y$. For $\mu\in\mathscr{P}^2(Y)$, taking a sum of

$$\begin{aligned}\frac{k}{2}d^2(y_i,o_i) &\le F_{y_i}(o_i)\\ &= \int_{Y_i} d_i^2(o_i,x_i)\mu_i(dx_i) - \int_{Y_i} d_i^2(y_i,x_i)\mu_i(dx_i)\\ &\le \int_{Y_i} d_i^2(o_i,x_i)\mu_i(dx_i)\end{aligned}$$

over I, we get

$$\begin{aligned}\frac{k}{2}\sum_{i\in I} d_i^2(y_i,o_i)m_i &\le \sum_{i\in I}\int_{Y_i} d_i^2(o_i,x_i)\mu_i(dx_i)m_i\\ &= \sum_{i\in I}\int_{Y} d_i^2(o_i,x_i)\mu(dx)m_i\\ &= \int_Y d^2(o,x)\mu(dx)<\infty\end{aligned}$$

which implies $y\in Y$. On the other hand, for $z\in Y$, we have

$$\begin{aligned}Y\ni z\mapsto F_y(z) &= \int_Y (d^2(z,x)-d^2(y,x))\mu(dx)\\ &= \int_Y \sum_{i\in I} m_i(d_i^2(z_i,x_i)-d_i^2(y_i,x_i))\mu(dx)\\ &= \sum_{i\in I} m_i\int_{Y_i}(d_i^2(z_i,x_i)-d_i^2(y_i,x_i))\mu_i(dx_i)\\ &= \sum_{i\in I} m_i F_{y_i}(z_i)\ge \sum_{i\in I} m_i\frac{k}{2}d_i^2(z_i,y_i)\\ &= \frac{k}{2}d^2(z,y)\end{aligned}$$

which implies $y=b(\mu)$ in view of Proposition 2.1. □

3. JENSEN'S INEQUALTY

Theorem 3.1 (Jensen's Inequality over CAT(κ)-space). *Let (Y,d) be complete CAT(κ)-space with diameter strictly less than $R_\kappa/2$, and φ a lower semi continuous convex function on Y. Suppose that $\kappa \le 0$ or φ is upper bounded on Y. Then, for any $\mu \in \mathscr{P}^1(Y)$,*

$$\varphi(b(\mu)) \le \int_Y \varphi d\mu \tag{3.6}$$

provided the right hand side makes sense, in particular, we have

$$d(b(\mu), z) \le \int_Y d(x,z)\mu(dx), \quad z \in Y. \tag{3.7}$$

Remark 3.1.

(1) In Appendix of [8], we assert that Jensen's inequality holds over general k-convex spaces, but its proof has a gap. So the assertion is only true for CAT(κ)-space with diameter strictly less than $R_\kappa/2$ for the moment.

(2) If the CAT(κ)-space (Y,d) with diameter strictly less than $R_\kappa/2$ admits the convex geometry in the sense of Kendall [6], then we can omit the upper boundedness of φ for the case $\kappa > 0$ in Theorem 3.1. Here (Y,d) is said to admit the convex geometry in the sense of [6] if there exists a bounded non-negative continuous convex function $\Phi : Y \times Y \to \mathbb{R}$ vanishing only on the diagonal. Indeed, for any lower semi continuous convex function $\varphi : Y \to \mathbb{R}$, $\varphi_n(x) := \inf_{y\in Y}\{\varphi(y) + n\Phi(x,y)\}$ satisfies $|\varphi_n(x) - \varphi_n(y)| \le n\Phi(x,y)$, which yields the continuity and boundedness of φ_n. The lower semi continuity of φ yields $\varphi = \uparrow \lim_{n\to\infty}\varphi_n$. We can see the convexity of φ_n in the following way: Let $\gamma = (\gamma_t)_{t\in[0,1]}$ be a minimal geodesic and $\{y_k\}$, $\{z_k\}$ the minimizing sequence for $\varphi_n(\gamma_0)$, $\varphi_n(\gamma_1)$ respectively, and $\eta = (\eta_t)_{t\in[0,1]}$ the minimal geodesic joining $\eta_0 := y_k$ and $\eta_1 := z_k$. Then

$$\begin{aligned}
\varphi_n(\gamma_t) &\le \varphi(\eta_t) + n\Phi(\gamma_t,\eta_t) \\
&\le (1-t)(\varphi(\eta_0) + n\Phi(\gamma_0,\eta_0)) + t(\varphi(\eta_1) + n\Phi(\gamma_1,\eta_1)) \\
&= (1-t)(\varphi(y_k) + n\Phi(\gamma_0,y_k)) + t(\varphi(z_k) + n\Phi(\gamma_1,z_k)) \\
&\to (1-t)\varphi_n(\gamma_0) + t\varphi_n(\gamma_1), \text{ as } k \to \infty.
\end{aligned}$$

(3) Let (M,g) be a complete smooth Riemannian manifold with sectional curvature bounded above by $\kappa > 0$ and d its Riemannian distance. Let $\bar{B}_r(o)$ be a closed ball in (M,g) with center $o \in M$ and radius $r < R_\kappa/2$ having no intersection with the cut-locus of o. Such a ball is called a *regular geodesic ball* (see [6],[7]). Then $(\bar{B}_r(o), d)$ is a CAT(κ)-space with diameter strictly less than R_κ. Kendall [6] proved that any probability measure over a regular geodesic ball admits barycenter and Jensen's inequality holds for it. His proof is based on the convex geometry and completely different from ours. Theorem 3.1 does not cover (but, is not included in) the Jensen's inequality in [6].

Proof of Theorem 3.1. (3.7) follows from (3.6) because the distance function $x \mapsto d(x,z)$ is convex for each $z \in Y$ (see [14] for $\kappa = 0$ and Chapter II Exercises 2.3 (1) of [1] for $\kappa > 0$).

Since a CAT(κ)-space is always a CAT(0)-space for $\kappa \le 0$, (3.6) is proved by [2], [14] for this case. So we take $\kappa > 0$ and assume the upper boundedness of φ. Moreover we may assume that φ is bounded. Indeed, it holds that $\int_Y \varphi d\mu = \downarrow \lim_{k\to\infty}\int_Y \varphi \vee (-k)d\mu$ and $\varphi \vee (-k)$ is a bounded convex function. By scaling, we may assume that φ is non-negative. Take an $\varepsilon_1 \in]0, R_\kappa/2[$ such that $\mathrm{diam}(Y) \le (R_\kappa/2) - \varepsilon_1$. Then (Y,d) is a k_1-convex space with $k_1 := (\pi - 2\sqrt{\kappa}\varepsilon_1)\tan\sqrt{\kappa}\varepsilon_1$. For $T > 0$, we set $\varphi_T(x) := \varphi(x)/T$ and $Z_T := [0, \sup_{x\in Y}\varphi_T(x)]$. Taking a large T enough, $\hat{Y}_T := Y \times Z_T$ is a complete CAT(κ)-space with diameter strictly less than $R_\kappa/2$. Here we use the fact that for $\kappa \ge 0$ the Euclidean direct product of CAT(κ)-spaces is again a CAT(κ)-space (see Chapter II Exercises 1.16(2) in [1]). Therefore, for sufficiently large $T > 0$, there exists an $\varepsilon_2 \in]0, R_\kappa/2[$ such that $\mathrm{diam}(\hat{Y}_T) \le (R_\kappa/2) - \varepsilon_2$. Let $Y_{\varphi_T} := \{(x,t) \in \hat{Y}_T \mid \varphi_T(x) \le t\}$. Then it is a complete k_2-convex space for $k_2 := (\pi - 2\sqrt{\kappa}\varepsilon_2)\tan\sqrt{\kappa}\varepsilon_2$. Put $k := \min\{k_1, k_2\}$. Then Y and $\hat{Y}_T$ are complete k-convex spaces. Set $\widehat{\varphi_T}(x) := (x, \varphi_T(x))$, $\hat{\mu} := (\widehat{\varphi_T})_*\mu$ and $\nu := (\varphi_T)_*\mu$. Then $\hat{\mu} \in \mathscr{P}^1(\hat{Y}_T)$ and $\nu \in \mathscr{P}^1(Z_T)$. Indeed, for $\hat{z} = (z,t) \in \hat{Y}_T$, $s \in Z_T$,

$$
\begin{aligned}
\int_{\hat{Y}_T} d(\hat{z},\hat{x})\hat{\mu}(d\hat{x}) &= \int_Y d(\hat{z},(x,\varphi_T(x)))\mu(dx) \\
&= \int_Y \sqrt{d^2(z,x) + |t - \varphi_T(x)|^2}\mu(dx) \\
&\le \int_Y (d(z,x) + |t - \varphi_T(x)|)\mu(dx) < \infty, \\
\int_{Z_T} |s-t|\nu(dt) &= \int_Y |s - \varphi_T(x)|\mu(dx) < \infty.
\end{aligned}
$$

On the other hand, μ (resp. ν) is the marginal of $\hat{\mu}$ to Y, (resp. to Z_T). Indeed, for $A \in \mathcal{B}(Y)$ and $B \in \mathcal{B}(Z_T)$, we see

$$
\begin{aligned}
\hat{\mu}(A \times Z_T) &= \int_{\hat{Y}_T} 1_{A\times Z_T}(x,t)\hat{\mu}(dxdt) \\
&= \int_Y 1_{A\times Z_T}(x,\varphi_T(x))\mu(dx) = \mu(A), \\
\hat{\mu}(Y \times B) &= \int_{\hat{Y}} 1_{Y\times B}(x,t)\hat{\mu}(dxdt) \\
&= \int_Y 1_{Y\times B}(x,\varphi_T(x))\mu(dx) \\
&= \int_Y 1_B(\varphi_T(x))\mu(dx) = \nu(B).
\end{aligned}
$$

For $u \in Z_T$, the minimizer of

$$
\begin{aligned}
Z_T \ni s \mapsto F_u(s) :&= \int_{Z_T} ((s-t)^2 - (u-t)^2)\nu(dt) \\
&= \int_Y ((s-\varphi_T(x))^2 - (u - \varphi_T(x))^2)\mu(dx)
\end{aligned}
$$

is $\frac{1}{T}\int_Y \varphi d\mu$, hence $b(\nu) = \frac{1}{T}\int_Y \varphi d\mu$, because Z_T is a complete CAT(0)-space. By Proposition 2.2, we have

$$
b(\hat{\mu}) = (b(\mu), b(\nu)) = \left(b(\mu), \frac{1}{T}\int_Y \varphi d\mu\right).
$$

Finally we claim $\mathrm{supp}[\hat{\mu}] \subset Y_{\varphi_T}$. Indeed,

$$\begin{aligned}\hat{\mu}(\hat{Y} \setminus Y_{\varphi_T}) &= \int_{\hat{Y}_T} 1_{\hat{Y}_T \setminus Y_{\varphi_T}}(x,t)\hat{\mu}(dxdt)\\ &= \int_Y 1_{\hat{Y}_T \setminus Y_{\varphi_T}}(x,\varphi_T(x))\mu(dx)\\ &= \int_Y 1_{\{\varphi_T(x)>\varphi_T(x)\}}\mu(dx) = 0.\end{aligned}$$

By Proposition 1.5 in [14], Y_{φ_T} is a closed convex subset of $\hat{Y}_T$. Therefore, we obtain $(b(\mu), \frac{1}{T}\int_Y \varphi d\mu) = b(\hat{\mu}) \in C(\mathrm{supp}[\hat{\mu}]) \subset Y_{\varphi_T}$ in view of Remark 2.1(2), which implies $\varphi(b(\mu)) \le \int_Y \varphi d\mu$. □

Example 3.1. Let H be a complete separable real Hilbert space with inner product $\langle x, y\rangle_H$ and $S^\infty := \{x \in H \mid \langle x, x\rangle_H = 1\}$ the *unit sphere in H* endowed with the distance d defined by $d(x, y) := \cos^{-1}\langle x, y\rangle$, $x, y \in S^\infty$. Then (S^∞, d) is a complete CAT(1)-space with diameter less than π. Consider a closed ball $\bar{B}_r(o) := \{x \in S^\infty \mid d(o,x) \le r\}$ with $r < \pi/4$, $o \in S^\infty$. Then $(\bar{B}_r(o), d)$ is a complete CAT(1)-space with diameter strictly less than $\pi/2$. Any $\mu \in \mathscr{P}^1(\bar{B}_r(o))$ admits its barycenter $b(\mu)$ and Jensen's inequality holds for it as in Theorem 3.1.

ACKNOWLEDGMENTS

This note is based on a part of the lecture at Graduate School of Informatics, Kyoto University. The author thanks to Professor Masanori Hino, who organizes this lecture, and to Professor Kazumasa Kuwada and Dr. Naotaka Kajino for their valuable comments.

The author was partially supported by a Grant-in-Aid for Scientific Research (C) No. 19540220 from Japan Society for the Promotion of Science.

REFERENCES

[1] M. R. Bridson and A. Haefliger, *Metric Spaces of Non-Positive Curvature* (Fundamental Principles of Mathematical Sciences **319**). Springer-Verlag, Berlin, 1999.
[2] J. Jr. Eells and B. Fuglede, *Harmonic Maps between Riemannian Polyhedra* (Cambridge Tracts in Mathematics **142**). Cambridge University Press, Cambridge, 2001.
[3] B. Fuglede, The Dirichlet problem for harmonic maps from Riemannian polyhedra to spaces of upper bounded curvature. *Trans. Amer. Math. Soc.* **357** (2005), no. 2, 757-792 (electronic).
[4] B. Fuglede, Dirichlet problems for harmonic maps from regular domains. *Proc. London Math. Soc.* (3) **91** (2005), no. 1, 249-272.
[5] W. S. Kendall, Martingales on manifolds and harmonic maps. In R.Durret and M.Pinsky (Eds.), *The Geometry of Random Motion*, pp. 121-157. Contemp. Math. **73**, Amer. Math. Soc., Providence, RI, 1988.
[6] W. S. Kendall, Probability, convexity, and harmonic maps with small image I: Uniqueness and fine existence. *Proc. London Math. Soc.* **61** (1990), no. 2, 371-406.
[7] W. S. Kendall, From stochastic parallel transport to harmonic maps. In: J. Jost et al. (Eds.) *New Directions in Dirichlet Forms* (*AMS/IP Stud. Adv. Math.* **8**), pp. 49-115. Amer. Math. Soc., Providence, RI, 1998.
[8] K. Kuwae and K.-Th. Sturm, On a Liouville type theorem for harmonic maps to convex spaces via Markov chains. In: *Proceedings of German-Japanese Symposium in Kyoto 2006*, pp. 177-192. RIMS Kôkyûroku Bessatsu B6, 2008.
[9] S.-I. Ohta, Convexities of metric spaces. *Geom. Dedicata* **125** (2007), no. 1, 225-250.

[10] S.-I. Ohta, Extending Lipschitz and Hölder maps between metric spaces, preprint, 2007, *Positivity* **13** (2009), no. 2, 407-425.

[11] K.-Th. Sturm, Nonlinear Markov operators associated with symmetric Markov kernels and energy minimizing maps between singular spaces. *Calc. Var. Partial Differential Equations* **12** (2001), no. 4, 317–357.

[12] K.-Th. Sturm, Nonlinear Markov operators, discrete heat flow, and harmonic maps between singular spaces. *Potential Anal.* **16** (2002), no. 4, 305–340.

[13] K.-Th. Sturm, Nonlinear martingale theory for processes with values in metric spaces of nonpositive curvature. *Ann. Probab.* **30** (2002), no. 3, 1195–1222.

[14] K.-Th. Sturm, Probability measures on metric spaces of nonpositive curvature. In: P. Auscher, T. Coulhon, A. Grigor´yan (Eds.), *Heat Kernels and Analysis on Manifolds, Graphs, and Metric Spaces* (Contemporary Mathematics **338**), pp. 357-390. Amer. Math. Soc., Providence, RI, 2003.

DEPARTMENT OF MATHEMATICS AND ENGINEERING, GRADUATE SCHOOL OF SCIENCE AND TECHNOLOGY, KUMAMOTO UNIVERSITY, KUMAMOTO 860–8500, JAPAN

E-mail address: kuwae@gpo.kumamoto-u.ac.jp

Potential Theory and Stochastics in Albac, 183–192
Aurel Cornea Memorial Volume

RICH MEASURE SPACES

PETER A. LOEB

*This article is dedicated to the memory of Aurel Cornea;
his courage, insight, and love of life will always be an example for me.*

ABSTRACT. We outline the construction of the rich measure spaces that are now called Loeb measure spaces in the literature. We then review some recent work by Yeneng Sun and the author. Sun's work shows that there are results, some used for decades without a rigorous foundation, that are only true for spaces with the rich structure of these measure spaces. His joint work with the author uses that structure to extend an important result on the purification of measure valued maps and to obtain a Fatou-type result for the Gelfand integral. With the appropriate map, results obtained using these measure spaces as prototype can also be shown to be valid for more general "nowhere countably generated" measure spaces.

1. INTRODUCTION

In 1975 [15], the author constructed a class of standard measure spaces formed on nonstandard models. These spaces, now called "Loeb spaces" in the literature, are very close to the underlying internal spaces, and are rich in structure. This brief survey is an invitation to read and not a substitute for the many articles presenting the numerous applications of these spaces. In particular, see the author's three chapters and Osswald's two chapters in [23] for more extensive background. We will begin here with the essentials of nonstandard analysis and the construction of these measure spaces. We will then discuss some recent work by Yeneng Sun that uses these measure spaces. Sun has shown that there are results, some used in applications for decades without a rigorous foundation, that are only true for spaces with the rich structure of Loeb measure spaces. We will follow the description of Sun's work by reviewing the recent result in [20] by Yeneng Sun and the author on the purification of measure-valued maps. A counterexample shows that the result we give is false if the Loeb space is replaced by the unit interval with Lebesgue measure. We also briefly describe a Fatou-type result for the Gelfand integral from [21]. Finally we note that with the appropriate map, results obtained using Loeb measure spaces as prototype can also be shown to be valid for more general measure spaces having the property that the relative measure algebra for any subset of positive measure cannot be countably generated.

2000 *Mathematics Subject Classification.* Primary 03H05, 28E05. Secondary 60G05. 91A60.
Key words and phrases. Nonstandard analysis, Loeb probability measures, purification of measure valued maps.

2. Basic Nonstandard Analysis

One thinks of a standard mathematical model as a world that exists in some sense. For example, we think of the real numbers as having an existence independent of what we may know about them. Theorems in a formal language form correct statements about such a model. It is important to recognize the distinction between objects in a standard model and the statements about those objects in a formal language. The reason for this distinction is that there is another mathematical object, called a nonstandard model, for which all theorems have a meaning and are correct. Informally, what we have are two worlds, the standard and the nonstandard, and the theorems about the first are also correct statements about the second. In Robinson's nonstandard model [26], the number system contains infinitely large and infinitely small positive numbers together with all of the numbers in the original real number system.

One way to explain Robinson's result is to invoke a theorem of Kurt Gödel. Take a name not use for anything in the standard number system – for example, Bach. To the theorems about the standard real number system add new statements: "Bach is bigger than 1", "Bach is bigger than 2", etc. Add one such statement for each natural number. The standard number system is not a model for the collection of theorems augmented by these statements about Bach. There is no number simultaneously bigger than $1,2,3$, etc. The standard number system is, however, a model for any finite subset of the augmented collection of statements. To see this, fix a finite subset of the augmented collection. Find the biggest number named in these statements, and let Bach be the name of a number that is even bigger. Since every finite subset of our augmented collection of statements has a model, it follows from a result of Gödel that the entire augmented collection of statements has a model. That is there is a number system extending the standard one and a number in that system, call it Bach, that is bigger than $1,2,3$, etc. Bach's reciprocal, 1 divided by Bach, is then an infinitesimal number.

Another approach to understanding Robinson's result is to construct a simple number system with infinitesimals using sequences of real numbers and a free ultrafilter $\mathscr{U}$ on the natural numbers. Two sequences represent the same nonstandard number if they agree on a set $U \in \mathscr{U}$. A constant sequence represents an ordinary number. For example, the sequence $5,5,5,\ldots$, represents the number 5. On the other hand, a sequence of positive real numbers tending to zero such as the sequence $1,1/2,1/3,1/4,\ldots$, becomes a positive infinitesimal. A sequence of real numbers increasing to infinity such as the sequence $1,2,3,\ldots.$, becomes an infinitely large number.

Rather than thinking of constructions, however, it is better in practice to work with the properties of the extended number system. In doing so, one should keep in mind the fact that any theorem for the ordinary numbers is also a theorem, when properly interpreted, for the enlarged number system.

What is meant by saying "when properly interpreted"? Briefly, when we say "all" subsets of a given set, we can't formally specify what we mean. Even for the set of natural numbers, the idea of all subsets cannot be formalized. This inability to formalize the notion of "all subsets" means that when interpreting theorems in the nonstandard model, we can cheat. We don't interpret the word "all" to really mean "all". We work instead with what are called **internal** sets, and interpret "all sets" to mean all internal sets. If A is a set in the standard model, then ${}^*A$ denotes the set in the nonstandard model with the same name and formal properties; it is internal. Any object that can be described using only the names of known internal objects is also internal. An object that is not internal is called **external**.

Important for applications is the fact that the set of natural numbers $\mathbb{N} = \{1,2,3,\cdots\}$ has been extended along with the real numbers. The theorem that says every positive real number is within distance one of a natural number is still valid for the extended number system. Therefore, we now have infinitely large natural numbers. We use ${}^*\mathbb{N}$ to denote the extended system of natural numbers, and ${}^*\mathbb{N}_\infty$ to denote the new members of that system, all of which are larger than any ordinary natural number. The set ${}^*\mathbb{N}_\infty$ is clearly external, since every nonempty internal subset of ${}^*\mathbb{N}$ must have a first element; subtracting 1 from a first element of ${}^*\mathbb{N}_\infty$ would yield the last ordinary natural number. It follows that the set of ordinary natural numbers is also external since ${}^*\mathbb{N}_\infty$ can be described as the complement of that set in ${}^*\mathbb{N}$.

It is also important to know that if ρ is an element of the extended real number system, which we denote by ${}^*\mathbb{R}$, and $|\rho| \leq n$ for some ordinary $n \in \mathbb{N}$, then there is a unique real-number r such that $\rho = r + \varepsilon$ where ε is infinitesimal, i.e., $\forall n \in \mathbb{N}, |\varepsilon| < 1/n$. We call r the standard part of ρ and write $r = \mathrm{st}(\rho)$. If ρ has a standard part, we will say it is **limited**; otherwise we will say it is **unlimited**.

Let us consider the problem of finding the area under the graph of the curve $y = x^2$ between 0 and 1. If we choose a natural number n and divide the interval between 0 and 1 into n intervals of length $1/n$, then the sum $(1/n)^2 \cdot 1/n + (2/n)^2 \cdot 1/n + \cdots + 1^2 \cdot 1/n$ is an approximation to the area under the curve; the larger the value of n, the better the approximation. If we take an unlimited natural number n, we have a sum that equals the area except for an infinitesimal error.

Here however, we come to the question "what is meant by the sum?" First let's consider the question "what is meant by a finite set?" In the standard world a set is finite if it can be enumerated with natural numbers finishing with a largest natural number. Having extended the real numbers and the natural numbers we have infinitely large, i.e., unlimited natural numbers. If a set can be internally enumerated with natural numbers up to such an unlimited natural number, then the set is called a hyperfinite set. Hyperfinite sets are infinite sets, but they have all of the formal properties of finite sets. In particular, since we can sum any finite set of real numbers, the summing function in the nonstandard model also gives an answer for any hyperfinite set in ${}^*\mathbb{R}$.

Hyperfinite sets play a central role in the applications of nonstandard analysis to many areas of mathematics beyond the calculus. In probability, for example, it is easy to analyze a finite coin toss. It is much harder to analyze an infinite coin toss. Any particular outcome has zero probability. On the other hand, in our nonstandard real number system there are unlimited natural numbers. We can choose such a number, again call it Bach, and, at least in our imagination, we can perform a hyperfinite coin toss. That is, we toss the coin Bach times. Now any particular outcome has probability equal to one divided by two raised to the power Bach. Moreover, this hyperfinite coin tossing space contains all standard infinite coin tosses.

The author showed in [15] how to make this nonstandard experiment, with probabilities given by nonstandard real numbers, into an ordinary probability space with real probabilities. The pair of spaces, one with nonstandard probabilities and the other with ordinary probabilities, is an example of what are called Loeb spaces in the literature. One can parametrize ordinary probability experiments with these spaces. The underlying space of points is a set in a nonstandard model, and the simple nonstandard space and the corresponding standard one are very close.

As shown by Anderson [1], hyperfinite coin tosses can be used to form a model for Brownian motion. One divides time up into infinitesimal intervals, and at the beginning of each interval tosses a coin. If the toss is a head, one moves to the right; if the toss is a tail, one moves to the left. The step size is the square root of the time change. These random walks, one for each hyperfinite coin tossing sequence, form a good underlying probability space for Brownian motion. Anderson also showed that the corresponding Loeb space generates Wiener measure.

Anderson's construction of Brownian motion can be viewed as a nonstandard formulation of Donsker's theorem. (See, for example, [4].) Other nonstandard treatments of weak convergence of measures can be found in [2], [16], and [17]. In the latter article, the author uses nonstandard analysis to produce a new, standard limit procedure for obtaining representing measures for positive harmonic functions.

Hyperfinite sets and the corresponding Loeb spaces play an important role in mathematical economics. A central problem in this subject is to study equilibria in economies with a vary large number of individuals when each individual has only a negligible influence on the economy. For this, it is quite natural to consider an economy with a hyperfinite number of individuals, each individual having only an infinitesimal influence on the economy.

3. More on Nonstandard Measure Theory

We next give a more detailed description of the construction of Loeb measure spaces. The principal device used is $\aleph_1$-saturation. This means that any ordinary sequence taken from an internal set is the beginning of an internal sequence from that set.

Working in an $\aleph_1$-saturated nonstandard model, we can construct a hyperfinite set Λ as the set of elementary outcomes in a conceptual experiment in the "nonstandard world". For coin tossing, for example, Λ can be the set of internal sequences of -1's and 1's of length $\eta \in {}^*\mathbb{N}_\infty$. Given such a hyperfinite Λ, we can let $\mathscr{C}$ consist of all internal subsets of Λ. The collection $\mathscr{C}$ is an internal σ-algebra, but it is also an algebra in the ordinary sense. Suppose P is an internal probability measure on $(\Lambda, \mathscr{C})$. For the coin tossing experiment, for example, each internal set A with internal cardinality $|A|$, would be given the probability $P(A) = |A|/2^\eta$ in ${}^*[0,1]$. In the general case, we can form a finitely additive real-valued measure $\widehat{P}$ on $(\Lambda, \mathscr{C})$ with values in the real interval $[0,1]$ by setting $\widehat{P}(A) = \operatorname{st}(P(A))$. The question is, "Can we extend $\widehat{P}$ to a countably additive measure on $\sigma(\mathscr{C})$, i.e., the σ-algebra generated by $\mathscr{C}$?" This is the question we consider next.

We start with an arbitrary internal measure space $\left(\Lambda, \mathscr{C}, \mu\right)$. We will assume here that $\mu(\Lambda)$ is limited. When Λ is a hyperfinite set one usually sets $\mathscr{C}$ equal to the collection of all internal subsets of Λ, and μ is usually an internal probability measure. In general, $\mathscr{C}$ is an internal σ-algebra as well as an algebra in the usual sense in Λ. Let $\widehat{\mu}$ be the finitely additive, extended real valued measure on $(\Lambda, \mathscr{C})$ defined at each $A \in \mathscr{C}$ by setting $\widehat{\mu}(A) = \mathrm{st}(\mu(A))$. Let $\sigma(\mathscr{C})$ denote the σ-algebra in the ordinary sense generated by $\mathscr{C}$. The collection $\sigma(\mathscr{C})$ is, in general, an external collection of subsets of Λ. We can extend $\widehat{\mu}$ to a measure μ_L defined on the measure completion $L_\mu(\mathscr{C})$ of $\sigma(\mathscr{C})$, and thus obtain a standard measure space $(\Lambda, L_\mu(\mathscr{C}), \mu_L)$ on Λ, by using the standard Carathéodory Extension Theorem.

To apply this technique, we note that when a sequence $\langle A_i : i \in \mathbb{N}\rangle$, indexed by the ordinary natural numbers, consists of pairwise disjoint elements of $\mathscr{C}$ and the union A is also in $\mathscr{C}$ then A is actually a finite union since all but a finite number of the A_i's are empty. Here is the simple proof: Using $\aleph_1$-saturation, we extend the sequence $\langle A_i : i \in \mathbb{N}\rangle$ to an internal sequence $\langle A_i : i \in {}^*\mathbb{N}\rangle$; the set

$$\{m \in {}^*\mathbb{N} : A \subseteq \cup_{1 \le i \le m} A_i\}$$

is internal and contains ${}^*\mathbb{N}_\infty$, so it must contain some standard natural number since ${}^*\mathbb{N}_\infty$ is external.

It now follows that $\widehat{\mu}(A) = \sum_{i \in N} \widehat{\mu}(A_i)$. By the Carathéodory Extension Theorem, the finitely additive measure $\widehat{\mu}$ has a unique σ-additive extension μ_L defined on the completion $L_\mu(\mathscr{C})$ of the σ-algebra $\sigma(\mathscr{C})$.

When used in probability theory, the above general construction allows one to tackle problems of continuous parameter stochastic processes using the combinatorial tools available for discrete parameter processes. Examples are the author's construction of Poisson processes in [15] and Anderson's representation of Brownian Motion and the Itô integral in [1]. As demonstrated by the work of Keisler , Fajardo and Keisler , Sun, and others, these measure spaces have special closure properties not shared by even Lebesgue measure spaces. The above construction has yielded new standard-analysis results by many researchers in areas such as probability theory, potential theory, mathematical economics and mathematical physics. Here are five examples: Keisler's new existence theorem for stochastic differential equations [11], Perkins' award winning research (Rollo Davidson Prize in Probability Theory) on the theory of local time [24], Arkeryd's results on gas kinetics [3], Cutland and Ng's work on the Wiener sphere and Wiener measure [5], and Sun's work in probability and economics, some of which we discuss next.

4. Recent Work of Yeneng Sun

We briefly discuss next some work in the late 1990's of Yeneng Sun ([28], [29], [30], and Chapter 7 of [23]). To set the background we consider the following question: Does it make sense to speak of an infinite number of independent individuals or random variables indexed by points in a uniform probability space? Can one, for example, reasonably consider an infinite number of independent tosses of a fair coin with the tosses indexed by a uniform probability space, and if so, does it make sense to say that half the tosses should be heads? Of course, there is no problem in speaking of independent random variables indexed by the first n integers with each integer having probability $1/n$. An infinite index set with a uniform probability measure, however, must be uncountable; there is no uniform probability measure on the full set of natural numbers.

The problem thus evolves to finding the possible meaning of a continuum of independent random variables. As Doob indicated in 1937, [6], whatever way one approaches this problem, the usual measure-theoretic tools fail.

One approach to representing a continuum of independent random variables is to consider a function $f(i,\omega)$ (i.e., a process) where i is an index from an uncountable probability space, the parameter space (the probability measure need not be uniform) and ω is taken from a second probability space, the sample space. The question then is whether it makes sense to work with the usual product of these two probability spaces. Sun has shown in Proposition 7.33 of [23] that no matter what kind of measure spaces, even Loeb measure spaces, one might take as the parameter space and sample space of a process, independence and joint measurability with respect to the classical measure-theoretic product, i.e., formed using measurable rectangles as in [27], are never compatible with each other except for a trivial case.

In the articles cited at the beginning of this section, Sun has shown that a construction overcoming these measure-theoretic problems is obtained by forming the internal product of internal factors, and then taking not just the Loeb space of each factor, but also the Loeb space of the internal product. This new σ-algebra, call it the big product σ-algebra, forms a rich extension of the usual product σ-algebra of the Loeb factors, one that still has the Fubini Property equating the integral over the product space to the iterated integrals over the factor spaces forming the product. In his articles, cited above, Sun shows that to work with independence in a continuum setting, as has been attempted in informal mathematical applications without rigor, and for decades, one needs a rich product σ-algebra such as the one just described. It therefore makes sense to use the big product space as the underlying product space for an infinite number of equally weighted, independent random variables or agents.

As part of his work with large product spaces, Sun has extended the law of large numbers. The usual strong law states that if random variables X_i, $i \in \mathbb{N}$, are independent with the same distribution and finite mean m, then $\frac{1}{n}\Sigma_{i=1}^{n} X_i$ tends almost surely to the constant random variable m. That is, for almost all samples ω, the value of the sequence at ω tends to the constant m.

Theorem 4.1 (Sun). *Let f be a real-valued integrable process on the big product space. Suppose f is jointly measurable on the product probability space, and outside an appropriate set of measure* 0 *of the product of the index set with itself, the random variables $f_t := f(t,\cdot)$ are pairwise independent for all remaining pairs of indices. Then for almost all samples $\omega \in \Omega$, the mean of the sample function $f_\omega := f(\cdot,\omega)$ on the parameter space T is the mean of f viewed as a random variable on the big product space. There is no requirement of identical distributions.*

Another facet of this same work deals with independence. It is well known that for a finite collection of random variables, pairwise independence is strictly weaker than mutual independence. Sun has shown that for processes on big product spaces, pairwise independent and mutual independent coalesce, and they coalesce with other notions of independence that are distinct for a finite number of random variables. This implies asymptotic results for finite families of random variables where the families are ordered by containment and have increasing cardinality.

5. PURIFICATION OF MEASURE-VALUED MAPS

We now turn to a special case of the joint work of Yeneng Sun with the author in [20]. That special case generalizes the following celebrated theorem of Dvoretzky, Wald and Wolfowitz (see [7], [8], [9]).

Theorem 5.1. *Let A be a finite set and $\mathscr{M}(A)$ the space of probability measures on A. Let $(T,\mathscr{T})$ be a measurable space, and μ_k, $k=1,\cdots,m$, finite, atomless signed measures on $(T,\mathscr{T})$. Given $f:T\to\mathscr{M}(A)$ so that for each $a\in A$, $f(\cdot)(\{a\})$ is $\mathscr{T}$-measurable, there is a $\mathscr{T}$-measurable map $g:T\to A$ so that for each $a\in A$,*

$$\int_T f(t)(\{a\})d\mu_k(t)=\mu_k(\{t\in T:g(t)=a\}).$$

This theorem justifies the elimination, i.e., purification, of randomness in some settings. For example, in games, T represents information available to the players, and A represents the actions players may choose, given $t\in T$. Every player's objective is to maximize his/her own expected payoff, but that payoff depends on the actions chosen by all the players.

For each player, a mapping from T to A is called a pure strategy. A mapping from T to $\mathscr{M}(A)$ is called a mixed strategy; in this case, each player chooses a "lottery on A". A Nash equilibrium is achieved if every player is satisfied with his/her own choice of strategy given the choices of the other players.

In quite general settings, such an equilibrium can be achieved using mixed strategies. When results such as Theorem 5.1 apply, those mixed strategies can be purified yielding an equilibrium with the same expected payoffs. We refer the reader to [20] for more information on the game-theoretic consequences of Theorem 5.1. In [20], the DWW Theorem is extended from a finite set A to a complete, separable metric space. Since there always exists a Borel bijection from such a space to a compact metric space, it suffices to let A be compact metric space. For simplicity here, we state the result for a finite set of measures. We refer the reader to [18] for the proof and to [20] for the more general result and its proof.

Theorem 5.2. *Let K be a finite set, and let A be a compact metric space. For each $k\in K$, let μ_k be a nonatomic, finite, signed Loeb measure on a Loeb measurable space $(T,\mathscr{T})$. If f is a $\mathscr{T}$-measurable mapping from T to $\mathscr{M}(A)$, then there is a $\mathscr{T}$-measurable mapping g from T to A such that for each $k\in K$ and for all Borel sets B in A, $\int_T f(t)(B)\mu_k(dt)=\mu_k\left(g^{-1}[B]\right)$. This is equivalent to the condition that for each $k\in K$ and for any bounded Borel measurable function θ on A,*

$$\int_T\int_A\theta(a)f(t)(da)\mu_k(dt)=\int_T\theta(g(t))\mu_k(dt).$$

Example 5.3. Sun and the author show by a counter example in [20] that the Loeb measures of the theorem cannot be replaced with measures absolutely continuous with respect to Lebesgue measure λ on $[0,1]$. In their example, the compact set A is the interval $[-1,1]$, and the measure-valued map is given by $f(t):=(\delta_t+\delta_{-t})/2$ for each $t\in[0,1]$. The example uses just two measures, $\mu_1=\lambda$ on $[0,1]$, and $\mu_2=2t\lambda$ on $[0,1]$. If there is a function g that works, then it is not hard to see using the functions on A

given by $\theta(a) = |a|$ and $\theta(a) = a^2$ that $\int_0^1 \left(t - |g(t)|\right)^2 \lambda(dt) = 0$. That is, $g(t)$ must take the value t or $-t$ λ-a.e. on $[0,1]$. It is also not hard to see that this is impossible. The moral is that a Lebesgue measurable g can't switch values on $[0,1]$ fast enough to work.

6. Further Results

The main result in [21] establishes a general Fatou type result for a sequence of Gelfand integrable functions from a vector Loeb measure space to the dual of a separable Banach lattice. Such a result fails when the Loeb measure space is replaced by a Lebesgue measure space. To be exact, Sun and the author consider a sequence $\{g_n\}_{n=1}^{\infty}$ of Gelfand integrable, non-negative functions mapping a measurable space $(\Omega, \mathscr{A})$ supplied with countably many Loeb measures μ_j, $j \in J$, into a dual Banach lattice X. In some cases, a weak*-tightness condition on the sequence $\{g_n\}_{n=1}^{\infty}$ is needed. Forming a control measure $\bar{\mu}$ from the measures μ_j, we show that if the Gelfand integrals $\int_\Omega g_n d\mu_j$ have a weak* limit $a_j \in X$ as n goes to infinity, then there is a function g from $(\Omega, \mathscr{A})$ to X such that for $\bar{\mu}$-a.e. $\omega \in \Omega$, $g(\omega)$ is a weak* limit point of $\{g_n(\omega)\}_{n=1}^{\infty}$, and g is Gelfand μ_j-integrable with $\int_\Omega g d\mu_j \le a_j$ for each $j \in J$. If in addition, the sequence $\{g_n\}_{n=1}^{\infty}$ satisfies a strong condition, that of uniform integrability, then one can claim $\int_\Omega g d\mu_j = a_j$ for each $j \in J$ without imposing a lattice structure on X.

We note that another result needing rich measure spaces can by found in the article [19] by Osswald, Sun, Zhang and the author. It relates uncorrelatedness and orthogonality for vector-valued processes.

Some results established for Loeb measure spaces can be extended with a technique established by D. N. Hoover and H. J. Keisler in [10]. That is, certain results valid for an atomless Loeb measure space are also valid for measure spaces having the property that we shall refer to as being nowhere countably generated. A measure space has this property if for any set of positive measure, the relative σ-algebra of measurable subsets (modulo sets of measure 0) cannot be countably generated. In addition to establishing important properties for such spaces, the Hoover-Keisler paper [10] contains a number of far-reaching ideas on the equivalence and convergence of processes with filtrations.

The purification result in [20] has been extended to such spaces by Konrad Podczeck in [25]. His extension is a consequence of new results in functional analysis produced for that purpose. In [22], Sun and the author apply the Hoover-Keisler technique to demonstrate that with the choice of appropriate mappings, results of this type can easily be validated for nowhere countably generated measure spaces once they have been established for Loeb measure spaces, or even just such a space formed on a hyperfinite set. This generalization technique is analogous to the extension of certain results established for the Lebesgue unit interval to more general atomless spaces. Just as the Lebesgue unit interval forms the prototype and simplest such space, hyperfinite Loeb counting spaces play a similar role in the class of nowhere countably generated measure spaces.

REFERENCES

[1] R. M. Anderson, A nonstandard representation of Brownian motion and Itô integration. *Israel J. Math.* **25** (1976), 15-46.

[2] R. M. Anderson and S. Rashid, A nonstandard characterization of weak convergence. *Proc. Amer. Math. Soc.* **69** (1978), 327-332.

[3] L. Arkeryd, Loeb solutions of the Boltzmann equation. *Arch. Rational Mech. Anal* **86** (1984), 85-97.

[4] P. Billingsley, *Convergence of Probability Measures*, 3rd Edition. Wiley, New York, 1999.

[5] N. Cutland and S-A. Ng, The Wiener sphere and Wiener measure. *Ann. Probability* **21** (1993), 1-13.

[6] J. L. Doob, Stochastic processes depending on a continuous parameter. *Trans. Amer. Math. Soc.* **42** (1937), 107-140.

[7] A. Dvoretsky, A. Wald and J. Wolfowitz, Elimination of randomization in certain problems of statistics and of the theory of games. *Proc. Nat. Acad. Sci. USA* **36** (1950), 256-260.

[8] A. Dvoretsky, A. Wald and J. Wolfowitz, Relations among certain ranges of vector measures. *Pac. J. Math.* **1** (1951), 59-74.

[9] A. Dvoretsky, A. Wald and J. Wolfowitz, Elimination of randomization in certain statistical decision problems in certain statistical decision procedures and zero-sum two-person games. *Ann. Math. Stat.* **22** (1951), 1-21.

[10] D. N. Hoover and H. J. Keisler. Adapted probability distributions. *Trans. Amer. Math. Soc.* **286** (1984), 159-201.

[11] H. J. Keisler, An infinitesimal approach to stochastic analysis. *Memoirs Amer. Math. Soc.* **48** (1984).

[12] H. J. Keisler and Y. N. Sun, A metric on probabilities, and products of Loeb spaces. *Jour. London Math. Soc.* **69** (2004), 258-272.

[13] M. A. Khan, K. P. Rath and Y. N. Sun, The Dvoretzky-Wald-Wolfowitz theorem and purification in atomless finite-action games, submitted.

[14] M. A. Khan and Y. N. Sun, Non-cooperative games on hyperfinite Loeb spaces. *J. Math. Econ.* **31** (1999), 455-492.

[15] P. A. Loeb, Conversion from nonstandard to standard measure spaces and applications in probability theory. *Trans. Amer. Math. Soc.* **211** (1975), 113-122.

[16] P. A. Loeb, Weak limits of measures and the standard part map. *Proc. Amer. Math. Soc.* **77** (1979), 128-135.

[17] P. A. Loeb, A generalization of the Riesz-Herglotz theorem on representing measures. *Proc. Amer. Math. Soc.* **71** (1978), 65-68.

[18] P. A. Loeb, Applications of rich measure spaces formed from nonstandard models. In: I. van den Berg and V. Neves (Eds.) *The Strength of Nonstandard Analysis*, pp. 206-210. Springer-Verlag Wien–New York, 2007.

[19] P. A. Loeb, H. Osswald, Y. Sun, and Z. Zhang, Uncorrelatedness and orthogonality for vector-valued processes. *Trans. Amer. Math. Soc.* **356** (2004), 3209-3225.

[20] P. A. Loeb and Y. Sun, Purification of measure-valued maps. *Doob Memorial Volume* of the *Illinois Journal of Mathematics*, **50** (2006), 747-762.

[21] P. A. Loeb and Y. Sun, A general Fatou lemma. *Advances in Mathematics*, **213** (2007), 741-762.

[22] P. A. Loeb and Y. Sun, Purification and Saturation, *Proc. Amer. Math. Soc.*, to appear.

[23] P. A. Loeb and M. Wolff, *Nonstandard Analysis for the Working Mathematician*. Kluwer Acad. Publish., Springer, 2000.

[24] E. Perkins, A global intrinsic characterization of local time. *Ann. Probability* **9** (1981), 800-817.

[25] K. Podczeck, On purification of measure-valued maps. *Economic Theory* **38** (2009), 399-418.

[26] A. Robinson, *Non-standard Analysis*. North-Holland, Amsterdam, 1966.

[27] H. L. Royden, *Real Analysis*, third edition. Macmillan, New York, 1988.

[28] Y. Sun, Hyperfinite law of large numbers. *Bull. Symbolic Logic* **2** (1996), 189-198.

[29] Y. Sun, A theory of hyperfinite processes: the complete removal of individual uncertainty via exact LLN. *J. Math. Econ.* **29** (1998), 419-503.

[30] Y. Sun, The almost equivalence of pairwise and mutual independence and the duality with exchangeability. *Probab. Theory and Relat. Fields* **112** (1998), 425-456.

DEPARTMENT OF MATHEMATICS, UNIVERSITY OF ILLINOIS, URBANA, IL 61801, USA
E-mail address: loeb@math.uiuc.edu

Potential Theory and Stochastics in Albac, 193–202
Aurel Cornea Memorial Volume

INTEGRABILITY OF MAXIMAL FUNCTIONS FOR GENERALIZED LEBESGUE SPACES $L^{p(\cdot)}(\log L)^{q(\cdot)}$

YOSHIHIRO MIZUTA, TAKAO OHNO, AND TETSU SHIMOMURA

ABSTRACT. We study the integrability of maximal functions for generalized Lebesgue spaces $L^{p(\cdot)}(\log L)^{q(\cdot)}$, where the exponent $p(\cdot)$ approaches 1 on some part of the domain. Our integrability results depend on the shape of that part and the speed of the exponent approaching 1.

1. INTRODUCTION

A crucial tool in the development of the function space theory is the boundedness of Hardy-Littlewood maximal operator. For a locally integrable function f on $\mathbf{R}^n$, the maximal function is defined by

$$Mf(x) = \sup_{r>0} \frac{1}{|B(x,r)|} \int_{B(x,r)} |f(y)|dy,$$

where $|B(x,r)|$ denotes the volume of the open ball $B(x,r)$ centered at x of radius $r > 0$.

In the classical Lebesgue L^p spaces, we know (see the book by Stein [14, Chapter 1]) that if $p > 1$, then

$$\|Mf\|_{L^p(\mathbf{R}^n)} \le C\|f\|_{L^p(\mathbf{R}^n)}$$

for all $f \in L^p(\mathbf{R}^n)$. Unfortunately, this is not true when $p = 1$ even if we are restricted to a bounded domain G in $\mathbf{R}^n$. Thus we need to consider the space $L\log L(G)$ of measurable functions f on G whose norm

$$\|f\|_{L\log L(G)} = \inf\left\{\lambda > 0 : \int_G \frac{|f(y)|}{\lambda} \log\left(e + \frac{|f(y)|}{\lambda}\right) dy \le 1\right\}$$

is finite. It is known from Stein [13] and [14, Sections I.1 and I.5] that $Mf \in L^1(G)$ if and only if $f \in L\log L(G)$.

Following Orlicz [11], Kováčik-Rákosník [6] and Musielak [10], we consider a positive continuous function $p(\cdot)$ on G whose value is not less than 1, and the space of all measurable functions f on G satisfying

$$\int_G \left|\frac{f(y)}{\lambda}\right|^{p(y)} dy < \infty$$

2000 *Mathematics Subject Classification.* 46E30,42B25.

Key words and phrases. Maximal functions, Lebesgue spaces with variable exponent, Minkowski content.

for some $\lambda > 0$. We define the norm on this space by

$$\|f\|_{L^{p(\cdot)}(G)} = \inf\left\{\lambda > 0 : \int_G \left|\frac{f(y)}{\lambda}\right|^{p(y)} dy \le 1\right\}.$$

In recent years, the generalized Lebesgue spaces $L^{p(\cdot)}$ have attracted more and more attention, in connection with the study of elasticity, fluid mechanics and differential equations with $p(\cdot)$-growth; see R̊užička [12].

In this paper we study the integrability of the maximal operator in the Lebesgue space of variable exponent approaching 1 in some part of G. Recently, Hästö [5] has obtained an interesting class of variable exponents approaching 1 in some part of G such that $Mf \in L^1(G)$ for all $f \in L^{p(\cdot)}(G)$. As in Hästö [5], letting $\log_{(1)} t = \log t$ and $\log_{(m+1)} t = \log(\log_{(m)} t)$ for $m = 1, 2, \ldots$, we consider a function ω on the interval $[0, \infty)$ such that $\omega(0) = 0$ and

$$\omega(r) = \frac{a_1}{\log(1/r)} + \frac{a_2 \log_{(2)}(1/r)}{\log(1/r)} + \frac{a_3 \log_{(3)}(1/r)}{\log(1/r)}$$

for $0 < r \le r_1$, where the numbers a_1, a_2, a_3 and $0 < r_1 < 1/e$ are chosen so that $\omega(r)$ is nondecreasing on $[0, r_1]$; set $\omega(r) = \omega(r_1)$ for $r > r_1$.

THEOREM A (cf. Hästö [5]). *Let $x_0 \in G$ and consider a variable exponent $p(\cdot)$ such that*

$$p(x) = 1 + \omega(|x - x_0|) \tag{1}$$

for $x \in G$. If $a_2 > 1/n$, then the maximal operator $\mathcal{M} : f \to Mf$ is bounded from $L^{p(\cdot)}(G)$ into $L^1(G)$, that is,

$$\|Mf\|_{L^1(G)} \le C\|f\|_{L^{p(\cdot)}(G)}.$$

Futamura and Mizuta [4] have proved that the conclusion is still valid when $a_2 = 1/n$ and $a_3 \ge 0$.

In this paper, following Cruz-Uribe, Fiorenza and Neugebauer [2], we further consider a function $q(\cdot)$ such that $q(0) = q_0 \in \mathbf{R}$ and

$$q(x) = q_0 + \eta(|x - x_0|) \tag{2}$$

for $x \in G$, where $\eta(r) = \dfrac{b_1}{\log_{(2)}(1/r)} + \dfrac{b_2 \log_{(3)}(1/r)}{\log_{(2)}(1/r)}$ for $0 < r \le r_2$, and $\eta(r) = \eta(r_2)$ for $r > r_2$. Here the numbers b_1, b_2 and $0 < r_2 < 1/e$ are chosen so that $\eta(r)$ is nondecreasing on $(0, \infty)$.

Now we set

$$\Phi(x, t) = t^{p(x)}(\log(c_0 + t))^{q(x)},$$

where c_0 is chosen that $\Phi(\cdot, t)$ is a convex function of t. Define the norm

$$\|f\|_{L^{p(\cdot)}(\log L)^{q(\cdot)}(G)} = \inf\left\{\lambda > 0 : \int_G \Phi(x, |f(x)|/\lambda)\, dx \le 1\right\}$$

for a measurable function f on G. We denote by $L^{p(\cdot)}(\log L)^{q(\cdot)}(G)$ the family of all measurable functions f on G such that $\|f\|_{L^{p(\cdot)}(\log L)^{q(\cdot)}(G)} < \infty$.

Our aim in this paper is to establish the following result, as an extension of the recent result by Cruz-Uribe, Fiorenza and Neugebauer [2].

THEOREM B. *Let $p(\cdot)$ and $q(\cdot)$ be of the form* (1) *and* (2).

(i) *If $na_2 + q_0 > 0$, then*

$$\int_G Mf(x)(\log(1+Mf(x)))^{na_2+q_0-1}(\log(1+(\log(1+Mf(x)))))^{na_3+b_2}\,dx \le C;$$

(ii) *if $na_2 + q_0 = 0$ and $na_3 + b_2 > 0$, then*

$$\int_G Mf(x)(\log(1+Mf(x)))^{-1}(\log(1+(\log(1+Mf(x)))))^{na_3+b_2-1}\,dx \le C$$

for all measurable functions f on G with $\|f\|_{L^{p(\cdot)}(\log L)^{q(\cdot)}(G)} \le 1$, where C denotes a positive constant independent of f.

This also gives extensions of Theorem A by Hästö [5], Futamura-Mizuta [4] and Mizuta-Ohno-Shimomura [9].

2. VARIABLE EXPONENTS

Throughout this paper, let C denote various positive constants independent of the variables in question. Further let G denote a bounded open set in $\mathbf{R}^n$.

We say that a positive nondecreasing function φ on the interval $[0,\infty)$ satisfies (φ) if there exist $\varepsilon_1 > 0$ and $0 < r_1 < 1$ such that

(φ) $(\log(1/r))^{-\varepsilon_1}\varphi(1/r)$ is nondecreasing on $(0, r_1)$.

Similarly, a positive nondecreasing function ψ on the interval $[0,\infty)$ is said to satisfy (ψ) if there exist $\varepsilon_2 > 0$ and $0 < r_2 < 1/e$ such that

(ψ) $(\log_{(2)}(1/r))^{-\varepsilon_2}\psi(1/r)$ is nondecreasing on $(0, r_2)$.

Consider positive nondecreasing functions φ and ψ satisfying (φ) and (ψ), respectively. Set

$$\varepsilon_0 = \max\{\varepsilon_1, \varepsilon_2\}.$$

For the sake of convenience, we assume that

(φ') $\varphi(t) \ge e^{\varepsilon_0}$ for all $t > 0$,

(ψ') $\psi(t) \ge e^{\varepsilon_0}$ for all $t > 0$.

First we give the following results, which can be derived by conditions (φ) and (φ').

Lemma 1 ([8, Lemma 3.1, Section 5.3], [9, Lemmas 2.1 and 2.2]). (i) *$\varphi(r)$ is of log-type, that is, there exists $C > 0$ such that*

$$C^{-1}\varphi(r) \le \varphi(r^2) \le C\varphi(r) \qquad \text{whenever } r > 0. \tag{3}$$

(ii) *For $\gamma > 0$, there exists $C > 0$ such that*

$$t^{-\gamma}\varphi(t) \le Cs^{-\gamma}\varphi(s) \qquad \text{whenever } t \ge s > 0.$$

(iii) *There exists $0 < \tilde{r}_1 < r_1$ such that $\omega(r) = \log\varphi(1/r)/\log(1/r)$ is nondecreasing on $[0, \tilde{r}_1]$.*

Further, we see from (ψ) and (ψ') that ψ satisfies (i), (ii) and

(iv) there exists $0 < \tilde{r}_2 < r_2$ such that $\eta(r) = \log\psi(1/r)/\log(\log(1/r))$ is nondecreasing on $[0, \tilde{r}_2]$.

Condition (3) implies the doubling condition on φ, that is, there exists a constant $C > 1$ such that

$$\varphi(r) \le \varphi(2r) \le C\varphi(r) \qquad \text{whenever } r > 0.$$

Our typical example of φ is of the form

$$\varphi(r) = a_1 (\log r)^{a_2} (\log_{(2)} r)^{a_3}$$

for large r, where $a_1 > 0$, $a_2 \ge 0$ and $a_3 \in \mathbf{R}$; similarly, that of ψ is of the form

$$\psi(r) = b_1 (\log_{(2)} r)^{b_2}$$

for large r, where $b_1 > 0$ and $b_2 \ge 0$.

For simplicity, set

$$r_0 = \min\{\tilde{r}_1, \tilde{r}_2\}.$$

Consider

$$\omega(r) = \frac{\log \varphi(1/r)}{\log(1/r)}$$

for $0 < r \le r_0$; set $\omega(r) = \omega(r_0)$ for $r > r_0$. We further consider

$$\eta(r) = \frac{\log \psi(1/r)}{\log_{(2)}(1/r)}$$

for $0 < r \le r_0$; set $\eta(r) = \eta(r_0)$ for $r > r_0$.

For a compact set K in $\mathbf{R}^n$, we define

$$K(r) = \{x \in \mathbf{R}^n : \delta_K(x) \le r\}$$

for $r \ge 0$, where $\delta_K(x)$ denotes the distance of x from K. For $0 < \alpha \le n$, we say that the $(n-\alpha)$-dimensional upper Minkowski content of K is finite if

$$|K(r)| \le C r^{\alpha} \qquad \text{for small } r > 0;$$

see the book by Mattila [7]. Note here that if K is a singleton, then its 0-dimensional upper Minkowski content is finite, and if K is a spherical surface, then its $(n-1)$-dimensional upper Minkowski content is finite. As examples of K, we may consider fractal type sets like Cantor sets or Koch curves.

Now we define variable exponents $p(\cdot)$ and $q(\cdot)$ by

$$p(x) = 1 + \omega(\delta_K(x))$$

and

$$q(x) = q_0 + \eta(\delta_K(x))$$

for $x \in \mathbf{R}^n$ and $q_0 \in \mathbf{R}$.

We here remark the following easy result.

Lemma 2 (cf. [9, Lemma 2.2]). *$p(\cdot)$ and $q(\cdot)$ are continuous on $\mathbf{R}^n$.*

3. Integrability results

Let us begin with the following result.

Lemma 3 (cf. [9, Lemma 2.3]). *Let K be a compact set in G whose $(n-\alpha)$-dimensional upper Minkowski content is finite. Then*

$$\int_G \delta_K(x)^{-\alpha}(\log(1+\delta_K(x)^{-1}))^{-\beta}\,dx < \infty$$

for every $\beta > 1$.

Lemma 4 (cf. [4, Lemma 2.3], [9, Lemma 2.4]). *Suppose the $(n-\alpha)$-dimensional upper Minkowski content of K is finite. If f is a measurable function on G with $\|f\|_{L^{p(\cdot)}(\log L)^{q(\cdot)}(G)} \le 1$, then*

$$\int_G |f(x)|(\log(e+|f(y)|))^{q_0}\varphi(|f(x)|)^{\alpha}\psi(|f(x)|)\,dx \le C.$$

Proof. Consider the set

$$G' = \{x \in K(r_0)\cap G : |f(x)| < \delta(x)^{-\alpha}(\log(1/\delta(x)))^{-\beta}\},$$

where the constant β is determined later and $\delta(x) = \delta_K(x)$ for simplicity. If $x \in G'$, then we have by (φ) and (ψ)

$$\begin{aligned}
&|f(x)|(\log(e+|f(x)|))^{q_0}\varphi(|f(x)|)^{\alpha}\psi(|f(x)|)\\
&\le C\delta(x)^{-\alpha}(\log(1/\delta(x)))^{-\beta}(\log(1/\delta(x)))^{q_0}\varphi(1/\delta(x))^{\alpha}\psi(1/\delta(x))\\
&\le C\delta(x)^{-\alpha}(\log(1/\delta(x)))^{-\beta+q_0+\varepsilon_3},
\end{aligned}$$

where $\varepsilon_3 > \varepsilon_1\alpha$. If we take β so large that $\beta > 1+q_0+\varepsilon_3$, then it follows from Lemma 3 that

$$\int_{G'} |f(x)|(\log(e+|f(x)|))^{q_0}\varphi(|f(x)|)^{\alpha}\psi(|f(x)|)\,dx \le C.$$

If $x \notin G'$ and $\delta(x) < r_0$, then $|f(x)| \ge \delta(x)^{-\alpha}(\log(1/\delta(x)))^{-\beta}$, so that

$$\delta(x) \ge C|f(x)|^{-1/\alpha}(\log|f(x)|)^{-\beta/\alpha}.$$

Hence, in view of Lemma 1, we see that

$$\begin{aligned}
\frac{\log\varphi(1/\delta(x))}{\log(1/\delta(x))}\log|f(x)| &\ge \frac{\log\varphi(C|f(x)|^{1/\alpha}(\log|f(x)|)^{\beta/\alpha})}{\log(C|f(x)|^{1/\alpha}(\log|f(x)|)^{\beta/\alpha})}\log|f(x)|\\
&\ge \frac{\alpha\log(C\varphi(|f(x)|))}{\log|f(x)|+C\log(C\log|f(x)|)}\log|f(x)|\\
&= \alpha\log(C\varphi(|f(x)|))\left(1-\frac{C\log(C\log|f(x)|)}{\log|f(x)|+C\log(C\log|f(x)|)}\right)\\
&\ge \alpha\log\varphi(|f(x)|) - C,
\end{aligned}$$

which yields

$$\begin{aligned}
|f(x)|^{p(x)-1} &= \exp\left(\frac{\log\varphi(1/\delta(x))}{\log(1/\delta(x))}\log|f(x)|\right) \ge \exp\left(\alpha\log\varphi(|f(x)|)-C\right)\\
&= C\varphi(|f(x)|)^{\alpha}.
\end{aligned}$$

Similarly, we have

$$\frac{\log \psi(1/\delta(x))}{\log_{(2)}(1/\delta(x))} \log_{(2)} |f(x)| \geq \log \psi(|f(x)|) - C,$$

which yields

$$(\log |f(x)|)^{q(x)-q_0} \geq C\psi(|f(x)|).$$

Thus it follows that

$$\int_{K(r_0)\setminus G'} |f(x)|(\log(e+|f(x)|))^{q_0} \varphi(|f(x)|)^{\alpha} \psi(|f(x)|)dx$$
$$\leq C \int_G |f(x)|^{p(x)} (\log(c_0 + |f(x)|))^{q(x)} \, dx \leq C.$$

Finally, since $p(x) \geq p_1 > 1$ when $\delta(x) \geq r_0$, we find

$$\int_{G\setminus K(r_0)} |f(x)|(\log(e+|f(x)|))^{q_0} \varphi(|f(x)|)^{\alpha} \psi(|f(x)|)dx$$
$$\leq C \int_G |f(x)|^{p(x)} (\log(c_0 + |f(x)|))^{q(x)} \, dx + C \leq C.$$

The required assertion is now proved. □

For simplicity, set

$$\lambda(t) = t(\log(e+t))^{q_0} \varphi(t)^{\alpha} \psi(t)$$

and consider a continuous function Λ such that

$$C^{-1} \frac{\lambda(t)}{t} \leq \int_1^t \frac{\Lambda(s)}{s^2} ds \leq C \frac{\lambda(t)}{t} \qquad \text{whenever } t > 2.$$

The next lemma is an extension of Stein [14, Chapter 1], whose proof will be done along the same lines as in Stein [14, Chapter 1] (see [9]); for another proof, see Cianchi [1].

Lemma 5 ([9, Lemma 2.5]). *For a locally integrable function f on G, the following are equivalent:*

(i) $\int_G \lambda(|f(x)|) \, dx \leq A_1$;

(ii) $\int_G \Lambda(Mf(x)) \, dx \leq A_2$,

where $A_1, A_2 > 1$ are constants (independent of f) such that $C^{-1} A_1 \leq A_2 \leq C A_1$.

4. Proof of Theorem B

By Lemmas 4 and 5, we have the following result.

Theorem 1. *Let λ and Λ be as before Lemma 5. Suppose the $(n-\alpha)$-dimensional upper Minkowski content of K is finite. If $\|f\|_{L^{p(\cdot)}(\log L)^{q(\cdot)}(G)} \leq 1$, then*

$$\int_G \lambda(|f(x)|) \, dx \leq C,$$

or equivalently,

$$\int_G \Lambda(Mf(x)) \, dx \leq C.$$

Remark 1. Let $\varphi(r) = A(\log r)^{a_2}(\log_{(2)} r)^{a_3}$ for large r, where A, a_2 and a_3 are chosen so that φ is nondecreasing. Moreover, let $\psi(r) = B(\log_{(2)} r)^{b_2}$ for large r, where $B > 0$ and $b_2 \geq 0$. In this case, we can take

$$\lambda(r) = r(\log r)^{a_2\alpha+q_0}(\log_{(2)} r)^{a_3\alpha+b_2}$$

for large r.

(i) If $a_2\alpha + q_0 > 0$, then we can take

$$\Lambda(r) = r(\log r)^{a_2\alpha+q_0-1}(\log_{(2)} r)^{a_3\alpha+b_2}$$

for large r.

(ii) If $a_2\alpha + q_0 = 0$ and $a_3\alpha + b_2 > 0$, then we can take

$$\Lambda(r) = r(\log r)^{-1}(\log_{(2)} r)^{a_3\alpha+b_2-1}$$

for large r.

Thus Theorem 1 gives Theorem B in the Introduction and the following corollary.

Corollary 1. *For $\alpha > 0$, let K be a compact subset of G whose $(n-\alpha)$-dimensional upper Minkowski content is finite. Consider the exponents*

$$p(x) = 1 + \frac{a_1}{\log(1/\delta_K(x))} + \frac{a_2 \log_{(2)}(1/\delta_K(x))}{\log(1/\delta_K(x))}$$

and

$$q(x) = q_0 + \frac{b_1}{\log_{(2)}(1/\delta_K(x))} + \frac{b_2 \log_{(3)}(1/\delta_K(x))}{\log_{(2)}(1/\delta_K(x))}$$

when $\delta(x) \leq r_0$. If $a_2, b_2 > 0$ and $a_1, b_1, q_0 \in \mathbf{R}$ are chosen so that $a_2\alpha + q_0 > 0$,

$$\omega(r) = a_1/\log(1/r) + a_2(\log_{(2)}(1/r))/\log(1/r)$$

and

$$\eta(r) = b_1/\log_{(2)}(1/r) + b_2(\log_{(3)}(1/r))/\log_{(2)}(1/r)$$

are nondecreasing on $(0, r_0]$, then

$$\int_G |f(x)|(\log(1+|f(x)|))^{a_2\alpha+q_0}(\log_{(2)}(1+|f(x)|))^{b_2}\,dx < \infty, \tag{4}$$

or equivalently,

$$\int_G Mf(x)(\log(1+Mf(x)))^{a_2\alpha+q_0-1}(\log_{(2)}(1+Mf(x)))^{b_2}\,dx < \infty$$

for all $f \in L^{p(\cdot)}(\log L)^{q(\cdot)}(G)$.

5. Further remarks

In this section, we discuss the sharpness of Theorem 1 (Corollary 1).

If K is a compact subset of G, then we write

$$\delta(x) = \delta_K(x)$$

for simplicity.

Remark 2. We show that the exponents in (4) are sharp under the condition that

$$C^{-1}r^{\alpha} \le |K(r)| \le Cr^{\alpha} \qquad \text{for } 0 < r < r_0.$$

In fact, if we replace $q(x)$ by

$$\tilde{q}(x) = q_0 + \frac{b_1}{\log_{(2)}(1/\delta(x))} + \frac{b_2 \log_{(3)}(1/\delta(x))}{\log_{(2)}(1/\delta(x))} - \frac{b_3 \log_{(4)}(1/\delta(x))}{\log_{(2)}(1/\delta(x))}$$

with $b_3 > 0$, then we can find $f \in L^{p(\cdot)}(\log L)^{\tilde{q}(\cdot)}(G)$ such that

$$\int_G |f(x)|(\log(1+|f(x)|))^{a_2\alpha+q_0}(\log_{(2)}(1+|f(x)|))^{b_2}\,dx = \infty.$$

To show this, we consider the function

$$f(x) = \delta(x)^{-\alpha}(\log(1/\delta(x)))^{-a_2\alpha-q_0-1}(\log_{(2)}(1/\delta(x)))^{-b_2-1}(\log_{(3)}(1/\delta(x)))^{-1}$$

for $x \in G$ with $\delta(x) \le r_0$; set $f(x) = 0$ when $\delta(x) > r_0$. Then

$$\begin{aligned} &\int_G f(x)(\log(1+f(x)))^{a_2\alpha+q_0}(\log_{(2)}(1+f(x)))^{b_2}\,dx \\ \ge\ & C\int_0^{r_0} t^{-1}(\log(1/t))^{-1}(\log_{(2)}(1/t))^{-1}(\log_{(3)}(1/t))^{-1}dt = \infty. \end{aligned}$$

Further, since $f(x) \le \delta(x)^{-\alpha}$, we have for $t = \delta(x) \le r_0$

$$f(x)^{p(x)-1} \le \exp\left(\alpha\log(1/t)\frac{a_1 + a_2\log_{(2)}(1/t)}{\log(1/t)}\right) = C(\log(1/t))^{a_2\alpha}$$

and

$$\begin{aligned} (\log f(x))^{\tilde{q}(x)-q_0} &\le C\exp\left(\log_{(2)}(1/t)\frac{b_1 + b_2\log_{(3)}(1/t) - b_3\log_{(4)}(1/t)}{\log_{(2)}(1/t)}\right) \\ &= C(\log_{(2)}(1/t))^{b_2}(\log_{(3)}(1/t))^{-b_3}, \end{aligned}$$

so that

$$\begin{aligned} &\int_G f(x)^{p(x)}(\log(c_0+f(x)))^{\tilde{q}(x)}\,dx \\ &\le C\int_0^{r_0} t^{-1}(\log(1/t))^{-1}(\log_{(2)}(1/t))^{-1}(\log_{(3)}(1/t))^{-b_3-1}dt < \infty. \end{aligned}$$

Remark 3. Let

$$p(x) = 1 + \frac{a_1}{\log(1/|x|)} + \frac{a_2\log_{(2)}(1/|x|)}{\log(1/|x|)} + \frac{a_3\log_{(3)}(1/|x|)}{\log(1/|x|)}$$

and

$$q(x) = q_0 + \frac{b_1}{\log_{(2)}(1/|x|)} + \frac{b_2\log_{(3)}(1/|x|)}{\log_{(2)}(1/|x|)}$$

for $x \in B_0 = B(0, r_0)$ with $0 < r_0 < 1/e$, where $a_2, a_3, b_2 > 0$, $a_1, b_1, q_0 \in \mathbf{R}$, $a_2 n + q_0 > 0$,

$$\omega(r) = \{a_1 + a_2\log_{(2)}(1/r) + a_3\log_{(3)}(1/r)\}/\log(1/r)$$

and

$$\eta(r) = \{b_1 + b_2\log_{(3)}(1/r)\}/\log_{(2)}(1/r)$$

are nondecreasing on $(0, r_0]$. Then we can find $f \in L^{p(\cdot)}(\log L)^{q(\cdot)}(B_0)$ such that Mf fails to belong to $L^{p(\cdot)}(\log L)^{q(\cdot)}(B_0)$.

For this purpose, consider

$$f(y) = |y|^{-n}(\log(1/|y|))^{-\beta}$$

for $y \in B_0$ and β such that $a_2 n + q_0 + 1 < \beta < a_2 n + q_0 + 2$ or $\beta = a_2 n + q_0 + 1$ and $a_3 n + b_2 < -1$. Then we see that

$$|f(y)|^{p(y)-1} \le C(\log(1/|y|))^{a_2 n}(\log_{(2)}(1/|y|))^{a_3 n}$$

and

$$(\log|f(y)|)^{q(y)-q_0} \le C(\log_{(2)}(1/|y|))^{b_2}$$

for $y \in B_0$, so that

$$\begin{aligned} & \int_{B_0} |f(y)|^{p(y)}(\log(c_0 + |f(y)|))^{q(y)}\, dy \\ \le\ & C\int_0^{1/e} t^{-1}(\log(1/t))^{-\beta + a_2 n + q_0}(\log_{(2)}(1/t))^{a_3 n + b_2}\, dt < \infty \end{aligned}$$

by our assumption on β.

On the other hand,

$$Mf(x) \ge \frac{1}{|B(x, 2|x|)|}\int_{B(x,2|x|)} |y|^{-n}(\log(1/|y|))^{-\beta} dy \ge C|x|^{-n}(\log(1/|x|))^{-\beta+1}$$

since $\beta > 1$. Hence

$$\begin{aligned} & \int_{B_0} Mf(x)^{p(x)}(\log(c_0 + Mf(x)))^{q(x)}\, dx \\ \ge\ & C\int_0^{1/e} t^{-1}(\log(1/t))^{-\beta + a_2 n + q_0 + 1}(\log_{(2)}(1/t))^{a_3 n + b_2}\, dt = \infty, \end{aligned}$$

which implies that $Mf \notin L^{p(\cdot)}(\log L)^{q(\cdot)}(B_0)$.

Remark 4. Let

$$p(x) = p_0 + \frac{a_1}{\log(1/|x|)}$$

and

$$q(x) = q_0 + \frac{b_1}{\log_{(2)}(1/|x|)}$$

for $x \in B_0 = B(0, r_0)$ with $0 < r_0 < 1/e$, where $a_1, b_1, q_0 \in \mathbf{R}$, $p_0 > 1$ and $\inf_{x \in B_0} p(x) > 1$. Then, in view of Diening [3], we see that $f \in L^{p(\cdot)}(\log L)^{q(\cdot)}(B_0)$ if and only if $Mf \in L^{p(\cdot)}(\log L)^{q(\cdot)}(B_0)$.

References

[1] A. Cianchi, Strong and weak type inequalities for some classical operators in Orlicz spaces. *J. London Math. Soc.* **60** (1999), 187-202.

[2] D. Cruz-Uribe, SFO, A. Fiorenza and C. J. Neugebauer, $L \log L$ results for the maximal operator in variable L^p spaces. to appear in *Trans. Amer. Math. Soc.*

[3] L. Diening, Maximal functions in generalized $L^{p(\cdot)}$ spaces. *Math. Inequal. Appl.* **7** (2004), 245-254.

[4] T. Futamura and Y. Mizuta, Maximal functions for Lebesgue spaces with variable exponent approaching 1. *Hiroshima Math. J.* **36** (2006), no. 1, 23-28.

[5] P. Hästö, The maximal operator in Lebesgue spaces with variable exponent near 1. *Math. Nachr.* **280** (2007), no. 1-2, 74-82.
[6] O. Kováčik and J. Rákosník, On spaces $L^{p(x)}$ and $W^{k,p(x)}$. *Czechoslovak Math. J.* **41** (1991), 592-618.
[7] P. Mattila, *Geometry of Sets and Measures in Euclidean Spaces* (Cambridge Studies in Advanced Mathematics **44**). Cambridge University Press, Cambridge, 1995.
[8] Y. Mizuta, *Potential theory in Euclidean spaces.* Gakkōtosyo, Tokyo, 1996.
[9] Y. Mizuta, T. Ohno and T. Shimomura, Integrability of maximal functions for generalized Lebesgue spaces with variable exponent. *Math. Nachr.* **281** (2008), no. 3, 386-395.
[10] J. Musielak, *Orlicz Spaces and Modular Spaces* (Lecture Notes in Mathematics **1034**). Springer-Verlag, Berlin, 1983.
[11] W. Orlicz, Über konjugierte Exponentenfolgen. *Studia Math.* **3** (1931), 200-211.
[12] M. Růžička, *Electrorheological fluids : Modeling and Mathematical Theory* (Lecture Notes in Math. **1748**). Springer, 2000.
[13] E. M. Stein, Note on the class $L\log L$. *Studia Math.* **32** (1969), 305-310.
[14] E. M. Stein, *Singular Integrals and Differentiability Properties of Functions.* Princeton Univ. Press, Princeton, 1970.

THE DIVISION OF MATHEMATICAL AND INFORMATION SCIENCES, FACULTY OF INTEGRATED ARTS AND SCIENCES, HIROSHIMA UNIVERSITY, HIGASHI-HIROSHIMA, 739-8521, JAPAN
E-mail address: yomizuta@hiroshima-u.ac.jp

GENERAL ARTS, HIROSHIMA NATIONAL COLLEGE OF MARITIME TECHNOLOGY, HIGASHINO, OOSAKIKAMIJIMA, TOYOTAGUN, 725-0231, JAPAN
E-mail address: ohno@hiroshima-cmt.ac.jp

DEPARTMENT OF MATHEMATICS, GRADUATE SCHOOL OF EDUCATION, HIROSHIMA UNIVERSITY, HIGASHI-HIROSHIMA, 739-8524, JAPAN
E-mail address: tshimo@hiroshima-u.ac.jp

Potential Theory and Stochastics in Albac, 203–213
Aurel Cornea Memorial Volume

PROBABILISTIC APPROACHES TO MONOTONICITY AND MAXIMUM PRINCIPLES

MIHAI N. PASCU

ABSTRACT. In this paper we present a survey of some recent results of the author related to certain monotonicity and maximum principles: the Hot Spots conjecture of Jeffrey Rauch and the Laugesen-Morpurgo conjecture. The methods used in the proofs are probabilistic in spirit, involving coupling of reflecting Brownian motions in smooth domains.

We conclude with a discussion on the possibility of obtaining new results, or simplified proofs of known results, for similar monotonicity and maximum principles.

1. INTRODUCTION

In this paper we present a survey of some recent results of the author regarding probabilistic approaches to monotonicity and maximum principles of solutions of certain differential equations.

The method used in the proofs involves representing the solution function as a certain functional of the reflecting Brownian motion, and then, using coupling techniques, showing that the this functional is strictly increasing along certain family of lines covering the domain. In particular, from the strict monotonicity we conclude that the solution function satisfies a strong maximum/minimum principle.

In general, a coupling is a pair of stochastic processes constructed on the same probability space, between which there is a certain (deterministic) dependence. In the case of reflecting Brownian motion, there are mainly two couplings in the literature: the mirror and the synchronous couplings. These couplings extend the ordinary couplings of unreflected (free) Brownian motion to the case of reflecting Brownian motion (see for example [BuKe]). We will also present a third coupling of reflecting Brownian motions introduced by the author, namely the scaling coupling (see [Pa]).

The structure of the paper is as follows: in Section 2 we present the classical maximum principle for harmonic functions and the Hopf maximum principle, and, as an extension of the later, we present the Hot Spots conjecture of Jeffrey Rauch. Using the properties of the scaling coupling of reflecting Brownian motion, we present a proof of the conjecture for a certain classes of convex planar domains.

In Section 3, we present the mirror coupling of reflecting Brownian motions in the particular case of the unit disk. As an application, using the properties of the coupling in the particular case of the unit ball in $\mathbb{R}^n$ ($n \in \mathbb{N}$), we present a recent result of the author concerning the validity of Laugesen-Morpurgo conjecture (see [PaGa]).

2000 *Mathematics Subject Classification*. Primary 60J65, 60J35. Secondary 35B50.

Key words and phrases. Monotonicity and maximum principles, Hot Spots conjecture, Laugesen-Morpurgo conjecture, Chavel's conjecture.

In Section 4 we conclude with some extensions and open directions of research concerning the Hot Spots conjecture and Chavel's conjecture.

2. MAXIMUM PRINCIPLES AND THE HOT SPOTS CONJECTURE

Maximum principles enable us to obtain information about the solutions of certain differential equations, without any explicit knowledge of the solutions themselves, with many applications in Differential Equations, Analysis, Physics, etc.

As an example, the classical maximum/minimum principle for harmonic functions can be stated as follows:

Theorem 1. *Let D be an open connected subset of $\mathbb{R}^n$ and let $u : D \subset \mathbb{R}^n \to \mathbb{R}$ be harmonic in D. If u attains its maximum (or minimum) in D at an interior point of D, then u reduces to a constant in D.*

The above maximum principle has been extended to more general operators beside the Laplaceian $\Delta = \sum_{i=1}^{n} \frac{\partial^2}{\partial x_i^2}$, for example to the case of linear elliptic operators of the form

$$L = \sum_{i=1}^{n} a_{ij}(x) \frac{\partial^2}{\partial x_i \partial x_j} + \sum_{i=1}^{n} b_i(x) \frac{\partial}{\partial x_i}, \tag{1}$$

where the matrix $A = \left(a_{ij}(x)\right)_{i,j}$ of the coefficients is locally uniformly positive definite in D and the coefficients $b_i(x)$ are locally bounded.

The extension, known as the Hopf maximum principle, can be stated as follows:

Theorem 2. *Let u be a $C^2(D)$ function satisfying the differential inequality*

$$Lu + cu \geq 0 \ (\leq 0) \tag{2}$$

in a domain D, where the coefficients of L satisfy the previous conditions, and $c = c(x)$ is a non-positive function on D. If u takes a positive maximum (negative minimum) value M in D, then $u \equiv M$ in D.

One can ask if the above maximum principle can be extended for example to the case when c is a positive (constant, for example) function in D, for example the case of the differential equation

$$\Delta u + \lambda u = 0. \tag{3}$$

Without additional conditions, the answer to the above question is in general negative, as it can be seen by considering the function $u(x, y) = \cos x \cos y$, which satisfies the differential equation

$$\Delta u + 2u = 0 \tag{4}$$

in $D = \left\{(x, y) \in \mathbb{R}^2 : -\frac{\pi}{2} < x, y < \frac{\pi}{2}\right\}$ and attains its maximum inside D at the origin, without reducing to a constant.

One extension of the maximum principle in this direction was formulated by Jeffrey Rauch at a conference at Tulane University in 1974, who considered the case of solutions of (3) satisfying Neumann boundary conditions on ∂D. More precisely, he proposed the following conjecture:

Conjecture 3. *Extrema of second Neumann eigenfunctions are attained solely at the boundary of the domain, that is:*

$$\inf_{x\in\partial D} \varphi_2(x) < \varphi_2(y) < \sup_{x\in\partial D} \varphi_2(x), \qquad y \in D, \tag{5}$$

for every second Neumann eigenfunction φ_2 of D.

Remark 4. It is known that the set of real numbers (eigenvalues) for which there exists a nontrivial solution of the differential equation (3) with either Neumann boundary conditions

$$\frac{\partial u}{\partial n} = 0 \text{ on } \partial D \tag{6}$$

or mixed Dirichlet-Neumann boundary conditions

$$\begin{cases} \left.\frac{\partial u}{\partial n}\right|_{\partial D} = 0 \text{ on } B \\ u = 0 \text{ on } \partial D - B \end{cases} \tag{7}$$

forms a discrete set, and we will order the corresponding eigenvalues as follows

$$0 = \lambda_1 < \lambda_2 < \lambda_3 < \dots \nearrow +\infty,$$

respectively

$$0 < \mu_1 < \mu_2 < \dots \nearrow +\infty.$$

Remark 5. We will refer to λ_1, φ_1, $\lambda_2, \varphi_2, \dots$ as the first, second,... Neumann eigenvalue, respectively eigenfunction for the Laplaceian on D, and similarly in the mixed Dirichlet-Neumann case.

Due to the physical interpretation in terms of the solution of the heat equation, the above conjecture is known in the literature as the "Hot Spots" conjecture, and there are several results in the literature concerning it. For example, B. Kawohl states ([Ka], 1985) that the conjecture holds for balls, annuli and parallelepipeds in $\mathbb{R}^n$, K. Burdzy and R. Bañuelos ([BaBu], 1999) showed that the conjecture holds for convex domains with two axes of symmetry, or just one axis of symmetry and some additional hypothesis (which insure that the second Neumann eigenspace is 1-dimensional, being spanned by a symmetric eigenfunction) and obtuse triangles; D. Jerisson and N. S. Nadirashvili ([JeNa], 2000) showed independently that the conjecture holds for convex domains with two orthogonal axes of symmetry.

Below, we will give a proof of the conjecture due to the author ([Pa], 2002) for anti-symmetric second Neumann eigenfunctions of convex domains with one axis of symmetry, which leads to a resolution of the Hot Spots conjecture in the case of convex domains with two orthogonal axes of symmetry.

On the negative side, K. Burdzy and W. Werner ([BuWe], 1999) proved that the conjecture does not hold for a certain non-convex domain, and R. Bass and K. Burdzy ([BaBu1], 2000) provided a stronger counterexample for the conjecture (they showed that for a certain non-convex domain both the maximum and the minimum of a second Neumann eigenfunction lie inside the domain). For some other recent results on the conjecture see [Bu] (the case of some doubly-connected domains) or [Mi] (for nearly circular domains, without symmetries).

For smooth planar domains, the reflecting Brownian motion (with normal reflection on the boundary of the domain) can be defined as follows:

Definition 6. Let $D \subset \mathbb{R}^2$ be a $C^{1,\alpha}$ $(0 < \alpha < 1)$ simply connected domain and let ν_D the inward unit normal vector field on ∂D. We define the *reflecting Brownian motion* in D starting at $x_0 \in \overline{D}$ as a solution of the differential equation

$$X_t = X_0 + B_t + \frac{1}{2}\int_0^t \nu_D(X_s)\,dL_s, \tag{8}$$

where B_t is a 2-dimensional Brownian motion started at 0, L_t is a continuous nondecreasing process which increases only when $X_t \in \partial D$, X_t is adapted with respect to the filtration $(\mathscr{F}_t^B)$ corresponding to the Brownian motion B_t, and almost surely $X_0 = x_0$ and $X_t \in \overline{D}$ for all $t \geq 0$.

Remark 7. For existence and pathwise uniqueness of the solution in the sense of the above definition, see for example [BaHs].

In [Pa], the author introduced the scaling coupling of reflecting Brownian motions in a smooth planar domain D, as follows. First, in the case when D is the unit disk $U = \{z \in \mathbb{C} : |z| < 1\}$, he showed that given a reflecting Brownian motion Z_t in U starting at $z_0 \in \overline{U} - \{0\}$ and $|z_0| \leq a \leq 1$, the process $\widetilde{Z}_t$ defined by

$$\tilde{Z}_t = \frac{1}{M_{\gamma_t}} Z_{\gamma_t}, \qquad t \geq 0, \tag{9}$$

where

$$M_t = a \vee \sup_{s \leq t} |Z_s|, \tag{10}$$

$$C_t = \int_0^t \frac{1}{M_s^2}\,ds, \tag{11}$$

$$\gamma_t = \inf\{s > 0 : C_s \geq t\}, \tag{12}$$

is a $(\mathscr{F}_{\gamma_t}^Z)$-adapted reflecting Brownian motion in U starting at $\tilde{z}_0 = \frac{z_0}{a}$, and called the pair $\left(Z_t, \tilde{Z}_t\right)$ constructed above a scaling coupling of reflecting Brownian motions in U starting at $z_0 \in \overline{U} - \{0\}$, respectively at $\tilde{z}_0 = \frac{z_0}{a}$ (see Figure 2).

Using conformal mapping arguments and Lévy's theorem on the conformal invariance of the reflecting Brownian motion, this construction is then extended to an arbitrary $C^{1,\alpha}$ $(0 < \alpha < 1)$ simply connected domains (see [Pa] for the details).

As an application to this construction, we present the proof of the Hot Spots conjecture in the case of the unit disk, since it is really simple and it contains the basic ideas needed for the extension to the more general case of planar convex domains with 1-axis of symmetry.

Theorem 8. *If φ is a second Neumann eigenfunction for U which is antisymmetric with respect to the horizontal axis, then φ is strictly increasing on any diameter of U which intersects non-trivially the horizontal axis. In particular, the hot spots conjecture (5) holds for φ.*

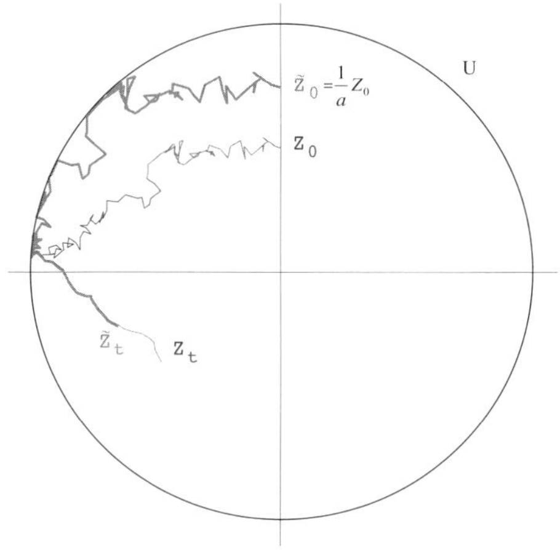

FIGURE 1. Scaling coupling of RBM in U

Proof. In order to prove the claim, since φ is antisymmetric with respect to the horizontal axis, the nodal line of φ is the line segment $[-1,1]$, and without loss of generality we may assume that $\varphi > 0$ in $U^+ = \{z \in U : \operatorname{Im} z > 0\}$. In order to prove the claim it suffices therefore to show that

$$\varphi\left(xe^{i\theta}\right) < \varphi\left(ye^{i\theta}\right) \tag{13}$$

for any $\theta \in (0,\pi)$ and $0 < x < y < 1$.

Also note that the restriction of φ to the half-disk U^+ is a mixed Dirichlet-Neumann eigenfunction (with Dirichlet conditions on $[-1,1]$, and Neumann conditions on the remaining part of the boundary of ∂U^+) with constant sign, and therefore it corresponds to the first Dirichlet-Neumann eigenvalue μ_1 of U^+. It follows that

$$P^x\left(\tau^x > t\right) = E^x\left(1; \tau^x > t\right) = \int_{U^+} p_{U^+}(t,x,y)\,dy,$$

where $\tau^x = \inf\{t > 0 : B_t \in [-1,1]\}$ is the lifetime of a reflecting Brownian motion in U^+ killed on $[-1,1]$, and $p_{U^+}\left(t,x,y\right)$ is the transition density of reflecting Brownian motion in U^+ killed on hitting $[-1,1]$. Using an eigenfunction expansion, it can be shown (see [Pa]) that for large t we have

$$P^x\left(\tau^x > t\right) \approx e^{-\lambda_2 t}\varphi(x)\int_{U^+} \varphi\left(y\right)dy,$$

hence, in order to prove the monotonicity in (13) it suffices to prove the corresponding monotonicity for the lifetime τ^x of a reflecting Brownian motion in U^+ starting at x, killed on $[-1,1]$.

For $x \in U^+$ and $|x| < a < 1$ arbitrarily fixed, consider $\left(Z_t, \tilde{Z}_t\right)$ a scaling coupling of reflecting Brownian motions in U starting at x and $\frac{1}{a}x$, respectively. By the construction (9) - (12) of the scaling coupling it follows that if $\tilde{Z}_t \in [-1,1]$ then $Z_{\gamma_t} \in [-1,1]$. Since $C_t \geq t$ we have $\gamma_t \leq t$, and therefore if $\tilde{Z}_t \in [-1,1]$, there exists $t' = \gamma_t \leq t$ such that $Z_{t'} \in [-1,1]$, which shows that $\tau^x \leq \tau^{\frac{1}{a}x}$ for any $|x| < a < 1$.

Combining with the preliminary remarks, this shows that φ is a radially increasing function in U^+. Being a solution of a differential equation with real analytic coefficients, φ cannot be constant on an open set without reducing to a constant, and therefore φ is in fact a strictly increasing function (see [Pa]), concluding the proof. □

The previous result extends by conformal mapping arguments to convex domains with a line of symmetry (see [Pa]), as follows:

Theorem 9. *Let D be a convex $C^{1,\alpha}$ domain ($0<\alpha<1$) which is symmetric with respect to the horizontal axis. If φ is a second Neumann eigenfunction for D which is antisymmetric with respect to the horizontal axis, then the hot spots conjecture holds for φ.*

From the above theorem, in [Pa], the author obtained the validity of the Hot Spots conjecture for convex domains with two orthogonal axes of symmetry

Theorem 10. *If D is a convex $C^{1,\alpha}$ domain ($0<\alpha<1$), symmetric with respect to both coordinate axes, then hot spots conjecture holds for D.*

More generally, using the same ideas, it can be shown the following (see [Pa]):

Theorem 11. *Let D be a convex $C^{1,\alpha}$ ($0<\alpha<1$) domain and let φ be a second Neumann eigenfunction for D. If the nodal line of φ is a hyperbolic line in D, then φ is strictly monotone on the family of hyperbolic lines in D which intersect nontrivially the nodal line of φ. In particular, the Hot Spots conjecture holds for φ.*

For some extensions of the Hot Spots conjecture to the case of the first mixed Dirichlet-Neumann eigenfunctions, see also [BaPaPa].

3. The Laugesen-Morpurgo conjecture

In connection with their work on conformal extremals on Riemann zeta function of Neumann eigenvalues, in [LaMo], R. Laugesen and C. Morpurgo advanced the following conjecture:

Conjecture 12. *Let $p_U(t,x,y)$ denote the heat kernel for the Laplaceian with Neumann boundary conditions on the unit disk $U=\{z\in\mathbb{C}:|z|<1\}$ in $\mathbb{C}$. For any $t>0$ arbitrarily fixed, the radial function $p_U(t,x,x)$, called the diagonal element of the heat kernel, is a strictly increasing function of $|x|$, that is,*

$$p_U(t,x,x)<p_U(t,y,y), \tag{14}$$

for any $t>0$ and $x,y\in U$ with $|x|<|y|$.

Until recently, despite of the simple nature of the conjecture and the fact that it is well known since 1994, the only known results on the validity of the conjecture were some partial results (see [PaPa1] and [PaPa2]). Recently, in [BaKuSi], Bañuelos et al. showed that the Laugesen-Morpurgo conjecture (14) holds for the transition density of the n-dimensional Bessel processes on $(0,1]$ reflected at 1 in the case $n>2$, and that this is false for $n=2$. Since the absolute value of a n-dimensional Brownian motion is a Bessel process of order n, this is equivalent to the monotonicity with respect to $x \in (0,1)$ of the integral mean

$$\int_0^{2\pi} p\left(t, x, xe^{i\theta}\right) d\theta.$$

More recently, in [PaNi] the authors showed that the 1-dimensional discrete version of the above conjecture holds, thus obtaining a new proof of the Laugesen-Morpurgo conjecture in the 1-dimensional case, and in [PaGa], the authors used coupling arguments to settle the conjecture.

The proof relies on properties of mirror coupling of reflecting Brownian motions, introduced by Burdzy et al. (see [BuKe] and [AtBu]).

In the case of the unit disk, the mirror coupling of reflecting Brownian motions in U with starting points $x, y \in U$, is defined as a pair $(X_t, Y_t)_{t\geq 0}$ of stochastic processes given by

$$\begin{cases} X_t = x + W_t + \int_0^t \nu(X_s)\, dL_s^X \\ Y_t = y + Z_t + \int_0^t \nu(Y_s)\, dL_s^Y \end{cases} \tag{15}$$

where W_t is a 2-dimensional Brownian motion starting at $W_0 = 0$, Z_t is the mirror image of the Brownian motion W_t with respect to the line of symmetry M_t between X_t and Y_t, that is

$$Z_t = W_t - 2\int_0^t \frac{X_s - Y_s}{\|X_s - Y_s\|^2} (X_s - Y_s) \cdot dW_s, \tag{16}$$

where L_t^X and L_t^Y denote the boundary local times of the reflecting Brownian motions X_t and respectively Y_t, and $\nu(z) = -z$, $z \in \partial U$, denotes the inward unit normal vector field on the boundary of U. The processes X_t and Y_t evolve according to (15) above for $t \leq \tau$, where τ is the (possibly infinite) coupling time

$$\tau = \inf\{t > 0 : X_t = Y_t\} \in \mathbb{R} \cup \{\infty\},$$

and the two processes evolve together after they have coupled.

Remark 13. Replacing B_t by a n-dimensional Brownian motion and ν by the inward unit normal $\nu_{\mathbb{U}}(x) = -x$, $x \in \mathbb{U}$, at the boundary of the unit ball $\mathbb{U}$ in $\mathbb{R}^n$, the same formulae give the mirror coupling of reflecting Brownian motions in the unit ball $\mathbb{U}$ in $\mathbb{R}^n$ in any dimension $n \geq 1$.

In [PaGa], the authors showed that the mirror coupling in the unit disk leaves invariant the left/right order of the starting points of the coupling, in the following sense

Lemma 14. *If X_t, Y_t is a mirror coupling of reflecting Brownian motions in U with starting points $X_0 = x + r$, $Y_0 = x + re^{i\theta}$, $x \in (0,1)$, $r \in (0, \min\{x, 1-x\})$ and $\theta \in [0, 2\pi)$, then for all times $t < \tau \wedge \tau_1$ where $\tau_1 = \inf\{t > 0 : 0 \in M_t\}$, the mirror M_t moves away from the point x, in such a way that a_t^1 and b_t^1 are non-increasing functions of t, where $A_t = \left(a_t^1, a_t^2\right)$ and $B_t = \left(b_t^1, b_t^2\right)$ are the points of intersection of M_t with the boundary of U (see Figure 2).*

From the above property it follows that:

$$E^{x+re^{i\theta}} 1_{B(x,\varepsilon)}(Y_t) \leq E^{x+r} 1_{B(x,\varepsilon)}(X_t), \qquad t \geq 0, \tag{17}$$

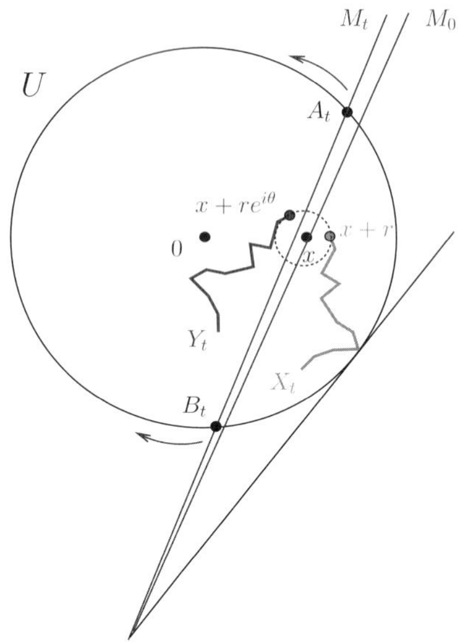

FIGURE 2. The mirror coupling of reflecting Brownian motions in U.

for any $x \in (0,1)$, $r, \varepsilon \in (0, \min\{x, 1-x\})$ and $\theta \in [0, 2\pi)$, where $B(x,\varepsilon)$ denotes the open disk with radius $\varepsilon > 0$ centered at x.

Dividing by $\pi\varepsilon^2$ and passing to the limit with $\varepsilon \to 0$, from the above inequality we obtain

$$p_U\left(t, x + re^{i\theta}, x\right) \le p_U(t, x+r, x), \qquad t \ge 0, \tag{18}$$

and therefore for all $t > 0$ we obtain the double inequality

$$\frac{1}{2\pi}\int_0^{2\pi} p_U\left(t, x + re^{i\theta}, x\right) d\theta \le p_U(t, x+r, x) \le p_U(t, x+r, x+r). \tag{19}$$

Using this result, in [PaGa] the authors proved the Laugesen-Morpurgo conjecture in the 2-dimensional case, as follows:

Theorem 15 (Laugesen-Morpurgo conjecture). *The diagonal element of the Neumann heat kernel of the unit disk is a radially increasing function, that is*

$$p_U(t, x, x) < p_U\left(t, y, y\right),$$

for any $t > 0$ and any $x, y \in (0,1)$ with $x < y$.

Proof. Using (19), we obtain

$$\begin{aligned}\frac{d}{dx}p_U(t,x,x) &= \lim_{r\searrow 0}\frac{p_U(t,x+r,x+r)-p_U(t,x,x)}{r}\\ &\geq \frac{1}{2\pi}\lim_{r\searrow 0}\int_0^{2\pi}\frac{p_U\left(t,x+re^{i\theta},x\right)-p_U(t,x,x)}{r}d\theta\\ &= \frac{1}{2\pi}\int_0^{2\pi}\nabla p_U(t,x,x)\cdot e^{i\theta}d\theta\\ &= 0.\end{aligned}$$

The strict monotonicity follows by using the fact that $p_U(t,x,x)$ is a real analytic function (see [PaGa] for the details). □

The above arguments extend naturally to the n-dimensional case (see [PaGa]), to give a resolution of the Laugesen-Morpurgo conjecture as follows:

Theorem 16. *The diagonal element $p_{\mathbb{U}}(t,x,x)$ of the Neumann heat kernel of the unit ball $\mathbb{U}$ in $\mathbb{R}^n$ $(n\in\mathbb{N}^*)$ is a radially increasing function, that is*

$$p_{\mathbb{U}}(t,x,x)<p_{\mathbb{U}}\left(t,y,y\right),$$

for any $t>0$ and any $x,y\in\mathbb{U}$ with $\|x\|<\left\|y\right\|$.

4. Extensions and concluding remarks

The ideas presented in the previous sections can be used for a possible new approach to the Hot Spots conjecture (the conjecture is still open at the present moment in its full generality). More precisely, by using an eigenfunction expansion, it can be seen that in the case of the unit disk the Hot Spots conjecture follows from the Laugesen-Morpurgo conjecture (see [PaGa] for the details). So by extending the Laugesen-Morpurgo conjecture to new classes of domains, we obtain by a similar reasoning the validity of the Hot Spots conjecture for these new classes of domains.

Another direction of research in which the coupling arguments presented above can be applied, is to Isaac Chavel conjecture on the domain monotonicity of the transition density of reflecting Brownian motion. More precisely, the conjecture states that for smooth bounded convex domains $D\subset\Omega\subset\mathbb{R}^n$ we have

$$p_D\left(t,x,y\right)\geq p_\Omega\left(t,x,y\right),\qquad \left(t,x,y\right)\in(0,\infty)\times D\times D. \tag{20}$$

It is known for example that the above conjecture holds if Ω is a ball centered at either x or y ([Ch]) or if D is ball centered at x or y ([Ke]). As shown in [BaBu2], without additional hypothesis Chavel's conjecture does not hold. In [CaZh] for example, the authors showed that the conjecture holds for all interior points $x,y\in D$ and all times t smaller than a constant depending on the distance of the points to the boundary ∂D.

Using the fact that the mirror coupling does not interchange the starting position of the points (see Lemma 14), the author showed that if either D or Ω is a ball with center x or y, then the inequality (20) holds, therefore obtaining a unitary proof of the results obtained by I. Chavel and W. S. Kendall. A more detaliate analysis of the mirror coupling could provide new classes of domains for which Chavel's conjecture holds, this research being currently in progress.

Acknowledgments

Work supported by CNCSIS grant PNII - IDEI 209.

References

[AtBu] R. Atar and K. Burdzy, Mirror couplings and Neumann eigenfunctions, preprint (available on http://www.arxiv.org)

[Ba] C. Bandle, *Isoperimetric Inequalities and Applications* (Monographs and Studies in Mathematics). Pitman Advanced Publishing Program, Boston · London · Melbourne, 1980.

[BaBu] R. Bañuelos and K. Burdzy, On the "hot spots" conjecture of J. Rauch. *J. of Funct. Ann.* **164** (1999), 1-33 .

[BaBu1] R. Bass and K. Burdzy, Fiber Brownian motion and the "hot spots" problem. *Duke Math J.* **105** (2000), No. 1, 25-58.

[BaBu2] R. Bass and K. Burdzy, On domain monotonicity of the Neumann heat kernel. *J. Funct. Anal.* **116** (1993), No. 1, 215-224.

[BaHs] R. F. Bass and E. P. Hsu, Pathwise uniqueness for reflecting Brownian motion in Euclidian domains. *Probab. Theory and Rel. Fields* **117** (2000), 183-200.

[BaKuSi] R. Bañuelos, T. Kulczycki and B. Siudeja, Neumann heat kernel monotonicity, preprint (available on http://www.arxiv.org)

[BaPaPa] R. Bañuelos, M. Pang and M. N. Pascu, Brownian motion with killing and reflection and the "hot-spots" problem. *Prob. Th. Rel. Fields* **130** (2004), No. 1, 56-68.

[Bu] K. Burdzy, The hot spots problem in planar domains with one hole. *Duke Math. J.* **129** (2005), No. 3, 481-502.

[BuKe] K. Burdzy and W. Kendall, Efficient Markovian couplings: Examples and counterexamples. *Ann. Appl. Probab.* **10** (2000), No. 2, 362-409.

[BuWe] K. Burdzy and W. Werner, A counterexample to the "hotspots" conjecture. *Ann. Math.* **149** (1999), No. 2, 309–317.

[CaZh] R. Carmona and W. Zheng, Reflecting Brownian motion and comparison theorems for Neumann heat kernels. *J. Funct. Anal.* **123** (1994), No. 1, 109-128.

[Ch] I. Chavel, Heat diffusion in insulated convex domains, *J. London Math. Soc.* **34** (1986), 473-478.

[Fr] P. Freitas, Closed nodal lines and interior hot spots of the second eigenfunction of the Laplacian on surfaces. *Indiana University Mathematics Journal* **51** (2002), 305-316.

[Hs] E. P. Hsu, A domain monotonicity property of the Neumann heat kernell. *Osaka J. Math.* **31** (1994), 215-223.

[JeNa] D. Jerison and N. Nadirashvili, The "hot spots" conjecture for domains with two axes of symmetry. *J. Amer. Math. Soc.* **13** (2000), 741-772.

[Ka] B. Kawohl, *Rearrangements and Convexity of Level Sets in PDE* (Lecture Notes in Math. **1150**). Springer, Berlin, 1985.

[Ke] W. S. Kendall, Coupled Brownian motions and partial domain monotonicity for the Neumann heat kernel. *J. Funct. Anal.* **86** (1989), 226-236.

[LaMo] R. Laugesen and C. Morpurgo, Extremals for eigenvalues of Laplacians under conformal mapping. *J. Funct. Anal.* **155** (1998), 64-108.

[Mi] Y. Miyamoto, The "Hot Spots" conjecture for nearly circular planar convex domains, RIMS (2007), preprint no. **1591** (online at http://www.kurims.kyoto-u.ac.jp)

[Pa] M. N. Pascu, Scaling coupling and the hot spots conjecture. *Trans. Amer. Math. Soc.* **354** (2002), No. 11, 4681-4702.

[PaGa] M. N. Pascu and M. E. Gageonea, On a conjecture of Laugesen and Morpurgo, preprint (available at http://arxiv.org/abs/0807.4726)

[PaNi] M. N. Pascu and A. Nicolaie, On a discrete version of the Laugesen-Morpurgo conjecture. *Statist. Probab. Lett.* **79** (2009), No. 6, 797-806.

[PaPa1] M. N. Pascu and N. R. Pascu, Monotonicity properties of reflecting Brownian motion. 21st Scientific Session on Mathematics and its Applications, Braşov, 2007. *Transilvania University Press*, 109-112.

[PaPa2] M. N. Pascu and N. R. Pascu, Brownian motion on the circle and applications. 22nd Scientific Session on Mathematics and its Applications, Braşov, May 2008. *Bull. Transilvania University Press* **1** (2008), No. 50, 469-478.

"TRANSILVANIA" UNIVERSITY OF BRAŞOV, FACULTY OF MATHEMATICS AND COMPUTER SCIENCE, STR. IULIU MANIU NR. 50, BRAŞOV – 500091, ROMANIA

E-mail address: mihai.pascu@unitbv.ro

Potential Theory and Stochastics in Albac, 215–224
Aurel Cornea Memorial Volume

POLYINVARIANT ELEMENTS IN A SEMI-DYNAMICAL SYSTEM

EUGEN POPA

ABSTRACT. Working in the abstract framework of Cornea–Licea [1], we are concerned with necessary and sufficient conditions in order that the following expression of the resolvent holds:

$$(*)\ \ V_\alpha x = \sum_{n=0}^{\infty} \frac{A^n x}{\alpha^{n+1}}$$

where $Ax := \lim_{\alpha \to +\infty} \alpha(\alpha V_\alpha x - x)$ is the generator of the resolvent.

When the resolvent is associated with a semidynamical system [7], we establish the connection between the generators of the resolvent and of the system.

Finally, we prove that the resolvent can be recovered, on a class of elements, form the generator, through the formula $(*)$.

This paper is the natural counterpart of [11].

1. PRELIMINARIES

Let X be an ordered convex cone of positive elements, in which any increasing and dominated sequence has a supremum (written as $x_n \nearrow x$).

$x \in X$ is called subtractible if $x + y \le x + y' \Rightarrow y \le y'$, or equivalently, if $\bigwedge_{n\ge 1} \frac{1}{n} x = 0$. In the usual way (see [1]) we can consider "differences" $y - x$, where x is subtractible. We denote by X_s the set of all such differences.

Let Y be another ordered convex cone of positive elements; we say that X and Y are in duality if there exists a map $[\cdot,\cdot] : X \times Y \longrightarrow [0, +\infty]$ for which the compatibility requirements hold:

1. $[x + x', y] = [x, y] + [x', y]$; $[x, y + y'] = [x, y] + [x, y']$.
2. $x \le x' \Longrightarrow [x, y] \le [x', y]$; $y \le y' \Longrightarrow [x, y] \le [x, y']$.
3. $x_n \nearrow x \Longrightarrow [x_n, y] \nearrow [x, y]$; $y_n \nearrow y \Longrightarrow [x, y_n] \nearrow [x, y]$.

We can consider the full dual, denoted by X^* and defined as the cone of all maps $\mu : X \to [0, +\infty]$, which has the properties:

$$\mu(x + y) = \mu(x) + \mu(y),$$

$$x \le y \Rightarrow \mu(x) \le \mu(y),$$

$$x_n \nearrow x \Rightarrow \mu(x) = \sup_n \mu(x_n).$$

However, in several situations, we shall be obliged to consider smaller cones.

The duality is called *separated* (resp. *strongly separated*) if $[x, y] = [x', y]\ \forall y \in Y \Rightarrow x = x'$ (resp. $[x, y] \le [x', y]\ \forall y \in Y \Rightarrow x \le x'$).

2000 *Mathematics Subject Classification.* 60J35, 31D05, 47D03.
Key words and phrases. Semi-group, resolvent, infinitesimal generator.

With each part $Y_0 \subseteq Y$ we may associate a $(\sigma(X, Y_0))$–weak topology, as the coarsest topology on X, such that the map $x \longmapsto [x, y]$ is continuous, for each $y \in Y_0$.

A kernel is a map $V : X \longrightarrow X$ with the properties:

- $V(x + y) = Vx + Vy$,
- $x \leq y \Longrightarrow Vx \leq Vy$,
- $x_n \nearrow x \Longrightarrow Vx_n \nearrow Vx$.

$V^* : Y \to Y$ is called adjoint kernel if $[Vx, y] = [x, V^* y]$, $\forall x \in X$, $y \in Y$. Obviously, if V has an adjoint, then it is weakly continuous.

Definition 1.1. A family $\mathcal{V} = (V_\alpha)_{\alpha>0}$ of kernels on $(X, \mathcal{X})$ is called a *resolvent* (of kernels) if the following relations hold:

$$V_\alpha \circ V_\beta = V_\beta \circ V_\alpha, \ \forall \alpha, \beta > 0,$$

$$V_\alpha = V_\beta + (\beta - \alpha).V_\alpha \circ V_\beta, \ \forall 0 < \alpha < \beta.$$

A. Cornea, G. Licea [1] defined and studied the potential theory associated with resolvents $(V_\alpha)_{\alpha>0}$ of kernels on an abstract ordered convex cone X.

Definition 1.2. For a subtractible $x \in X$, we say that $x \in D(A)$ if there exists an element $Ax \in X_s$ such that:

$$Ax := \lim_{\alpha \to +\infty} \alpha(\alpha V_\alpha x - x),$$

the limit being taken in the weak sense.

Definition 1.3. A *semidynamical system* is defined as a measurable map

$$\Phi : [0, +\infty) \times X \longrightarrow X$$

with the semigroup property

$$\Phi(t, \Phi(s, x)) = \Phi(s + t, x).$$

In the study of the generator, we accept that $\Phi(0, x) = x$, $\forall x \in X$.

A resolvent can be naturally associated with such a system through

$$V_\alpha x := \int_0^\infty e^{-\alpha t} \Phi(t, x) dt$$

(weak integral).

See [10] for the construction of the system associated with a resolvent.

Definition 1.4. For a subtractible $x \in X$, we say that $x \in D(A)$ if there exists an element $Ax \in X_s$ such that:

$$Ax = \lim_{\substack{t \to 0 \\ t > 0}} \frac{\Phi(t, x) - x}{t}$$

in the weak sense. In this way, a map $A : D(A) \to X_s$ is defined; obviously, A is additive, positively homogeneous;thus, it can be naturally extended to differences $D(A) - D(A)$.

The infinitesimal generator of a semigroup is well studied, both for the case of semigroups of linear operators on Banach spaces and of kernels on measurable spaces. In order to have the C_0 property for the dual semigroup, a special dual (the sun-dual) has to be considered (see [5]).

In the case of semigroups (more precisely, for resolvents) of kernels on measurable spaces, a thorough study was realized by Mokobodzki (see [4])

Let us cite also the papers [2], [3] where the potential theory is build from an axiomatic notion of generator (not related á priori with a semigroup or a resolvent).

Remark 1.5. The least one can ask would be: there exists the weak limit

$$\lim_{t\to 0}\Phi(t,x)=a_0(x).$$

If we suppose that each $\Phi(t,\cdot)$ is weakly continuous, then this requirement will imply the *right* continuity of the map $t\mapsto\Phi(t,x)$ on $(0,\infty)$, if we accept also $\Phi(t,x)=\Phi(t,a_0(x))$, $\forall t>0$.

It may happen that the map $t\mapsto[\Phi(t,x),y]$ is only right continuous: besides the translation semigroup on $\mathbb{R}$, where f may be not continuous (to the right), we may consider the translation semigroup on $(0,1)$:

$$P_t f(x) := \begin{cases} f(x+t) & ,x+t<1, \\ 0 & ,x+t\geq 1. \end{cases}$$

In this case, each P_t (for $t\in(0,1)$), produces discontinuous functions, independent of f.

At first, it seem natural to consider linear spaces and not cones of positive elements, since in general the derivative does not preserve positivity.

However, there exists natural cones of positive elements which are preserved under the infinitesimal generator. The advantage is that we do not need a convergence (as given for example by a duality), but we can limit ourselves to increasing families.

Definition 1.6. 1. $x\in X$ is called a *totally monotone element* if $x\in\bigcap_{n=1}^{\infty}D(A^n)$ and $A^n x\geq 0$, $\forall n$.

2. $x\in X$ is called a *polyinvariant element* if there exists $n\in\mathbb{N}$ such that $x\in\bigcap_{k=1}^{n}D(A^k)$ and $A^n x=0$.

2. Main results

Lemma 2.1. *Let A be the generator of a resolvent in duality and $x\in D(A)$. Then $V_\beta x\in D(A)$ and*

$$AV_\beta x=V_\beta(Ax)=\beta V_\beta x-x,\ \forall\beta>0.$$

In particular

$$\beta V_\beta(Ax)\longrightarrow Ax$$

as $\beta\to+\infty$.

Proof. Using the existence of the adjoint, it follows that each V_β is weakly continuous, hence we may write:

$$AV_\beta x = \lim_{\alpha\to+\infty} \alpha(\alpha V_\alpha(V_\beta x) - V_\beta x) = \lim_{\alpha\to+\infty} V_\beta[\alpha(\alpha V_\alpha x - x)] = V_\beta(Ax).$$

Using the resolvent equation, we may write:

$$\alpha\left[\alpha\frac{V_\alpha x - V_\beta x}{\beta-\alpha} - V_\beta x\right] = \frac{\alpha}{\beta-\alpha}\left[\alpha V_\alpha x - \beta V_\beta x\right] \longrightarrow \beta V_\beta x - x$$

for $\alpha \to +\infty$. □

Proposition 1. *If* $x \in \bigcap_{k=1}^{n} D(A^k)$, *then:*

$$\alpha^n V_\alpha^n x = x + C_n^1 V_\alpha A x + \cdots + C_n^{n-1} V_\alpha^{n-1} A^{n-1} x + V_\alpha^n A^n x.$$

Indeed, we apply successively αV_α in both members.
By induction, we get:

Proposition 2. *If* $x \in \bigcap_{k=1}^{n+1} D(A^k)$, *then:*

$$\alpha V_\alpha x = x + \frac{Ax}{\alpha} + \frac{A^2 x}{\alpha^2} + \cdots + \frac{A^n x}{\alpha^n} + \frac{1}{\alpha^n} V_\alpha A^{n+1} x.$$

It suffices to replace x with Ax and use the formula.
As a corollary:

Corollary 1. *Let x be a polyinvariant element of order $(n+1)$. Then:*

$$\alpha V_\alpha x = x + \frac{Ax}{\alpha} + \frac{A^2 x}{\alpha^2} + \cdots + \frac{A^n x}{\alpha^n}.$$

For totally monotone elements we have the following result:

Proposition 3. *Let x be a totally monotone element. Then the sequence*

$$x + \frac{Ax}{\alpha} + \frac{A^2 x}{\alpha^2} + \cdots + \frac{A^n x}{\alpha^n}$$

is increasing and dominated (hence convergent).

Theorem 2.2. *Let* $x \in \bigcap_{n=1}^{\infty} D(A^n)$ *and suppose that there exists $R \geq 0$ such that, for each $\alpha > R$ we have* $\dfrac{A^n x}{\alpha^n} \longrightarrow 0$ *weakly. Then*

$$\alpha V_\alpha x = \sum_{n=0}^{\infty} \frac{A^n x}{\alpha^n}.$$

3. Connections with the semidynamical system

For simplicity, we shall write x instead of $[x,y]$ in the next result.

Proposition 4. *Let Φ be a semidynamical system with generator A and let (V_α) be the resolvent associated with the system. If*

$$(*)\ \forall x \in X, y \in Y_0\ \exists M_{xy} \text{ such that } [\Phi(t,x), y] \le M_{xy},\ \forall t > 0,$$

then

$$\liminf_{t\to 0} \frac{\Phi(t,x)-x}{t} \le \liminf_{\alpha\to+\infty} \alpha\,(\alpha V_\alpha x - x) \le$$
$$\le \limsup_{\alpha\to+\infty} \alpha\,(\alpha V_\alpha x - x) \le \limsup_{t\to 0} \frac{\Phi(t,x)-x}{t}.$$

Proof. Let us denote

$$A_{xy} := \liminf_{t\to 0} \frac{[\Phi(t,x),y]-[x,y]}{t}.$$

The case $A_{xy} = -\infty$ is obvious, while $A_{xy} = +\infty$ is proved in a similar way as below.

$$\alpha\,(\alpha V_\alpha x - x) = \alpha\left[\alpha \int_0^\infty e^{-\alpha t}\Phi(t,x)dt - \alpha\int_0^\infty e^{-\alpha t} x dt\right]$$
$$= \alpha^2 \int_0^\infty e^{-\alpha t}\,[\Phi(t,x)-x]\,dt = \alpha^2\left[\int_0^\delta e^{-\alpha t}\Phi(t,x) - e^{-\alpha t}xdt\right.$$
$$\left. + \int_\delta^\infty e^{-\alpha t}\Phi(t,x) - e^{-\alpha t}xdt\right] = \alpha^2\int_0^\delta te^{-\alpha t}\frac{\Phi(t,x)-x}{t}dt +$$
$$+\alpha^2\int_\delta^\infty e^{-\alpha t}\,(\Phi(t,x)-x)\,dt \ge n[1-(1+\alpha\delta)e^{-\alpha\delta}] - \varepsilon.$$

Hence $\forall \varepsilon > 0\ \exists \delta > 0$ such that

$$\frac{[\Phi(t,x),y]-[x,y]}{t} \ge A_{xy} - \varepsilon, \forall 0 < t < \delta.$$

Now, we can write:

$$\alpha\,(\alpha V_\alpha x - x) - A_{xy} = \alpha\left[\alpha\int_0^\infty e^{-\alpha t}\Phi(t,x)dt - \alpha\int_0^\infty e^{-\alpha t}xdt\right] - \alpha\int_0^\infty e^{-\alpha t}A_{xy}dt$$
$$= \alpha\left[\int_0^\infty e^{-\alpha t}\left\{\alpha\Phi(t,x) - \alpha x - A_{xy}\right\}dt\right] = \alpha\left[\int_0^\delta e^{-\alpha t}\left\{\alpha\Phi(t,x)-\alpha x - A_{xy}\right\}dt\right]$$
$$+\alpha\left[\int_\delta^\infty e^{-\alpha t}\left\{\alpha\Phi(t,x) \quad \alpha x \quad A_{xy}\right\}dt\right].$$

We consider separately:

$$\alpha\left[\int_0^\delta e^{-\alpha t}\left\{\alpha\Phi(t,x)-\alpha x - A_{xy}\right\}dt\right] = \alpha\int_0^\delta e^{-\alpha t}\left\{[\alpha t\frac{\Phi(t,x)-x}{t} - \alpha t A_{xy}]\right.$$
$$\left. + (\alpha t - 1)A_{xy}\right\}dt = \alpha^2\int_0^\delta te^{-\alpha t}\left[\frac{\Phi(t,x)-x}{t} - A_{xy}\right]dt$$
$$+A_{xy}.\alpha\int_0^\delta(\alpha t-1)e^{-\alpha t}dt \ge -[1-(1+\alpha\delta)e^{-\alpha\delta}]\varepsilon - A_{xy}\alpha\delta e^{-\alpha\delta}.$$

For the second term:

$$\left|\int_\delta^\infty e^{-\alpha t}\{\alpha\Phi(t,x)-\alpha x-A_{xy}]\}\,dt\right|\le\alpha(2\alpha+1)M_{xy}\int_\delta^\infty e^{-\alpha t}dt=$$

$$=(2\alpha+1)M_{xy}e^{-\alpha\delta}\longrightarrow 0$$

for $\alpha\to+\infty$.

Hence, $\exists\alpha_\varepsilon$ such that $\forall\alpha\ge\alpha_\varepsilon$ we have:

$$\alpha\left(\alpha[V_\alpha x,y]-[x,y]\right)-A_{xy}\ge-\varepsilon.$$

□

Proposition 5. *Let Φ be a semidynamical system with generator A and let (V_α) be the resolvent associated with the system.*

Then, for $x\in D(A)$:

$$\lim_{\alpha\to+\infty}\alpha\,(\alpha V_\alpha x-x)=Ax$$

(in the weak sense).

Proof. Indeed:

$$\alpha\,(\alpha V_\alpha x-x)-Ax=\alpha\left[\alpha\int_0^\infty e^{-\alpha t}\Phi(t,x)dt-\alpha\int_0^\infty e^{-\alpha t}x\,dt\right]-\alpha\int_0^\infty e^{-\alpha t}Ax\,dt$$

$$=\alpha\left[\int_0^\infty e^{-\alpha t}\{\alpha\Phi(t,x)-\alpha x-Ax\}\,dt\right]=\alpha\left[\int_0^\delta e^{-\alpha t}\{\alpha\Phi(t,x)-\alpha x-Ax\}\,dt\right]$$

$$+\alpha\left[\int_\delta^\infty e^{-\alpha t}\{\alpha\Phi(t,x)-\alpha x-Ax\}\,dt\right].$$

We consider separately:

$$\alpha\int_0^\delta e^{-\alpha t}\left\{[\alpha t\frac{\Phi(t,x)-x}{t}-\alpha tAx]+(\alpha t-1)Ax\right\}dt$$

$$=\alpha^2\int_0^\delta te^{-\alpha t}\left[\frac{\Phi(t,x)-x}{t}-Ax\right]dt+Ax.\alpha\int_0^\delta(\alpha t-1)e^{-\alpha t}dt.$$

Let us denote

$$R_{xy}(t):=\left|\frac{[\Phi(t,x),y]-[x,y]}{t}-[Ax,y]\right|.$$

Hence, x,y being fixed, $\forall\varepsilon>0\;\exists\delta>0$ such that $R_{xy}(t)<\varepsilon$, $\forall t\in(0,\delta)$. Now:

$$\alpha^2\int_0^\delta te^{-\alpha t}R_{xy}(t)dt<\varepsilon$$

while

$$[Ax,y].\alpha\int_0^\delta(\alpha t-1)e^{-\alpha t}dt=[Ax,y].\alpha\left((\alpha-\delta)-\frac{1}{\alpha}\right)e^{-\alpha\delta}\longrightarrow 0$$

for $\alpha\to+\infty$.

The second term:

$$\left|\int_\delta^\infty e^{-\alpha t}\left\{\alpha[\Phi(t,x),y]-\alpha[x,y]-[Ax,y]\right\}dt\right|\le\alpha(2\alpha+1)M\int_\delta^\infty e^{-\alpha t}dt$$

$$=(2\alpha+1)Me^{-\alpha\delta}\longrightarrow 0\text{ for }\alpha\to+\infty.$$

□

Remark 3.1. Prop. 4. is weaker than prop. 5. The existence of the limit gives only the existence of a functional on Y_0. The fact that there exists an element from X which represents this functional is an additional assumption ("complete duality").

4. The system associated with a generator

Let $C \subseteq X$ and $C^* \subseteq Y$ be cones of positive elements or linear spaces in duality. No order relation or convergence is considered now.

For example, C may be one of: the linear space C_0^∞; the linear space of entire (real) functions; the cone of totally/completely monotone functions; while for C^* we may consider: the linear space of distributions; the cone of discrete measures.

We consider two additive and positively homogeneous maps

$$A : C \longrightarrow C \text{ and } A^* : C^* \longrightarrow C^*$$

such that $[Ax, y] = [x, A^* y]$, $\forall x \in C$, $y \in C^*$.

Let us stress that we do not suppose any monotony or continuity for A.

Proposition 6. *Let $A : C \longrightarrow C$ be given. Let*

$$X_0 := \{x \in C | \sum_{n=0}^{\infty} \frac{A^n x}{\alpha^{n+1}} \text{ is weakly convergent, } \forall \alpha > 0\}.$$

If we define, for $x \in X_0$:

$$V_\alpha x := \sum_{n=0}^{\infty} \frac{A^n x}{\alpha^{n+1}}$$

then the resolvent equation:

$$V_\alpha = V_\beta + (\beta - \alpha) V_\alpha \circ V_\beta, 0 < \alpha < \beta$$

holds. The generator of this resolvent is A.

Proof. Using the formula from Lemma 2.1., we can write (for $\alpha \neq \beta$):

$$(V_\alpha \circ V_\beta)x = \sum_{n=0}^{\infty} \frac{A^n V_\beta x}{\alpha^{n+1}} = \sum_{n=0}^{\infty} \frac{1}{\alpha^{n+1}} \sum_{m=n}^{\infty} \frac{A^{m+1} x}{\beta^{m+1-n}} = \sum_{m=0}^{\infty} \frac{A^{m+1} x}{\alpha \beta^{m+1}} \frac{1 - \frac{\beta^{m+1}}{\alpha^{m+1}}}{1 - \frac{\beta}{\alpha}}$$

$$= \frac{1}{\beta - \alpha} \left(V_\alpha x - V_\beta x \right).$$

Since:

$$\alpha(\alpha V_\alpha x - x) = Ax + \sum_{n=2}^{\infty} \frac{A^n x}{\alpha^{n-1}}$$

the assertion about the generator follows. □

Remark 4.1. (i) V_α is not a kernel: it remains to find out sufficient conditions such that V_α is monotone.

(ii) The condition for the definition of the resolvent is strictly stronger than that for the system. Let $f(x) := \sum_{n=1}^{\infty} \frac{n^{n/2}}{n!} x^n$. This is analytic on $\mathbb{R}$ and the series for the translation semigroup converges everywhere, for each $x \in \mathbb{R}$. However, the series giving the resolvent is nowhere convergent (at least for $x = 0$).

(iii) Under an additional assumption, this resolvent is associated with the system

$$\Phi(t,x) := \sum_{n=0}^{\infty} \frac{A^n x}{n!} t^n.$$

Indeed, as the series

$$\sum_{n=0}^{\infty} \frac{A^n x}{\alpha^{n+1}}$$

converges, it follows that $\limsup_{n\to+\infty} \sqrt[n]{|[A^n x, y]|} < +\infty$. Hence the series:

$$\sum_{n=0}^{\infty} \frac{A^n x}{n!} t^n$$

converges, for any $t > 0$. Thus, the system exists. Now, let us define:

$$\Phi^*(t,x) := \sum_{n=0}^{\infty} \frac{|[A^n x, y]|}{n!} t^n.$$

If we suppose that $\int_0^{\infty} e^{-\alpha t}\Phi^*(t,x)dt < +\infty$, then the dominated convergence theorem can be applied in order to prove that

$$V_\alpha x = \int_0^{\infty} e^{-\alpha t}\Phi(t,x)dt.$$

Some other situations when the resolvent is associated with the system are presented in the next propositions.

Proposition 7. *If*

$$\Phi(t,x) = \sum_{n=0}^{N} \frac{A^n x}{n!} t^n$$

then

$$\alpha V_\alpha x = \sum_{n=0}^{N} \frac{A^n x}{\alpha^n}.$$

Proof.

$$V_\alpha x = \int_0^{\infty} e^{-\alpha t}\left(\sum_{n=0}^{N} \frac{A^n x}{n!} t^n\right)dt = \sum_{n=0}^{N} \frac{A^n x}{n!} \int_0^{\infty} t^n e^{-\alpha t}dt = \sum_{n=0}^{N} \frac{A^n x}{\alpha^{n+1}}.$$ □

Proposition 8. *Let x be totally monotone. If*

$$\Phi(t,x) = \sum_{n=0}^{\infty} \frac{A^n x}{n!} t^n$$

then

$$\alpha V_\alpha x = \sum_{n=0}^{\infty} \frac{A^n x}{\alpha^n}.$$

Proof. Let us denote

$$x_N := \sum_{n=0}^{N} \frac{A^n x}{n!} t^n.$$

We have $x_N \nearrow \Phi(t,x)$. On the other hand:

$$\int_0^\infty e^{-\alpha t} x_N dt = \sum_0^N \frac{A^n x}{\alpha^{n+1}},$$

hence:

$$V_\alpha x = \int_0^\infty e^{-\alpha t}\Phi(t,x)dt = \int_0^\infty e^{-\alpha t}\left(\bigvee_N x_N\right)dt = \bigvee_N \left(\int_0^\infty e^{-\alpha t} x_N dt\right)$$
$$= \sum_{n=0}^{\infty} \frac{A^n x}{\alpha^{n+1}}.$$

□

Example 4.2. 1) Let $P_t f(x) := f(x+t)$ be the translation semigroup on $\mathbb{R}$. The infinitesimal generator is $Af(x) := f'(x)$, while the associated resolvent acts as:

$$V_\alpha f(x) = e^{\alpha x}\int_x^{+\infty} e^{-\alpha t} f(t)dt.$$

While the formula

$$P_t f(x) = \sum_{n=0}^{\infty} \frac{f^{(n)}(x)}{n!} t^n$$

holds for any entire function, the formula

$$V_\alpha f(x) = \sum_{n=0}^{\infty} \frac{f^{(n)}(x)}{\alpha^{n+1}}$$

holds only for some entire functions; among them: the polynomials (polyinvariant functions) and totally monotone functions.

2) Let

$$P_t f(x) := \frac{1}{\sqrt{2\pi t}}\int_{-\infty}^{\infty} f(y) e^{-\frac{(y-x)^2}{2t}} dy$$

be the brownian semigroup on $\mathbb{R}$. The infinitesimal generator is $Af(x) := \frac{1}{2} f''(x)$, while the associated resolvent acts as:

$$V_\alpha f(x) = \int_{-\infty}^{\infty} f(y)\frac{1}{\sqrt{2\alpha}} e^{-|y-x|\sqrt{2\alpha}} dy.$$

Now, the formula

$$P_t f(x) = \sum_{n=0}^{\infty} \frac{f^{(2n)}(x)}{2^n n!} t^n$$

holds for some entire functions [11], while the formula

$$V_\alpha f(x) = \sum_{n=0}^{\infty} \frac{f^{(2n)}(x)}{2^n \alpha^{n+1}}$$

holds only for a smaller class of entire functions; among them: the polynomials (polyinvariant functions) and totally convex functions.

ACKNOWLEDGMENTS

The author acknowledges support from the grant 2-CEx06-11-10/25.07.2006, partial support from grant CEx05-D11-23/2005 and from CNCSIS through grant GR17/2007.

The author gratefully acknowledges fruitful discussions with: Tusheng Zhanh, P. Gauthier, W. Stannat, E. Symeonidis during various stages of the preparation of the manuscript.

REFERENCES

[1] A. Cornea and G. Licea, *Order and Potential. Resolvent Families of Kernels* (Lecture Notes in Math. **494**). Springer-Verlag, 1975.

[2] C. Dellacherie, Théorie générale du potentiel. I. Hommage a P. A. Meyer et J. Neveu. *Astérisque* **236** (1996), 109-124.

[3] C. Dellacherie, Non linear Dirichlet problem and non linear integration. In: P. Picco et al. (Eds.), *From Classical to Modern Probability* (Progress in Probability, CIMPA summer school 2001), pp. 83-92 Birkhäuser, 2003.

[4] C. Dellacherie and P.-A. Meyer, *Probabilités et Potentiel.* Act. Sci. Ind. Hermann, Paris, 1987.

[5] J. van Neerven, *The Adjoint of a Semigroup of Operators* (Lecture Notes in Math. **1519**). Springer-Verlag, Berlin Heidelberg, 1992.

[6] R. S. Phillips, The adjoint semigroup. *Pac. J. Math.* **5** (1955), 269-283.

[7] E. Popa, Semi-dynamical systems on algebraic structures. *An. Ştiinţ. Univ. Al. I. Cuza Iaşi, Ser. Nouă, Mat.* **44** (1998) (Suppl.), 585-596.

[8] E. Popa, Excessive elements in semi-dynamical systems. *An. Stiint. Univ. Al. I. Cuza Iasi, Ser. Noua, Mat.* **45** (1999), No.2, 367-378.

[9] E. Popa, On Riesz' splitting property. *Libertas Math.* **22** (2002), 87-104.

[10] E. Popa, A Hille–Yosida type theorem in ordered convex cones. *Positivity* **10** (2006), No. 3, 555-571.

[11] E. Popa, The semi-dynamical system associated with a generator, to appear in: *Proceedings of The 5th International Congress of Romanian Mathematicians.*

[12] D.V. Widder, *The Laplace Transform.* Princeton University Press, 1946.

DEPARTMENT OF MATHEMATICS, UNIVERSITY ALEXANDRU IOAN CUZA, 700506 IAŞI, ROMANIA
E-mail address: epopa@uaic.ro

Potential Theory and Stochastics in Albac, 225–237
Aurel Cornea Memorial Volume

LIPSCHITZ CONTINUITY OF THE PSEUDO RESOLVENT OF THE STOCHASTIC BURGERS EQUATION

WILHELM STANNAT

ABSTRACT. The Kolmogorov operator $\tilde{L}$ of the stochastic Burgers equation with space-time white noise is studied in coordinates induced by the Cole-Hopf transformation. Based on gradient estimates obtained for the pseudo-resolvent associated with a general stochastic partial differential equation with multiplicative noise, we obtain new a priori estimates on the pseudo resolvent associated with $\tilde{L}$.

1. INTRODUCTION AND MAIN RESULTS

Consider the stochastic Burgers equation on [0,1],

$$dX_t = \left(\gamma\Delta X_t - X_t \cdot \partial_\xi X_t\right) dt + \sigma\, dW_t,\ X_0 = u \in H = L^2(0,1) \tag{1}$$

with Dirichlet boundary conditions, $\gamma > 0$, $\sigma \neq 0$ and space-time white noise. Equation (1) has been studied by many authors in the recent literature, because of its analogy with stochastic Navier-Stokes equations (see [7] and the references therein). It has been shown by Da Prato, Debussche and Teman in the case $\sigma = \gamma = 1$ in [3] that (1) has a unique solution in $C([0,T]; L^2(0,1))$ and that the solution admits a unique invariant measure ν. In the recent paper [1] by the first two authors, new a priori estimates on ν are given. Based on these a priori estimates and gradient estimates on the transition semigroup

$$T_t F(u) := E(F(X_t)|X_0 = u), \quad u \in H,\ t \geq 0,$$

induced by the unique solution of (1), the authors also prove that the associated Kolmogorov operator

$$LF(u) = \frac{\sigma^2}{2}\mathrm{Tr}(F''(u)) + \gamma\langle u, \Delta F'(u)\rangle + \frac{1}{2}\langle u^2, \partial_\xi F'(u)\rangle, \tag{2}$$

defined on the space $\mathrm{span}\{\exp(i\langle h, u\rangle) : h \in D(\Delta)\}$, is maximal dissipative in $L^2(\nu)$, i.e., its closure in $L^2(\nu)$ exists and generates a Markovian C_0-semigroup of contractions, which then must coincide with $(T_t)_{t\geq 0}$.

In this paper, we will consider the linearization of the deterministic Burgers equation

$$\partial_t u = \gamma\Delta u - u \cdot \partial_\xi u \tag{3}$$

2000 *Mathematics Subject Classification.* 60H15, 35Q35, 47D07, 76M35.
Key words and phrases. Stochastic Burgers equation, Kolmogorov operator, Cole-Hopf transformation, gradient estimates, bilinear stochastic partial differential equations.

given by the (nonlinear) Cole-Hopf transformation

$$\Phi(u)(x) = \exp\left(-\frac{1}{2\gamma}\int_0^x u(y)\,dy\right), x \in [0,1]$$

(see the monograph [6]), with inverse

$$u = \Psi(v) = \Phi^{-1}(v) = -2\gamma\frac{\partial_\xi v}{v}$$

and study some of its analytic implications on the stochastic Burgers equation and its associated Kolmogorov operator.

In particular, we are interested in obtaining new a priori estimates on the pseudo-resolvent $(R_\lambda)_{\lambda>0}$ associated with the transition semigroup of the stochastic Burgers equation. A priori estimates on the resolvent can be applied to obtain uniqueness results for the Kolmogorov operator L, but may be also used for the direct construction of full Markov processes associated with (1) (see [2, 8] for recent results on L^1-uniqueness of Kolmogorov operators associated with semilinear stochastic partial differential equations).

The change of variables induced by Φ

The following is well-known and easy to check: suppose that v is a solution of the heat equation

$$\partial_t v = \Delta_N v,$$

where Δ_N denotes the Neumann-Laplacian on $(0,1)$, it follows that $u = \Psi(v)$ is a solution of the Burgers equation (3) with Dirichlet boundary conditions (see Section 1.7 in [6]). The representation of the Kolmogorov operator in the new coordinates $v = \Phi(u)$ induced by Φ is provided in our first theorem.

Theorem 1.1. *Let $F \in C^2(H)$ with $F'(u) \in D(\Delta_N)$ for all $u \in H$. Let $u \in H_0^{1,2}(0,1)$. Then*

$$\begin{aligned} L(F\circ\Phi)(u) = &\frac{\sigma^2}{8\gamma^2}\int_0^1\int_0^1 x\wedge y\,\Phi(u)(x)\Phi(u)(y)\,F''(\Phi(u))(x,y)\,dx\,dy \\ &+\frac{\sigma^2}{8\gamma^2}\int_0^1 x\Phi(u)(x)F'(\Phi(u))(x)\,dx \\ &+\gamma\langle\Phi(u),\Delta_N F'(\Phi(u))\rangle. \end{aligned} \tag{4}$$

The proof of Theorem 1.1 is given in Section 2.

In the following, denote by

$$\begin{aligned} \tilde{L}F(\Phi(u)) = &\frac{\sigma^2}{8\gamma^2}\int_0^1\int_0^1 x\wedge y\,\Phi(u)(x)\Phi(u)(y)\,F''(\Phi(u))(x,y)\,dx\,dy \\ &+\frac{\sigma^2}{8\gamma^2}\int_0^1 x\Phi(u)(x)F'(\Phi(u))(x)\,dx \\ &+\gamma\langle\Phi(u),\Delta_N F'(\Phi(u))\rangle \end{aligned}$$

the Kolmogorov operator in this new coordinates.

Note that $v = \Phi(u)$ is strictly positive, bounded and has Neumann-boundary conditions. It is therefore reasonable to consider the space

$$E := H^{1,2}_+(0,1)$$

of nonnegative function in $H^{1,2}(0,1)$ as state space for $\tilde{L}$.

Duality for $\tilde{L}$

To simplify the notations in the following, let $I = (0,1)$ and define the polynomials

$$\varphi_f(v) := \langle f, v^m \rangle := \int_{I^m} f(x)\Pi_{k=1}^m v(x_k)\, dx .$$

Applying $\tilde{L}$ to φ_f we obtain that

$$\tilde{L}\varphi_f(v) = \langle \tilde{L}_m f, v^m \rangle = \varphi_{\tilde{L}_m f}(v), \qquad f \in \mathscr{B}_b(I^m) \tag{5}$$

with

$$\tilde{L}_m f := \gamma \Delta_N^m f + V_m f, \qquad V_m(x) = \frac{\sigma^2}{8\gamma^2} \sum_{i,j=1}^m x_i \wedge x_j .$$

Here, Δ_N^m denotes the m-fold tensor product of Δ_N acting on the space $L^2(I^m)$. The identity (5) establishes some duality between the Kolmogorov operator $\tilde{L}$ and the family of self-adjoint Schrödinger operators $\tilde{L}_m$ on $L^2(I^m)$, $m \ge 1$. This duality is reminiscent of the duality between the generator of a Fleming-Viot process, well-known in population genetics (see [5]), and an associated family of generators of Moran processes.

It is well-known that the symmetric semigroup $(e^{t\tilde{L}_m})_{t\ge 0}$, generated by $\tilde{L}_m$, is positivity preserving. We can therefore formally construct a semigroup $(\tilde{T}_t)$ on the space of polynomials

$$\text{span}\,\{\varphi_f : f \in \mathscr{B}_b(I^m), m \ge 1\}$$

by setting

$$\tilde{T}_t \varphi_f(v) := \varphi_{e^{t\tilde{L}_m} f}(v), f \in \mathscr{B}_b(I^m),$$

and extending by linearity. Unfortunately, it is not easy to specify a linear function space such that this explicit construction leads to a C_0-semigroup of bounded Markovian operators.

Remark 1.2. Note that the Dirac measure concentrated on $v = 0$ trivially is an infinitesimally invariant measure for $\tilde{L}$. To look further for nontrivial (infinitesimally) invariant (probability) measures, it is interesting to study the representation of $\tilde{L}$ on polynomials. Indeed, suppose that Π is an invariant measure for $\tilde{L}$ such that $\varphi_f \in L^1(\Pi)$ for all $f \in \mathscr{B}_b(I^m)$. Then we can define the measures ν_m on I^m by

$$\int f\, d\nu_m := \int \langle f, v^m \rangle\, \Pi(dv), \quad f \in \mathscr{B}_b(I^m).$$

Invariance of Π for $\tilde{L}$ then implies invariance of ν_m for the Schrödinger operator $\tilde{L}_m$, since

$$\int \tilde{L}_m f\, d\nu_m = \int \langle \tilde{L}_m f, v^m \rangle\, d\Pi(v) = \int \tilde{L}\varphi_f\, d\Pi = 0.$$

In the particular case $m = 1$, we obtain that

$$\int \left(\frac{\sigma^2}{8\gamma^2} x f + \gamma \Delta_N f \right) d\nu_1 = 0 \quad \forall f \in \mathscr{D}(\Delta_N).$$

It is easy to see that ν_1 must be absolutely continuous and its density is a nonnegative solution of the ordinary differential equation

$$\frac{\sigma^2}{8\gamma^2} x g(x) + \gamma g''(x) = 0 \qquad \text{on } (0,1). \tag{6}$$

It is well-known that the general solution of (6) can be represented as

$$c_1 \operatorname{Ai}\left(-\frac{\sigma^2}{8\gamma^3} x \right) + c_2 \operatorname{Bi}\left(-\frac{\sigma^2}{8\gamma^3} x \right), \tag{7}$$

where

$$\operatorname{Ai}(x) = \int_0^\infty \cos\left(\frac{t^3}{3} + xt \right) dt$$

$$\operatorname{Bi}(x) = \int_0^\infty e^{-\frac{t^3}{3} + xt} + \sin\left(\frac{t^3}{3} + xt \right) dt$$

are the Airy-functions. Also note that both solutions, Ai and Bi, are positive on (-1,0), hence (7) will be positive on [0,1] for positive constants c_1, c_2 and γ sufficiently large. On the contrary, for small γ, the solutions will have negative values on $(0,1)$. Hence, at least for small γ, there can be no nontrivial invariant measure Π for $\tilde{L}$ integrating the polynomials $\langle f, v \rangle$.

Our conjecture is, that also for large γ, any nontrivial invariant measure Π will integrate polynomials only up to some finite degree m. Unfortunately, we cannot prove it.

An SPDE with multiplicative noise

To construct the semigroup generated by $\tilde{L}$, we will use an associated stochastic partial differential equation. To this end note that $\tilde{L}$ can be realized as the Kolmogorov operator associated with the linear SPDE

$$dX_t = \left(\gamma \Delta_N X_t + \frac{\sigma^2}{8\gamma^2} x X_t \right) dt + \frac{\sigma}{2\gamma} B(X_t)\, dW_t, X_0 = \xi. \tag{8}$$

Here, the dispersion coefficient $B(X_t)$ is defined as the linear operator

$$B : H \to L(H), v \mapsto \left(h \mapsto \int_0^{\cdot} h(s)\, ds \cdot v \right).$$

Here, $L(H)$ denotes the space of bounded linear operators on H. To see that the second order part $\frac{\sigma^2}{8\gamma^2} \operatorname{Tr}\left(B^*(v) \varphi''(v) B(v) \right)$ of the Kolmogorov operator associated with (8) coincides with the second order part of $\tilde{L}$, note that $B^*(v)h(x) = \int_x^1 v(s)h(s)\, ds$, so that

$$\begin{aligned} \int_0^1 B(v) \circ B^*(v) e_k(\xi) e_l(\xi)\, d\xi &= \int_0^1 \int_\xi^1 v(s) e_k(s)\, ds \int_\xi^1 v(t) e_l(t)\, dt\, d\xi \\ &= \int_0^1 \int_0^1 s \wedge t\, v(s) v(t) e_k(s) e_l(t)\, ds\, dt, \end{aligned}$$

and consequently,

$$\begin{aligned}
&\frac{\sigma^2}{8\gamma^2}\operatorname{Tr}\left(B^*(v)\varphi''(v)B(v)\right)\\
&\quad=\frac{\sigma^2}{8\gamma^2}\sum_{k,l=1}^{\infty}\partial^2_{e_k e_l}\varphi(v)\int_0^1 B(v)\circ B^*(v)e_k(\xi)e_l(\xi)\,d\xi\\
&\quad=\frac{\sigma^2}{8\gamma^2}\int_0^1\int_0^1\varphi''(x,y)x\wedge y\,v(x)v(y)\,dx\,dy.
\end{aligned}$$

Note that the semigroup $(e^{t\gamma\Delta_N})_{t\ge 0}$ generated by the Neumann-Laplacian Δ_N is Hilbert-Schmidt with locally square integrable Hilbert-Schmidt norm $K(t)$, where

$$K(t)^2=\sum_{k=0}^{\infty}e^{-2t\gamma\pi^2k^2}\le\sum_{k=0}^{\infty}\frac{1}{1+2t\gamma\pi^2k^2}\le 1+\frac{1}{\sqrt{8\gamma t}}. \tag{9}$$

It follows from Theorem 5.3.1 in [4] that (8) has a unique mild solution for initial condition ξ satisfying $E(\|\xi\|_H^2)<\infty$. Denote by $X_t(x)$, $t\ge 0$, the unique mild solution with initial condition $\xi=x\in H$. Then

$$\tilde{T}_tF(x):=E(F(X_t(x)))\,,F\in B_b(H)$$

defines a semigroup of Markovian integral operators on $B_b(H)$. Moreover, $x\mapsto X_t(x)$ is continuous, so that $(\tilde{T}_t)$ is a Feller semigroup (see p. 69 in [4]).

Gradient estimates for SPDE with multiplicative noise

Throughout this subsection we will present a general result concerning gradient estimates for pseudo-resolvents of SPDE with multiplicative noise

$$dX_t=(AX_t+F(X_t))\,dt+B(X_t)\,dW_t \tag{10}$$

on a separable real Hilbert space H with cylindrical Wiener process (W_t). On the coefficients of the SPDE we make the following standard assumptions:

(H_1) A is a self-adjoint, negative semidefinite operator on H having compact resolvent. Moreover, e^{tA} is Hilbert-Schmidt for $t>0$.
In the following, we denote by $V_r=D((1-A)^r)$, $r\in\mathbb{R}$, the interpolation spaces with norm $|u|_r:=\left(\sum_{k=0}^{\infty}(1+\lambda_k)^{2r}\langle u,e_k\rangle^2\right)^{\frac{1}{2}}$, where $(e_k)_{k\ge 0}$ is an orthonormal basis of H consisting of eigenvectors of A with corresponding eigenvalues $-\lambda_k$, $k\ge 0$.

(H_2) $F:H\to H$ is Lipschitz continuous

(H_3) $B:H\mapsto L(H)$ is strongly continuous and there exists a locally square integrable function $K:[0,\infty[\to\mathbb{R}_+$ such that

$$\begin{aligned}
\|e^{tA}B(x)\|_{HS}&\le K(t)(1+|x|)\\
\|e^{tA}(B(x)-B(y))\|_{HS}&\le K(t)|x-y|.
\end{aligned}$$

(H_4) $\exists\,\alpha\in]0,\frac{1}{2}[$ such that $\int_0^1 t^{-2\alpha}K^2(t)\,dt<\infty$.

Theorem 5.3.1 in [4] implies that for any $x \in H$, a unique mild solution $X_t(x)$, $t \geq 0$, of (10) exists. Moreover, according to Remark 5.3.2 in [4], the unique mild solution takes values in V_α and is continuous there.

Theorem 1.3. *Let $r \in [0,\alpha]$ and suppose that there exist locally square integrable functions*

$$L_F, L_B : [0,\infty[\to \mathbb{R}_+$$

satisfying

(a) $|e^{tA}(F(x) - F(y))|_r \leq L_F(t)|x - y|_r$ *for all* $x, y \in V_r$

(b) $\sum_{k=0}^{\infty}(1+\lambda_k)^{2r} e^{-2t\lambda_k}|(B^*(x) - B^*(y))e_k|^2 \leq L_B^2(t)|x-y|_r^2$ *for all* $x, y \in V_r$.

Here, $B^*(x) \in L(H)$ denotes the adjoint linear operator of $B(x)$.

Suppose that there exists $\lambda_0 > 0$ such that

$$3\int_0^\infty e^{-\lambda_0 t}\left(e^t L_F^2(t) + L_B^2(t)\right) dt < 1.$$

Then

$$\int_0^\infty e^{-\lambda t} E\left(|X_t(x) - X_t(y)|_r^2\right) dt \leq C_\lambda \int_0^\infty e^{-\lambda t}|e^{tA}(x-y)|_r^2\, dt$$

for all $\lambda > \lambda_0$. Here

$$C_\lambda := 3\left(1 - 3\int_0^\infty e^{-\lambda t}\left(e^t L_F^2(t) + L_B^2(t)\right) dt\right)^{-1}. \tag{11}$$

Proof. To simplify notations in the following, define

$$\Delta_{x,y}(t) := X_t(x) - X_t(y), \quad t \geq 0.$$

Clearly,

$$\Delta_{x,y}(t) = e^{tA}\Delta_{x,y}(0) + \int_0^t e^{(t-s)A}(F(X_s(x)) - F(X_s(y)))\, ds$$
$$+ \int_0^t e^{(t-s)A}(B(X_s(x)) - B(X_s(y)))\, dW_s.$$

Since $X_\cdot(x)$ and $X_\cdot(y)$ take values in V_r and are continuous there, we can take the V_r norm in the last identity, to obtain that

$$|\Delta_{x,y}(t)|_r \leq |e^{tA}\Delta_{x,y}(0)|_r + \int_0^t |e^{(t-s)A}(F(X_s(x)) - F(X_s(y)))|_r\, ds$$
$$+ \left|\int_0^t e^{(t-s)A}(B(X_s(x)) - B(X_s(y)))\, dW_s\right|_r.$$

Note that

$$\int_0^t |e^{(t-s)A}(F(X_s(x)) - F(X_s(y)))|_r\, ds$$
$$\leq \left(\int_0^t e^{-(t-s)}\, ds\right)^{\frac{1}{2}} \left(\int_0^t e^{t-s}|e^{(t-s)A}(F(X_s(x)) - F(X_s(y)))|_r^2\, ds\right)^{\frac{1}{2}}$$
$$\leq \left(\int_0^t e^{t-s}|e^{(t-s)A}(F(X_s(x)) - F(X_s(y)))|_r^2\, ds\right)^{\frac{1}{2}},$$

and consequently

$$
\begin{aligned}
E\left(|\Delta_{x,y}(t)|_r^2\right) \le 3|e^{tA}\Delta_{x,y}(0)|_r^2 \\
+3\int_0^t e^{t-s} E\left(|e^{(t-s)A}(F(X_s(x)) - F(X_s(y)))|_r^2\right) ds \\
+3E\left(\left|\int_0^t e^{(t-s)A}(B(X_s(x)) - B(X_s(y)))\, dW_s\right|_r^2\right).
\end{aligned}
\tag{12}
$$

Using the assumptions on F and B, the second term on the right hand side can be estimated from above by

$$3\int_0^t e^{t-s} L_F^2(t-s) E\left(\left|\Delta_{x,y}(s)\right|_r^2\right) ds$$

and the third term can be estimated from above by

$$3\int_0^t L_B^2(t-s) E\left(\left|\Delta_{x,y}(s)\right|_r^2\right) ds,$$

since

$$
\begin{aligned}
&E\left(\left|\int_0^t e^{(t-s)A}\left(B(X_s(x)) - B(X_s(y))\right) dW_s\right|_r^2\right) \\
&= \sum_{k=0}^{\infty} (1+2\lambda_k)^{2r} E\left(\left(\langle \int_0^t e^{(t-s)A}(B(X_s(x)) - B(X_s(y)))\, dW_s, e_k\rangle\right)^2\right) \\
&= \sum_{k=0}^{\infty} (1+2\lambda_k)^{2r} \int_0^t e^{2(t-s)\lambda_k} E\left(\left|(B^*(X_s(x)) - B^*(X_s(y)))e_k\right|^2\right) ds \\
&\le \int_0^t L_B^2(t-s) E\left(\left|\Delta_{x,y}(s)\right|_r^2\right) ds.
\end{aligned}
$$

Inserting the upper bounds into (12), we obtain that

$$
\begin{aligned}
E\left(|\Delta_{x,y}(t)|_r^2\right) \le 3|e^{tA}\Delta_{x,y}(0)|_r^2 \\
+3\int_0^t \left(e^{t-s}L_F^2(t-s) + L_B^2(t-s)\right) E\left(|\Delta_{x,y}(s)|_r^2\right) ds.
\end{aligned}
\tag{13}
$$

Integrating the last inequality w.r.t. $e^{-\lambda t}\,dt$ yields

$$
\begin{aligned}
\int_0^{\infty} e^{-\lambda t} E\left(|\Delta_{x,y}(t)|_r^2\right) dt \le 3\int_0^{\infty} e^{-\lambda t}|e^{tA}\Delta_{x,y}(0)|_r^2\, dt \\
+3\left(\int_0^{\infty} e^{-\lambda t}\left(e^t L_F^2(t) + L_B^2(t)\right) dt\right)\int_0^{\infty} e^{-\lambda t} E\left(|\Delta_{x,y}(t)|_r^2\right) dt
\end{aligned}
$$

which implies the assertion, after rearranging the terms. □

For part (ii) of the next corollary, we need in addition to the previous hypotheses (H_1)-(H_4) the following assumption:

(H_5) F and B are Frechet differentiable with bounded and continuous differentials.

Theorem 5.4.1 in [4] implies that under this additional assumption, $x \mapsto X_t(x)$ is continuously differentiable. It follows that for $\varphi \in C_b^1(H)$, the function $x \mapsto \varphi(X_t(x))$ is continuously differentiable too.

Corollary 1.4. *Let*

$$R_\lambda\varphi(x) := \int_0^\infty e^{-\lambda t} T_t\varphi(x)\, dt,\ \lambda > 0,$$

be the pseudo-resolvent associated with the transition semigroup

$$T_t\varphi(x) := E\big(\varphi(X_t(x))\big)\,, \varphi \in \mathscr{B}_b(H)\,,$$

induced by (10).

(i) *Assume (H_1)-(H_4). Let $\varphi \in \mathscr{B}_b(H)$ be such that its restriction $\varphi : V_r \to \mathbb{R}$ is Lipschitz with Lipschitz constant L_φ. Then for $\lambda > \lambda_0$, $R_\lambda\varphi$ is Lipschitz too, with Lipschitz constant $L_{R_\lambda\varphi} \le \frac{L_\varphi}{\lambda}\sqrt{C_\lambda}$, where C_λ is as in Theorem 1.3.*

(ii) *Assume in addition to (i) that (H_5) holds and that $\varphi \in C_b^1(H)$. Then for $\lambda > \lambda_0$, $R_\lambda\varphi$ is continuously differentiable and*

$$\left|\partial_{e_k} R_\lambda\varphi(x)\right|^2 \le C_\lambda \frac{L_\varphi^2}{\lambda(\lambda+2\lambda_k)}(1+\lambda_k)^{2r} \quad \textit{for any } x \in H.$$

Proof. For the proof of (i) note that

$$\begin{aligned}
\left|R_\lambda\varphi(x) - R_\lambda\varphi(y)\right|^2 &\le \frac{1}{\lambda}\int_0^\infty e^{-\lambda t}\left|T_t\varphi(x) - T_t\varphi(y)\right|^2 dt \\
&\le \frac{1}{\lambda}\int_0^\infty e^{-\lambda t} L_\varphi^2 E\left(\left|X_t(x) - X_t(y)\right|_r^2\right) dt \qquad (14)\\
&\le C_\lambda \frac{L_\varphi^2}{\lambda}\int_0^\infty e^{-\lambda t}|e^{tA}(x-y)|_r^2\, dt.
\end{aligned}$$

Since A is negative semidefinite, $|e^{tA}x|_r \le |x|_r$, which implies (i).

For the proof of (ii) note that the assumptions on the coefficients of (10) imply that R_λ is continuously differentiable. In particular, for $y = x + he_k$, $h \in \mathbb{R}$, the last inequality (14) implies that

$$\begin{aligned}
\left|R_\lambda\varphi(x) - R_\lambda\varphi(y)\right|^2 &\le C_\lambda \frac{L_\varphi^2}{\lambda}\int_0^\infty e^{-\lambda t} e^{-2\lambda_k t}\, dt\, |h|^2 |e_k|_r^2 \\
&= C_\lambda \frac{L_\varphi^2}{\lambda(\lambda+2\lambda_k)}(1+\lambda_k)^{2r}\,|h|^2.
\end{aligned}$$

Taking the limit $h \downarrow 0$, we obtain the assertion. □

Estimates on the pseudo resolvent associated with (8)

In the case of the stochastic partial differential equation (8) associated with the Kolmogorov operator $\tilde{L}$, obtained as the image of the Kolmogorov operator L associated with the stochastic Burgers equation under the Cole-Hopf transformation, we can now apply the previous result on gradient estimates for the associated pseudo-resolvent.

Note that in this case A is the Neumann-Laplacian with eigenvalues $-\lambda_k = -\gamma\pi^2 k^2$, $k \ge 0$, $F(v)(\xi) = \frac{\sigma^2}{8\gamma^2}\xi v(\xi)$ and $B(v)h = \frac{\sigma}{2\gamma}\int_0^\cdot h(\xi)\, d\xi \cdot v$, so that the assumptions (H_1)-(H_5) are clearly satisfied.

Indeed, (H_1) and (H_2) are obvious, (H_3) follows from estimate (9), since

$$\|e^{tA}(B(v_1)-B(v_2))\|_{HS} \le \|e^{tA}\|_{HS}\|B(v_1-v_2)\|_{L(H)} \le K(t)\frac{\sigma}{2\gamma}|v_1-v_2|,$$

and (H_4) follows from estimate (9), too, since

$$\int_0^1 t^{-2\alpha}K^2(t)\,dt < \infty \quad \text{for } \alpha < \frac{1}{4}.$$

(H_5) is implied by the fact that F is a bounded linear operator and B can be seen as a bounded bilinear operator. The assumptions (a) and (b) of Theorem 1.3 are satisfied (in the case $r=0$) with

$$2L_F(t) = L_B^2(t) = \frac{\sigma^2}{4\gamma^2}\|e^{t\Delta_N}\|_{HS}^2 \le \frac{\sigma^2}{4\gamma^2}\left(1+\frac{1}{\sqrt{8\gamma t}}\right)$$

since

$$|e^{tA}(F(v_1)-F(v_2))| \le |(F(v_1)-F(v_2))| \le \frac{\sigma^2}{8\gamma^2}|v_1-v_2|$$

and

$$\sum_{k=0}^{\infty} e^{-2t\gamma\pi^2k^2}|B^*(v_1)e_k - B^*(v_2)e_k|^2$$

$$\le \frac{\sigma^2}{4\gamma^2}\sum_{k=0}^{\infty} e^{-2t\gamma\pi^2k^2}|v_1-v_2|^2 \le \frac{\sigma^2}{4\gamma^2}\left(1+\frac{1}{\sqrt{8\gamma t}}\right)$$

for all $v_1, v_2, h \in H = L^2(0,1)$, where $B^*(v)h = \int_\cdot^1 v(\xi)h(\xi)\,d\xi$ is the adjoint of $B(v)$ in H (see estimate (9)). In this case

$$I_\lambda := \frac{3\sigma^2}{4\gamma^2}\int_0^\infty e^{-\lambda t}\left(\frac{\sigma^2}{16\gamma^2}e^t+1\right)\left(1+\frac{1}{\sqrt{8\gamma t}}\right)dt \to 0, \lambda\to\infty,$$

so that we can find $\lambda_0 > 1$ such that the integral I_λ becomes smaller than 1 for $\lambda > \lambda_0$. Corollary 1.4 now implies:

Corollary 1.5. *Let $(\tilde{R}_\lambda)_{\lambda>0}$ be the pseudo-resolvent associated with the stochastic partial differential equation (8).*

(i) *Let $\varphi \in \mathscr{B}_b(H)$ be Lipschitz with Lipschitz constant L_φ. Then for $\lambda > \lambda_0$, $\tilde{R}_\lambda\varphi$ is Lipschitz too, with Lipschitz constant $L_{\tilde{R}_\lambda\varphi} \le \frac{L_\varphi}{\lambda}\sqrt{C_\lambda}$.*

(ii) *Assume in addition to (i) that $\varphi \in C_b^1(H)$. Then for $\lambda > \lambda_0$, $\tilde{R}_\lambda\varphi$ is continuously differentiable and*

$$\left|\partial_{e_k}\tilde{R}_\lambda\varphi(v)\right| \le \frac{C_\lambda L_\varphi^2}{\lambda(\lambda+2\pi^2k^2)}, k\ge 0.$$

Remark 1.6. In this remark let $U := H_0^{1,2}(0,1)$ and

$$V := \{v \in H^{2,2}(0,1) : v>0, \dot{v}(0)=\dot{v}(1)=0, \int_0^1 \left(\partial_\xi \log v\right)^2 d\xi < \infty\}.$$

Clearly, $\Phi : U \to V$ and is invertible with inverse

$$\Psi(v) = \Phi^{-1}(v) = -2\gamma\partial_\xi \log v.$$

Note that the mapping $B(v)$ is not Hilbert-Schmidt on H, so that it is not clear, whether for $\varphi \in C_b^2(H)$, $\tilde{T}_t\varphi$ is a strict solution of the Cauchy problem

$$\partial_t G(t,v) = \frac{1}{2}\operatorname{Tr}\left(B^*(v)G_{vv}(t,v)B(v)\right) + \langle \gamma\Delta_N v + F(v), G_v(t,v)\rangle,$$
$$t > 0, v \in D(\Delta_N)$$

with initial condition $G(0,v) = \varphi(v)$, $v \in H$.

However, assuming this is the case, it follows that for $\varphi \in C_b^2(H)$ $\tilde{R}_\lambda\varphi$ is a solution of the resolvent equation $(\lambda - \tilde{L})\tilde{R}_\lambda\varphi = \varphi$. In this case, a formal calculation yields that $R_\lambda\tilde{\varphi}(u) := (\tilde{R}_\lambda\varphi)(\Phi(u))$ solves the equation

$$(\lambda - L)R_\lambda\tilde{\varphi} = \tilde{\varphi} \quad \text{with} \quad \tilde{\varphi}(u) = \varphi(\Phi(u)),$$

because formally,

$$\begin{aligned} LR_\lambda\tilde{\varphi}(u) &= \tilde{L}(R_\lambda\tilde{\varphi}\circ\Phi^{-1})(\Phi(u)) = \tilde{L}(\tilde{R}_\lambda\varphi)(\Phi(u)) \\ &= \lambda\tilde{R}_\lambda\varphi(\Phi(u)) - \varphi(\Phi(u)) = \lambda R_\lambda\tilde{\varphi}(u) - \tilde{\varphi}(u). \end{aligned}$$

Hence, if φ is Lipschitz with Lipschitz constant L_φ and if $\lambda > \lambda_0$, it follows that the formal solution $R_\lambda\tilde{\varphi}$ to the resolvent equation associated with the Kolmogorov operator of the stochastic Burgers equation has the Lipschitz property

$$\begin{aligned} \left|R_\lambda\tilde{\varphi}(u_1) - R_\lambda\tilde{\varphi}(u_2)\right| &= \left|\tilde{R}_\lambda\varphi(\Phi(u_1)) - \tilde{R}_\lambda\varphi(\Phi(u_2))\right| \\ &\le \frac{L_\varphi}{\lambda}\sqrt{C_\lambda}\,|\Phi(u_1) - \Phi(u_2)|. \end{aligned}$$

2. Proof of Theorem 1.1

We will first need some preparations:

Lemma 2.1. *$\Phi : H \to H$ is Frechet differentiable with differential*

$$\Phi'(u)(v)(x) = -\frac{1}{2\gamma}\int_0^x v(y)\,dy\,\Phi(u)(x).$$

Proof. For the first differential we need to show that

$$\lim_{t\to 0}\left\|\frac{1}{t}(\Phi(u+tv) - \Phi(u)) + \frac{1}{2\gamma}\int_0^{\cdot} v(\xi)\,d\xi\,\Phi(u)\right\|_H = 0.$$

This follows from the fact that

$$\begin{aligned} &\left|\frac{1}{t}(\Phi(u+tv) - \Phi(u))(x) + \frac{1}{2\gamma}\int_0^{\cdot} v(\xi)\,d\xi\,\Phi(u)\right| \\ &\quad= \Phi(u)(x)\left|\frac{1}{t}(\Phi(tv)(x) - 1) + \frac{1}{2\gamma}\int_0^x v(y)\,dy\right| \\ &\quad\le \frac{t}{2}\Phi(u)(x)\left(\frac{1}{2\gamma}\right)^2\left(\int_0^x v(y)dy\right)^2\exp\left(\frac{1}{2\gamma}\|v\|\right) \\ &\quad\le \frac{t}{2}\exp\left(\frac{1}{\gamma}\|u\|_H\right)\frac{\|v\|_H^2}{4\gamma^2}. \end{aligned}$$

□

Lemma 2.2. *Let $F : H \to \mathbb{R}$ be two times Frechet differentiable. Then $G(u) := F(\Phi(u))$ is two times Frechet differentiable too, with differentials*

$$\begin{aligned} G'(u)(v) &= \int_0^1 F'(\Phi(u))(x)\Phi'(u)(v)(x)\,dx \\ &= -\frac{1}{2\gamma}\int_0^1 F'(\Phi(u))(x)\Phi(u)(x)\int_0^x v(y)\,dy\,dx \\ &= -\frac{1}{2\gamma}\int_0^1\int_y^1 F'(\Phi(u))(x)\Phi(u)(x)\,dx\,v(y)\,dy \end{aligned}$$

and (15)

$$\begin{aligned} G''(u)(v,w) = \frac{1}{4\gamma^2}&\int_0^1\int_0^1 F''(\Phi(u))(x,y)\Phi(u)(x)\Phi(u)(y)\cdot \\ &\cdot\int_0^x v(s)ds\int_0^y w(t)\,dt\,dx\,dy \\ &+\frac{1}{4\gamma^2}\int_0^1 F'(\Phi(u))(x)\Phi(u)(x)\int_0^x v(s)ds\int_0^x w(t)\,dt\,dx. \end{aligned}$$

Proof. The formula for the first Frechet differential is obvious due to Lemma 2.1. Note that $G'(u)(v)$ can be written as $H(\Phi(u))$, where

$$H(u) = -\frac{1}{2\gamma}\int_0^1 F'(u)(x)u(x)\int_0^x v(s)ds\,dx$$

is Frechet differentiable with differential

$$\begin{aligned} \langle H'(u), w\rangle = &-\frac{1}{2\gamma}\int_0^1\int_0^1 F''(u)(x,y)u(x)\int_0^x v(s)ds\,dx\,w(y)\,dy \\ &-\frac{1}{2\gamma}\int_0^1 F'(u)(x)\int_0^x v(s)ds\,w(x)\,dx. \end{aligned}$$

Inserting this formula for the Frechet differential of H into (15) yields

$$\begin{aligned} G''(u)(v,w) = \frac{1}{4\gamma^2}&\int_0^1\int_0^1 F''(\Phi(u))(x,y)\cdot\Phi(u)(x)\int_0^x v(s)\,ds \\ &\cdot\Phi(u)(y)\int_0^y w(t)dt\,dx\,dy \\ &+\frac{1}{4\gamma^2}\int_0^1 F'(\Phi(u))(x)\Phi(u)(x)\int_0^x v(s)ds\int_0^x w(t)dt\,dx. \end{aligned}$$ □

We are now ready to prove Theorem 1.1. First note that Lemma 2.2 implies

$$
\begin{aligned}
\frac{\sigma^2}{2}\operatorname{tr}((F\circ\Phi)''(u)) &= \frac{\sigma^2}{2}\sum_{k=1}^{\infty}(F\circ\Phi)''(e_k,e_k)\\
&= \frac{\sigma^2}{8\gamma^2}\sum_{k=1}^{\infty}\int_0^1\int_0^1 F''(\Phi(u))(x,y)\Phi(u)(x)\Phi(u)(y)\cdot\\
&\quad\cdot\int_0^x e_k(s)\,ds\int_0^y e_k(t)\,dt\,dx\,dy\\
&\quad+\frac{\sigma^2}{8\gamma^2}\sum_{k=1}^{\infty}\int_0^1 F'(\Phi(u))(x)\Phi(u)(x)\left(\int_0^x e_k(s)\,ds\right)^2
\end{aligned}
$$

which yields the first two terms in (4), using the fact that

$$
\sum_{k=1}^{\infty}\int_0^x e_k(s)\,ds\int_0^y e_k(t)\,dt = x\wedge y.
$$

The third term in (4) clearly follows from the fact that

$$
\gamma\Delta_N\Phi(u) = -\frac{1}{2}\partial_\xi u\cdot\Phi(u) \quad + \frac{1}{4\gamma}u^2\cdot\Phi(u),
$$

which implies, by the assumption $F'(\Phi(u))\in D(\Delta_N)$, that

$$
\begin{aligned}
\gamma\langle\Delta_N\Phi(u),F'(\Phi(u))\rangle &= -\frac{1}{2}\int_0^1\partial_\xi u\Phi(u)F'(\Phi(u))\,d\xi\\
&\quad+\frac{1}{4\gamma}\int_0^1 u^2\Phi(u)F'(\Phi(u))\,d\xi\\
&= +\frac{1}{2}\int_0^1 u(\xi)\partial_\xi\left(\Phi(u)F'(\Phi(u))\right)d\xi\\
&\quad-\frac{1}{4\gamma}\int_0^1 u^2(\xi)\partial_\xi\left(\int_\xi^1\Phi(u)(y)F'(\Phi(u))(y)\,dy\right)d\xi\\
&= \gamma\int_0^1 u(\xi)\Delta(F\circ\Phi)'(u)(\xi)\,d\xi + \frac{1}{2}\int_0^1 u^2(\xi)\partial_\xi(F\circ\Phi)'(u)(\xi)\,d\xi.
\end{aligned}
$$

References

[1] G. Da Prato and A. Debussche, m-dissipativity of Kolmogorov operators corresponding to Burgers equations with space-time white noise. *Potential Analysis* **26** (2007), 31-55.

[2] A. Es-Sarhir and W. Stannat, Invariant measures for semilinear SPDE's with local Lipschitz drift coefficients and applications. *J. Evol. Equ.* **8** (2008), 129-154.

[3] G. Da Prato, A. Debussche, and R. Teman, Stochastic Burgers equation. *NoDEA* **1** (1994), 389-402.

[4] G. Da Prato and J. Zabczyk, *Ergodicity for Infinite Dimensional Systems* (Lectures Note Series of the LMS **229**). Cambridge University Press, 1996.

[5] S. N. Ethier and T. G. Kurtz, Fleming-Viot processes in population genetics. *SIAM J. Control And Optimization* **31** (1993), 345-386.

[6] J. Kevorkian, *Partial Differential Equations: Analytical Solution Techniques* (Texts in Applied Mathematics **35**). Springer, 2000.

[7] M. Röckner and Z. Sobol, Kolmogorov equations in infinite dimensions: Well-posedness and regularity of solutions, with applications to stochastic generalized Burgers equations. *Annals of Probability* **34** (2006), 663-727.

[8] W. Stannat, A new a priori estimate for the Kolmogorov operator of a 2D-stochastic Navier-Stokes Equation. *IDAQP* **10** (2007), 1-15.

FACHBEREICH MATHEMATIK, TECHNISCHE UNIVERSITÄT DARMSTADT, SCHLOSSGARTENSTRASSE 7, 64289 DARMSTADT, GERMANY

E-mail address: stannat@mathematik.tu-darmstadt.de

Potential Theory and Stochastics in Albac, 239–246
Aurel Cornea Memorial Volume

EXCESSIVE FUNCTIONS WITH RESPECT TO MONGE-AMPÈRE RESOLVENTS AND CONCAVE FUNCTIONS

CORNELIU UDREA

ABSTRACT. We define a class of nonlinear operators associated with the non-homogeneous Dirichlet problem for the Monge-Ampère equation (on an open bounded strictly convex subset of $\mathbb{R}^n$), everyone of them being the initial operator of a nonlinear resolvent on the space of the essentially bounded functions (a such resolvent is called Monge-Ampère resolvent).

Afterwards we shall study the excessive functions with respect to these resolvents and we shall prove that these functions are the lower bounded concave functions.

Therefore we continue the study of some problems of the nonlinear potential theory as in [1], [2], [7], [8], [9], [10], [11], and we generalize and apply the results of the author.

1. INTRODUCTION

We shall study the excessive functions which are defined with respect to the resolvents associated with the solutions of the Dirichlet problem for the Monge-Ampère equation and we shall compare these excessive functions to the concave functions. The above resolvents are defined in [18] and they are called (on shortly) the Monge-Ampère resolvents. The results of this work generalize the similar ones what are presented in [14] or [16].

In this section we fix the notations which are used in this work, and we present some results of [20] concerning the study of the supermedian functions with respect to a nonlinear resolvent.

Throughout this work $U \subset \mathbb{R}^k$ is a non void open bounded and strictly convex set and $\mathcal{U}(U) := \{u \in \mathbb{R}^U :$ u is a convex function $\}$ and for all $\varphi \in C(\partial U)$ we denote

$$\mathcal{U}_\varphi(U) := \{u \in \mathbb{R}^{\overline{U}} : u \in C(\overline{U}) \text{ and u is convex function such that } u_{|\partial U} = \varphi\}.$$

We shall use the following notations, if λ is the Lebesgue measure on U, then:

$$\mathcal{M}_l(U) := \{f \in \overline{\mathbb{R}}^U : \text{ f is a } \lambda\text{-measurable function and } m_\infty(f) > -\infty\},$$

where $m_\infty(f)$ is the lower essential bound of f,

$$L^\infty(U) := \{f \in \overline{\mathbb{R}}^U : \text{ f is a } \lambda\text{-measurable function and } \lambda\text{-essentially bounded}\},$$

$$L^1(U) := \{f \in \overline{\mathbb{R}}^U : \text{ f is a } \lambda\text{-measurable function and } \lambda\text{-integrable function}\}.$$

2000 *Mathematics Subject Classification.* 31C45, 34G20.
Key words and phrases. Excessive functions, Monge-Ampère equation.

Also we shall consider two (nonlinear) operators $T,N : L^\infty(U) \to L^\infty(U)$, i.e. T,N are increasing maps ([7]); if $(I+T)(I-N) = I = (I-N)(I+T)$ then we shall say that the (T,N) is a pair of conjugated operators.

Definition 1.1. Let $u \in \mathcal{M}_l(U)$.

(i) (a) We shall say that u is *a weak T-dominant function* iff for all $f \in L^\infty(U)$ such that $f + Tf \le u$ on the set

$$\{f > 0\} := \{x \in E : f(x) > 0\}$$

it will result that $Tf \le u$.

(b) We shall use the notation

$$\mathcal{D}_T^w := \{u \in \mathcal{M}_l(U) : \text{ u is a weak T-dominant function}\}.$$

(c) We shall say that T satisfies *the weak maximum principle* iff

$$\mathbb{R}_+ = [0,+\infty) \subset \mathcal{D}_T^w.$$

(ii) (a) We shall say that u is a *weak complete T- dominant function* iff $\forall f,g \in L^\infty(U)$ such that $f + Tf \le g + Tg + u$ on the set $\{f > g\} := \{x \in E : f(x) > g(x)\}$ it will result that $Tf \le Tg + u$.

(b) We shall use the notation:

$$\mathcal{D}_T^{w,c} := \{u \in \mathcal{M}_l(U) : \text{ u is a weak complete T- dominant function}\}.$$

(c) We shall say that T satisfies the *weak complete maximum principle* iff $\mathbb{R}_+ \subset \mathcal{D}_T^{w,c}$ (according to [7]).

(iii) (a) We say that $u \in \mathcal{M}_l(U)$ is a *T-dominant function* (respectively a *complete T-dominant function*) iff for all $f \in L^\infty(U)$ (respectively for all $f,g \in L^\infty(U)$) such that $Tf \le u$ on $\{f > 0\}$ (respectively $Tf \le Tg + u$ on $\{f > g\}$) we have that $Tf \le u$ (respectively $Tf \le Tg + u$).

(b) We shall use the following notations:

$$\mathcal{D}_T := \{u \in \mathcal{M}_l(U) : \text{ u is a T-dominant function }\},$$

$$\mathcal{D}_T^c := \{u \in \mathcal{M}_l(U) : \text{ u is a complete T-dominant function }\}.$$

(c) If $\mathbb{R}_+ \subset \mathcal{D}_T$ (respectively $\mathbb{R}_+ \subset \mathcal{D}_T^c$), then we shall say that T satisfies *the maximum principle* (respectively *the complete maximum principle*).

Remark 1.2. Moreover $\overline{T}$ (respectively) $\overline{N}$ will denote the natural extension of T (respectively N) to the set $\mathcal{M}_l(U)$, that is for all $f \in \mathcal{M}_l(U)$:

$$\overline{T}(f) = \sup\{Tg : g \in L^\infty(U) \text{ and } g \le f\} = \sup\{T(\inf(f,n)) : n \in \mathbb{N}\}.$$

Obviously $\overline{T}$ and also $\overline{N}$ are increasingly continuous (on $\mathcal{M}_l(U)$) and $\overline{T}$ (respectively $\overline{N}$) will be denoted again by T (respectively N).

Definition 1.3. (i) Let $u \in \mathcal{M}_l(U)$.

(a) The function u is called *N-supermedian function* iff $Nu \le u$.

(b) We shall use the following notation

$$\mathcal{S}_N = \mathcal{S}(N) := \{u \in \mathcal{M}_l(U) : u \text{ is an } N\text{-supermedian function }\}.$$

(ii) (a) We shall say that a function $u \in \mathcal{M}_l(U)$ is a *strictly N-supermedian function* iff $\forall f,g \in L^\infty(U)$ such that $f \le g + u$, then $Nf \le Ng + u$.

(b) We shall use the following notation

$$\mathscr{S}_N^s := \{u \in \mathscr{M}_l(U) : u \text{ is a strictly } N\text{-supermedian function}\}.$$

Theorem 1.4 ([20]). *If (T,N) is a pair of conjugated operators then:*

(i) $\mathscr{S}_N = \mathscr{D}_T^w$.

(ii) $\mathscr{S}_N^s = \mathscr{D}_T^{w,c}$.

Definition 1.5 (see [7] and also [9] or [14]).

(i) (a) $\mathscr{V} := (V_p)_{p\in(0,\infty)}$ is called (nonlinear) resolvent (on $L^\infty(U)$) iff for all $p,q \in (0,\infty)$ we have that:

$$(I-(p-q)V_p)(I+(p-q)V_q) = I.$$

(b) If, for all $p \in (0,\infty)$,

$$(I-pV_p)(I+pV_0) = I = (I+pV_0)(I-pV_p),$$

then either V_0 will be called *the initial operator* of the resolvent $\mathscr{V}$, or $\mathscr{V}$ is called *the resolvent associated with* V_0.

(c) The resolvent $\mathscr{V}$ is called a *sub-Markov resolvent* if pV_p is a sub-Markov operator on $L^\infty(U)$ (i. e. $\|pV_pf - pV_pg\| \le \|f-g\|_\infty$, $\forall f \in L^\infty(U)$), $\forall p \in (0,\infty)$. For all $p \in (0,\infty)$, we shall denote by $L_p : \mathscr{M}_l(U) \to \mathscr{M}_l(U)$ the operator $L_pf := V_p(pf)$, $\mathscr{S}_p := \mathscr{S}_{L_p}$ and $\mathscr{S}_p^s := \mathscr{S}_{L_p}^s$.

(ii) (a) All functions $u \in \bigcap\limits_{p\in(0,\infty)} \mathscr{S}_p$ are called $\mathscr{V}$*-supermedian functions* and $\mathscr{S}_\mathscr{V} := \mathscr{S}(\mathscr{V}) := \bigcap\limits_{p\in(0,\infty)} \mathscr{S}_p$.

(b) The functions of the set $\bigcap\limits_{p\in(0,\infty)} \mathscr{S}_p^s$ are called *strictly* $\mathscr{V}$*-supermedian* functions and $\mathscr{S}_\mathscr{V}^s := \mathscr{S}^s(\mathscr{V}) := \bigcap\limits_{p\in(0,\infty)} \mathscr{S}_p^s$.

Theorem 1.6 ([20]). *We have the following assertions:*

(i) $\mathscr{S}_\mathscr{V} = \mathscr{D}_{V_0}$.

(ii) $\mathscr{S}_\mathscr{V}^s = \mathscr{D}_{V_0}^c$.

2. THE MONGE-AMPÈRE RESOLVENTS

In this section there are presented some useful notions and results about the convex functions, the Monge-Ampère resolvents (see [8], [13], [18], [20]) and the dominant and the supermedian functions.

Remark 2.1. We shall use the following results of the convex functions' theory:

(i) If $u, v \in \mathscr{U}(U)$ and $D \subset U$ is a non-void open and bounded set such that $u_{|D} \le v_{|D}$ and $(sci_D u)_{|\partial D} \ge (sci_D v)_{|\partial D}$, then $\partial_v(D) \subset \partial_u(D)$, (where $sci_D u : \overline{D} \to \overline{\mathbb{R}}$, $sci_D u(x) = \liminf\limits_{y\in D, y\to x} u(y)$, and $\partial_u(D) = \bigcup\limits_{x\in D} \partial_u(x)$, $\partial_u(x)$ being the subdifferential of u in x.)([8]).

(ii) For all $u, v \in \mathscr{U}(U)$ and $\alpha \in \mathbb{R}_+$ we have that $\nu_{u+v} \ge \nu_u + \nu_v$ and $\nu_{\alpha u} = \alpha^k \nu_u$, where ν_u is the curvature measure of the map u on U ([8]).

(iii) (The minimum principle for locally convex functions [8], [18].) Let $G \subset \mathbb{R}^k$ be a non-void open bounded set and $f, g : \overline{G} \to \mathbb{R}$ be continuous and locally convex functions on G such that $\nu_f \le \nu_g$ and $f_{|\partial G} \ge g_{|\partial G}$. We have that $f \ge g$ (where for all $D \subset G$ convex set $(\nu_f)_{|D} = \nu(f_{|D})$).

Definition 2.2 (according to [18]). (i) $\forall\, \varphi \in C(\partial U), \mathcal{V}^{\varphi} := (V_p^{\varphi})_{p\in(0,\infty)}$, where $V_p^{\varphi} : \mathcal{M}_l(U) \to \mathcal{M}_l(U), \forall\, p \in [0,\infty), \mathcal{V}^{\varphi}$ is a sub-Markov resolvent on the space $L^{\infty}(U)$, associated with the operator V_0^{φ}.

(ii) By the definition, for all $f \in L^{\infty}(U)$ and $p \in [0,\infty), V_p^{\varphi} f : \overline{U} \to \mathbb{R}$ is a concave continuous function such that

$$(V_p^{\varphi} f)_{|\partial U} = \varphi \text{ and } \nu_{-V_p^{\varphi} f} = (f - pV_p^{\varphi} f)^+ \cdot \lambda.$$

(iii) $\mathcal{V}^{\varphi}$ is called *the Monge-Ampère resolvent* for all $\varphi \in C(\partial U)$.

Remark 2.3. We shall present some useful results from [18].

(i) For all $\varphi \in C(\partial U)$ and $p \in [0,\infty)$ we have the following properties:

(a) $V_p^{\varphi} : L^{\infty}(U) \to -\mathcal{U}_{-\varphi}(U)$ is a compact, $\frac{1}{k}-$Hölder operator, satisfies the weak complete maximum principle (moreover V_0^{φ} satisfies the complete maximum principle) and is increasingly continuous.

(b) For all $f_1, f_2 \in L^{\infty}(U)$, we have:

$$V_p^{\varphi}(f_1 + f_2) \le V_p^{\varphi} f_1 + V_p^0 f_2.$$

(ii) For all $f \in \mathcal{M}_l(U)$ the following assertions are equivalent:

(a) $\{V_p^{\varphi} f = \infty\}$ is the void set for all $p \in (0,\infty)$.

(b) $(V_p^{\varphi} f)_{|U} \in -\mathcal{U}(U)$ for all $p \in (0,\infty)$.

(c) $(V_0^{\varphi} f)_{|U} \in -\mathcal{U}(U)$.

(d) There exists $h \in -\mathcal{U}(U)$ such that $\nu_{-h} = f^+.\lambda$ and $(scs_U h)_{|\partial U} \ge \varphi$, where $scs_U := -sci_U(-h)$.

Moreover if one of the previous sentences is valid we have that for all $p \in (0,\infty)$ $\nu_{-V_p^{\varphi} f} = (f - pV_p^{\varphi} f)^+ \cdot \lambda$ and f^+ is $\lambda-$locally integrable on U.

(iii) For all $p \in (0,\infty)$, the sequence $(V_p^{\varphi_n} f_n)_n$ converges uniformly on $\overline{U}$ to the map $V_p^{\varphi} f$ provided that:

(a) $(\varphi_n)_n$ converges uniformly to φ in $(C(\partial U), \|.\|_{\infty})$,

(b) $(f_n)_n \subset L^{\infty}(U)$ is a bounded sequence such that $(f_n)_n$ converges to f either in $(L^{\infty}(U), \|.\|_{\infty})$, or $\lambda - a.e.$ on U.

(iv) For all $u \in -\mathcal{U}(U), u \ge 0$ we have that u satisfies the complete $V_0^{\varphi}-$dominant principle.

Remark 2.4. For all $p \in (0,\infty)$ and $\varphi \in C(\partial U)$ we shall use the following notations:

(i) (a) $\mathcal{D}_p^w(\varphi) := \mathcal{D}^w_{pV_0^{\varphi}}; \mathcal{D}_p^{w,c}(\varphi) := \mathcal{D}^{w,c}_{pV_0^{\varphi}}$.

(b) $\mathcal{D}(\varphi) := \mathcal{D}_{V_0^{\varphi}}$ and $\mathcal{D}^c(\varphi) := \mathcal{D}^c_{V_0^{\varphi}}$.

(ii) In view of (a), Theorems 1.4 and 1.6, if

$$\mathcal{S}_p(\varphi) := \mathcal{S}(L_p^{\varphi}),\ \ \mathcal{S}_p^s(\varphi) := \mathcal{S}^s(L_p^{\varphi}),$$

then

$$\mathcal{S}_p(\varphi) = \frac{1}{p}\mathcal{D}_p^w(\varphi),\ \ \mathcal{S}_p^s(\varphi) = \frac{1}{p}\mathcal{D}_p^{w,c}(\varphi).$$

3. SUPERMEDIAN AND EXCESSIVE FUNCTION WITH RESPECT TO THE MONGE-AMPÈRE RESOLVENTS

In this section we shall apply the previous results to determine the class of the supermedian (respectively excessive) functions with respect to the Monge-Ampère resolvents.

Lemma 3.1. *Suppose that $\varphi, \psi \in C(\partial U)$ are such that $\varphi \leq \psi$. For all $f \in \mathcal{M}_l(U)$ and $p \in [0,\infty)$ we have that $V_p^{\varphi} f \leq V_p^{\psi} f$.*

Proof. If $f \in L^{\infty}(U)$ and $D := \{V_p^{\varphi} f > V_p^{\psi} f\}$, then we shall find that $D \subset U$ is an open set, $(V_p^{\varphi} f)_{|\partial D} \leq (V_p^{\psi} f)_{|\partial D}$ and $(\nu_{-V_p^{\varphi} f})_{|D} \leq (\nu_{-V p^{\nu} f})_{|D}$. According to the Remark 2.1.(iii) we shall have that $V_p^{\varphi} f \leq V_p^{\phi} f$ on D, hence $D = \emptyset$. □

Corollary 3.2. *For all $p \in (0,\infty)$ and $\varphi, \psi \in C(\partial U)$ such that $\varphi \leq \psi$ we shall have that $\mathcal{S}_p(\psi) \subset \mathcal{S}_p(\varphi)$.*

Proof. It is obvious. □

Theorem 3.3. *For all $\varphi \in C(\partial U)$ and $p \in (0,\infty)$ we have the following assertions:*
(i) $\{u \in -\mathcal{U}(U) : u \geq 0\} \subset \mathcal{S}_p^s(\varphi)$.
(ii) $\{u \in -\mathcal{U}(U) : \varphi \leq (scs_U u)_{|\partial U}\} \subset \mathcal{S}_p(\varphi)$.

Proof. (i) Since

$$\mathcal{S}_p^s(\varphi) = \frac{1}{p}\mathcal{D}_p^{w,c}(\varphi), \mathcal{D}_p^{w,c}(\varphi) \supset \mathcal{D}_p^c(\varphi),\ \frac{1}{p}\mathcal{D}^c_{pV_0\varphi} = \mathcal{D}^c(\varphi) \text{ and}$$

$$\{u \in -\mathcal{U}(U) : u \geq 0\} \subset \mathcal{D}_p^c(\varphi)$$

(Remark 2.4.(ii)), we have the sentence.

(ii) If $u \in -\mathcal{U}, \varphi \leq (scs_U u)_{|\partial U}(U)$ and $D := \{V_p^{\varphi}(pu) > u\}$, then $D \subset U$ is an open set such that $V_p^{\varphi}(pu) \leq scs_D u$ on ∂D; we have that $\partial_{-u}(D) \subset \partial_{-V_p^{\varphi}(pu)}(D)$ (in conformity with Remark 2.1.(i)).

On the other hand $\nu_{-V_p^{\varphi}(pu)} = p(u - pV_p^{\varphi}(pu))^+ \cdot \lambda$, and so that $\nu_{-V_p^{\varphi}(pu)} = 0$ on the set D and accordingly $\nu_{-u}(D) = 0$. Therefore $V_p^{\varphi}(pu) \leq u$ on D and $D = \emptyset$. □

Remark 3.4. For all $\varphi \in C(\partial U)$ we have that $\mathbb{R}_+ \subset \mathcal{S}_p^s(\varphi), \forall p \in (0,\infty)$. By the previous theorem we have also that $\{\alpha \in \mathbb{R} : \alpha \geq \varphi\} \subset \mathcal{S}_p(\varphi), \forall p \in (0,\infty)$.

Lemma 3.5. *For all $p \in (0,\infty)$ we have that:*
(i) *If $\varphi \in C(\partial U)$, then $\mathcal{S}_p(0) \subset \mathcal{S}_p^s(\varphi)$.*
(ii) $\mathcal{S}_p(0) = \mathcal{S}_p^s(0)$.

Proof. (i) If $u \in \mathcal{S}_p(0)$ and $g \in L^{\infty}(U)$, then, according to the Remark 2.3.(i)(b), $V_p^{\varphi}(g + pu) \leq V_p^{\varphi} g + V_p^0(pu)$ and so that $u \in \mathcal{S}_p^s(\varphi)$.

(ii) We have $V_p^0 0 = 0$, hence $\mathcal{S}_p^s(0) \subset \mathcal{S}_p(0)$ and by the previous point we shall find the equality. □

Definition 3.6. Let $\varphi \in C(\partial U)$.

(i) $\mathscr{S}(\varphi) := \mathscr{S}_{\mathcal{V}^\varphi} := \bigcap_{p\in(0,\infty)} \mathscr{S}_p(\varphi)$ and if $u \in \mathscr{S}(\varphi)$ then u will be called $\mathcal{V}^\varphi$*-supermedian function.*

(ii) A function $u \in \mathscr{M}_l(U)$ is $\mathcal{V}^\varphi$–supermedian if and only if, for all $p \in (0,\infty)$, $V_p^\varphi(pu) \leq u$. Moreover for all $u \in \mathscr{S}(\varphi)$, $u \geq V_0^\varphi 0$, and $V_0^\varphi 0 \in \mathscr{S}(\varphi)$.

Lemma 3.7. *For all* $\varphi \in C(\partial U)$ *we have that:*

(i) $\{u \in -\mathscr{U}(U) : \varphi \leq (scs_U u)_{|\partial U}\} \subset \mathscr{S}(\varphi)$. *In particular all real numbers* α *such that* $\varphi \leq \alpha$ *are* $\mathcal{V}^\varphi$*-supermedian functions.*

(ii) $\mathscr{S}(\varphi) = \mathscr{D}(\varphi)$.

Proof. (i) It is a consequence of the Theorem 3.(ii).

(ii) It is the result of the Theorem 1.6.(i). □

Definition 3.8. If $\varphi \in C(\partial U)$, then we shall define:

(i) $\mathscr{S}^s(\varphi) := \mathscr{S}^s_{\mathcal{V}^\varphi} := \bigcap_{p\in(0,\infty)} \mathscr{S}^s_p(\varphi)$ and if $u \in \mathscr{S}^s(\varphi)$, u will be called *strictly* $\mathcal{V}^\varphi$*-supermedian function.*

(ii) A function $u \in \mathscr{M}_l(U)$ is strictly $\mathcal{V}^\varphi$-supermedian if and only if for all

$$f \in L^\infty(U) \text{ and } p \in (0,\infty),\ V_p^\varphi(f + pu) \leq V_p^\varphi f + u.$$

Moreover, for all $u \in \mathscr{S}^s(\varphi)$ we have that $u \geq 0$.

Lemma 3.9. *For all* $\varphi \in C(\partial U)$, *we have that:*

(i) $\{u \in -\mathscr{U}(U) : u \geq 0\} \subset \mathscr{S}^s(\varphi)$, *and so that* $\mathbb{R}_+ \subset \mathscr{S}^s(\varphi)$.

(ii) $\mathscr{S}^s(\varphi) = \mathscr{D}^c_{V_0^\varphi}(\varphi)$.

(iii) (a) $\mathscr{S}^s(\varphi) + \mathscr{S}^s(\varphi) \subset \mathscr{S}^s(\varphi)$. (b) $\mathscr{S}^s(\varphi) + \mathscr{S}(\varphi) \subset \mathscr{S}(\varphi)$.

Proof. (i) It is a consequence of the Theorem 3.(i).

(ii) It is the result of the Theorem 1.6.(ii).

(iii) They are obvious. □

Definition 3.10. (i) For all $\varphi \in C(\partial U)$ and $u \in \mathscr{S}(\varphi)$ we shall define $\hat{u}_\varphi := \sup_{p\in(0,\infty)} V_p^\varphi(pu)$ that is the $\mathcal{V}^\varphi$ *- regularization of* u.

(ii) Obviously the above function $\hat{u}_\varphi$ has the properties:

(a) $\hat{u}_\varphi : \overline{U} \to (-\infty,\infty]$ and $\hat{u}_\varphi$ is a concave lower semicontinuous function such that $\hat{u}_\varphi|_{\partial U} = \varphi$.

(b) $\hat{u}_\varphi \in \mathscr{S}(\varphi)$ and $\hat{u}_\varphi \leq u$.

(iii) $\mathscr{E}(\varphi) := \mathscr{E}_{\mathcal{V}^\varphi} := \{u \in \mathscr{S}(\varphi) : u = \hat{u}_\varphi$ and $u < \infty\}$, and every $u \in \mathscr{E}(\varphi)$ will be called $\mathcal{V}^\varphi$*-excessive functions.*

Proposition 3.11. *For all* $u \in \mathscr{S}(\varphi)$ *we have that* $\hat{u}_\varphi = u$ *(obviously* λ *a.e.) on* U.

Proof. If $\{\hat{u}_\varphi = \infty\} \neq \emptyset$, then since $\hat{u}_\varphi$ is a concave function on U, we have that $\{\hat{u}_\varphi = \infty\} = U$ and $u = \hat{u}_\varphi = \infty$ everywhere on U.

Let $\{\hat{u}_\varphi = \infty\} = \emptyset$. Since $V_p^\varphi(pu) \le \hat{u}_\varphi$ on U, by Definition 2.2.(ii), we have that $\nu_{-V_p^\varphi(pu)} = p(u - V_p^\varphi(pu))\lambda, \forall p \in (0,\infty)$. Moreover, for all $p \in (0,\infty)$, we find that:

$$\frac{1}{p}\nu_{-V_p^\varphi(pu)} = \nu_{-\frac{1}{\sqrt[k]{p}}V_p^\varphi(pu)} \text{ and } \frac{1}{\sqrt[k]{p}} inf\varphi \le \frac{1}{\sqrt[k]{p}} V_p^\varphi(pu) \le \frac{1}{\sqrt[k]{p}}\hat{u}_\varphi,$$

hence $\lim_{p\to\infty} \frac{1}{\sqrt[k]{p}} V_p^\varphi(pu) = 0$, uniformly on the compact sets of U. Therefore $(\frac{1}{p}\nu_{-V_p^\varphi(pu)})_p$ converges vaguely to the zero measure on U when p converges to ∞.

By the previous relations and by the Beppo-Levi theorem we shall have that, for all $f \in C_c(U)$:

$$\int_U f(u-\hat{u}_\varphi)d\lambda = \lim_{p\to\infty}\int_U f(u - V_p^\varphi(pu))d\lambda = \lim_{p\to\infty}\int_U f.\frac{1}{p}\nu_{-V_p^\varphi(pu)} = 0,$$

that is

$$\int_U f(u-\hat{u}_\varphi)d\lambda = 0,$$

and so that $u = \hat{u}_\varphi$ $\lambda - a.e.$ on U. □

Corollary 3.12. *For all $\varphi, \psi \in C(\partial U)$ we have that:*

(i) *If $u \in \mathscr{S}(\varphi) \cap \mathscr{S}(\psi)$, then $\hat{u}_\varphi = \hat{u}_\psi$.*

(ii) *If $\varphi \le \psi$ then $\mathscr{E}(\psi) \subset \mathscr{E}(\varphi)$.*

Proof. (i) It is an immediate consequence of the previous proposition.

(ii) If $u \in \mathscr{E}(\psi) \subset \mathscr{S}(\psi) \subset \mathscr{S}(\varphi)$, then by the previous point it is obvious that $u \in \mathscr{E}(\varphi)$. □

Proposition 3.13. *The set $\mathscr{S}(0)$ is a convex cone of functions.*

Proof. By the Lemma 5.(ii) and Lemma 9.(iii)(a) we have that $\mathscr{S}(0) = \mathscr{S}^s(0)$ and $\mathscr{S}(0) + \mathscr{S}(0) \subset \mathscr{S}(0)$ so that it remains to prove that $\alpha\mathscr{S}(0) \subset \mathscr{S}(0), \forall \alpha \in (0,\infty)$.

Let $u \in \mathscr{S}(0)$ be a bounded map, $\alpha, p \in (0,\infty)$ and $D := \{V_p(p\alpha u) > \alpha\hat{u}\}$ (where $\hat{u} = \hat{u}_0$ and $V_p = V_p^0$). We shall find that $D \subset U$, D is an open set, $V_p(p\alpha u), \alpha scs_D\hat{u}$ are continuous functions on $\overline{D}$ and $V_p(p\alpha u) \le \alpha scs_D\hat{u}$ on ∂D. By the Remark 2.1.(i) $\nu_{-\alpha\hat{u}}(D) \le \nu_{-V_p(p\alpha u)}(D)$. Since on the set D the measure $\nu_{-V_p(p\alpha u)}$ is zero measure, we can apply the minimum principle for the locally convex functions (Remark 1.(iii)) and we shall find that $V_p(p\alpha u) \le \alpha\hat{u}$ on D and $D = \emptyset$. Therefore $V_p(p\alpha u) \le \alpha\hat{u} \le \alpha u$ on U and $\alpha u \in \mathscr{S}(0)$. □

Theorem 3.14. *For all $u \in \mathscr{S}(\varphi)$, where $\varphi \in C(\partial U)$, such that u is lower bounded the following statements are equivalent:*

(i) *If φ is such that $\varphi \le (scs_U u)_{|\partial U}$, then $u \in \mathscr{E}(\varphi)$.*

(ii) *The map u is a concave real function on U.*

Proof. (i)⇒ (ii) By the hypothesis we have that for all $\varphi \in C(\partial U)$ such that $\varphi \le (scs_U u)_{|\partial U}$, $u = \sup_{p\in[0,\infty)} V_p^\varphi(pu) < \infty$, and hence $u \in -\mathscr{U}(U)$.

(ii) ⇒ (i) By the Lemma 7.(i) $u \in \mathscr{S}(\varphi)$, for all $\varphi \in C(\partial U)$ such that $\varphi \le (scs_U u_{|\partial U})$ and by the Proposition 11. we have that $u = \hat{u}_\varphi$ on U for all such φ. Since $u < \infty$, then $u \in \mathscr{E}(\varphi)$. □

Corollary 3.15. *We have the equality:*

$$\{u \in -\mathcal{U}(U) : \ u \text{ is bounded}\} = \bigcup_{\varphi \in C(\partial U)} \mathcal{E}(\varphi).$$

Acknowledgments

Research supported by NASR of Government of Romania: CEEX-D11-23, 05.10.2005.

References

[1] E.M.J. Bertin, Fonctions convexes et théorie du potentiel. Preprint **89** (1978), Univ. Utrecht.

[2] E.M.J. Bertin, Convex potential theory. Preprint **489** (1987), Univ. Utrecht.

[3] N. Boboc and Gh. Bucur, *Măsură şi capacitate.* Ed. Şt. şi Enciclop., Bucureşti, 1985.

[4] N. Boboc, Gh. Bucur, and A. Cornea, *Order and Convexity in Potential Theory: H-Cones* (Lecture Notes in Math. **853**). Springer, Berlin, 1981.

[5] C. Constantinescu and A. Cornea, *Potential Theory and Harmonic Spaces.* Springer, Berlin, 1972.

[6] K. Deimling, *Nonlinear Functional Analysis.* Springer-Verlag, Berlin, 1985.

[7] C. Dellacherie, Une version non linéaire du théorème de Hunt. In: *Proc. of ICPT'90,* pp. 25-40. Nagoya, 1992.

[8] F. van Gool, Topics in non linear potential theory. Graduate thesis, Univ. Utrecht, 1992.

[9] F. van Gool, Nonlinear kernels and their resolvents. *Potential Anal.* **12** (2000), 203-210.

[10] H. Maagli, Perturbation semi-linéaire des resolvantes et des semi-groupes. In: *Proc. of ICPT'91,* pp. 61-89. Kluwer Acad. Publish., Dordrecht, 1994.

[11] H. Maagli, Perturbation semi-linéaire des resolvantes. In: *Proc. of ICPT'91,* pp. 247-259. Kluwer Acad. Publish., Dordrecht, 1994.

[12] P.-A. Meyer, *Probability and Potentials.* Blaisdell Publishing Company, London, 1966.

[13] T. Rockafellar, *Convex Analysis.* Princeton University Press, Princeton, 1970.

[14] C. Udrea, Nonlinear resolvents. *Rev. Roum. Math. Pures Appl.* **40** (1995), 691-712.

[15] C. Udrea, The Perron-Wiener-Brelot Method and the Monge-Ampère equation. In: *Proc. of XVI-th R. Nevanlinna Coll.,* pp. 279-287. Walter de Gruyters& Co, 1996.

[16] C. Udrea, Supermedian functions with respect to a nonlinear resolvent. *Math. Rep.* **49** (1997), 285-297.

[17] C. Udrea, On nonlinear operators: boundedness and maximum principle. *Rev. Roum. Math. Pures Appl.* **46** (2001), 125-136.

[18] C. Udrea, A family of nonlinear resolvents. *Southeast Asian Bull. Math.* **29** (2005) , 785-806, 2005.

[19] C. Udrea, Resolvents and nonlinear potential theory. In: *Current Trends in Potential Theory,* pp. 163-173. Theta Foundation Bucharest, 2005.

[20] C. Udrea, Dominant functions and supermedian functions. In: *Proc. of Sixth Cong. of Romanian Math.,* Vol 1. Romanian Acad. Publsh. House, Bucharest, 2009, (to appear).

Faculty of Mathematics and Computer Sciences, University of Piteşti, Str. Târgul din Vale nr. 1, Piteşti, Romania

E-mail address: corneliu_udrea@yahoo.com